Disaster Risk Reduction

Methods, Approaches and Practices

Series Editor

Rajib Shaw, Keio University, Shonan Fujisawa Campus, Fujisawa, Japan

Disaster risk reduction is a process that leads to the safety of communities and nations. After the 2005 World Conference on Disaster Reduction, held in Kobe, Japan, the Hyogo Framework for Action (HFA) was adopted as a framework for risk reduction. The academic research and higher education in disaster risk reduction has made, and continues to make, a gradual shift from pure basic research to applied, implementation-oriented research. More emphasis is being given to multi-stakeholder collaboration and multi-disciplinary research. Emerging university networks in Asia, Europe, Africa, and the Americas have urged process-oriented research in the disaster risk reduction field. With this in mind, this new series will promote the output of action research on disaster risk reduction, which will be useful for a wide range of stakeholders including academicians, professionals, practitioners, and students and researchers in related fields. The series will focus on emerging needs in the risk reduction field, starting from climate change adaptation, urban ecosystem, coastal risk reduction, education for sustainable development, community-based practices, risk communication, and human security, among other areas. Through academic review, this series will encourage young researchers and practitioners to analyze field practices and link them to theory and policies with logic, data, and evidence. In this way, the series will emphasize evidence-based risk reduction methods, approaches, and practices.

More information about this series at http://www.springer.com/series/11575

Riyanti Djalante · Mizan B. F. Bisri · Rajib Shaw
Editors

Integrated Research on Disaster Risks

Contributions from the IRDR Young Scientists Programme

Springer

Editors
Riyanti Djalante
Institute for the Advanced Study of Sustainability
United Nations University
Tokyo, Japan

Mizan B. F. Bisri
Institute for the Advanced Study of Sustainability
United Nations University
Tokyo, Japan

Rajib Shaw
Keio University
Fujisawa, Kanagawa
Japan

ISSN 2196-4106 ISSN 2196-4114 (electronic)
Disaster Risk Reduction
ISBN 978-3-030-55562-7 ISBN 978-3-030-55563-4 (eBook)
https://doi.org/10.1007/978-3-030-55563-4

This Springer imprint is published by the registered company Springer Nature Switzerland AG
The registered company address is: Gewerbestrasse 11, 6330 Cham, Switzerland

Contents

Chapter 1
Introduction: IRDR Young Scientists—Analysis of Researchers and Key Research Topics

Riyanti Djalante and Rajib Shaw

Abstract The Sendai Framework for Disaster Risk Reduction (SFDRR) is the global framework for disaster risk reduction adopted in 2015. The Sendai Framework for Disaster Risk Reduction 2015–2030 outlines seven clear targets and four priorities for action. Integrated Research on Disaster Risk (IRDR) is a decade-long research programme co-sponsored by the International Science Council and the United Nations Office for Disaster Risk Reduction (UNDRR). It is a global, multidisciplinary approach to dealing with the challenges brought by natural disasters, mitigating their impacts, and improving related policy-making mechanisms. This book presents the works of the IRDR Young Scientist Programme, showcasing works on characterization of hazard, vulnerability, and risk (Part 1), governance and management of disaster risks (Part 2), and emerging topics in DRR research such as post-disaster recovery and reconstruction, build-back-better approach, public health, role of young scientists, multi-stakeholder engagement, gender, and roles of indigenous knowledge (Part 3).

Keywords Disaster studies · Hazard · Risks · Vulnerability · Capacity building

1.1 Introduction

The Sendai Framework for Disaster Risk Reduction (SFDRR) is the global framework for disaster risk reduction adopted in 2015. The Sendai Framework for Disaster Risk Reduction 2015–2030 outlines seven clear targets and four priorities for action

R. Djalante (✉)
United Nations University – Institute for the Advanced Study for Sustainability (UNU-IAS), Tokyo, Japan
e-mail: djalante@unu.edu

R. Djalante · R. Shaw
Integrated Research on Disaster Risks (IRDR) Scientific Committee Member, Tokyo, Japan
e-mail: shaw@sfc.keio.ac.jp
URL: http://www.irdrinternational.org/

R. Shaw
Graduate School of Media and Governance, Keio University, Fujisawa, Japan

R. Djalante et al. (eds.), *Integrated Research on Disaster Risks*, Disaster Risk Reduction,
https://doi.org/10.1007/978-3-030-55563-4_1

to prevent new and reduce existing disaster risks: (i) Understanding disaster risk; (ii) Strengthening disaster risk governance to manage disaster risk; (iii) Investing in disaster reduction for resilience and; (iv) Enhancing disaster preparedness for effective response, and to "Build Back Better" in recovery, rehabilitation and reconstruction. It aims to achieve the substantial reduction of disaster risk and losses in lives, livelihoods, and health and in the economic, physical, social, cultural and environmental assets of persons, businesses, communities, and countries over the next 15 years. The SFDRR specifically calls for the strengthening of science to inform risk assessment and decision-making, including building capacity for research.

Integrated Research on Disaster Risk (IRDR) is a decade-long research programme co-sponsored by the International Council for Science (ICSU), the International Social Science Council (ISSC), and the United Nations Office for Disaster Risk Reduction (UNISDR). It is a global, multidisciplinary approach to dealing with the challenges brought by natural disasters, mitigating their impacts, and improving related policy-making mechanisms. The complexity of the task is such that it requires the full integration of research expertise from the natural, socio-economic, health, and engineering sciences as well as policy-making, coupled with an understanding of the role of communications, and public and political responses to reduce the risk. IRDR addresses technological and health-related events when these are consequences of natural hazards.

This book draws from research conducted by young scientists who are part of the Integrated Research in Disaster Risk (IRDR). It examines multidisciplinary in research and actions in disaster risk reduction, with case studies are taken globally. The Integrated Research on Disaster Risk (IRDR) Young Scientists programme is:

- A sub-program within IRDR which promote capacity building of young professionals and encourage them to undertake innovative and needs-based research which makes science-policy and science-practice linkages stronger.
- IRDR Young Scientists Programme was started in late 2016. Currently, it is a community of 115 young researchers from over 40 countries after 3 batches of application.
- IRDR network and partners provide academic advice and training courses, workshops and programmes for IRDR young scientists.
- IRDR young scientists contribute to innovative research in the field of disaster risk reduction and participate in conferences and/or social media as the ambassador of IRDR.

In terms of authorships, there are 20 IRDR Young Scientists with additional 32 co-authors, 19 females, and 33 males. The geographical distribution is presented in Fig. 1.1.

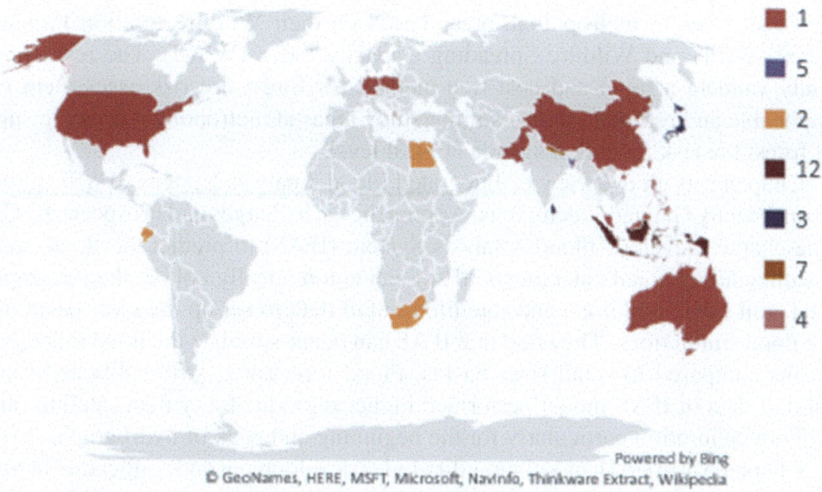

Fig. 1.1 Geographical distribution of authors

1.2 Structure of the Book and Summary of Book Chapters

The book is organized into three major parts which reflect the Sendai Framework for Disaster Risk Reduction. **Part 1 is on characterization of hazard, vulnerability, and risk. There are 7 chapters in this part, focusing on floods, tropical cyclones, heat, landslides, and earthquake and tsunamis**. Chapter 2 is titled Application of a Machine Learning Technique for Developing Short-Term Flood and Drought Forecasting Models in Tropical mountainous catchments, by Muñoz et al. Taking the case of tropical Andes in South America, they use Random Forest (RF) algorithm to develop short-term flood and drought forecasting models. They state that the applicability of this study is to assist authorities in flood and drought management to evaluate hazard risks and to find the basis for developing integrated action plans from a local and regional perspective.

Chapter 3 is titled Increasing trends in tropical cyclone induced surge impacts over North Indian Ocean by Sattar and Cheung. Tropical cyclone (TC) is a well-known natural disaster that can devastate much of a society, environment, economy, and result in people's deaths. The North Indian Ocean (NIO) is one ocean basin that is very prone to TC. TCs often cause huge human casualties in densely populated communities like Bangladesh, India, and Myanmar around the Bay of Bengal (BoB) region of the NIO. They found that while almost all intense TCs had historically made landfall around the BoB region, climate model projection simulated more intense TCs over the AS, which indicates larger and more destructive TCs especially around the coastal areas of the AS.

Chapter 4 is titled Forest Fire Risk Assessment in Metropolitan Areas using Wildfire Ignition Probability and Spreading Capacity Index: Case study of Tirana, Albania, by Hysa. The chapter develops a cost-free and rapid method for categorizing the forest surfaces in metropolitan areas based on their Wildfire Ignition Probability Index (WIPI) and Wildfire Spreading Capacity Index (WSCI). The results of the study validate a rapid and cost-free method for forest fire risk assessment being applicable and reproducible on similar study areas at metropolitan scale, in support of forest fire risk reduction agenda at local level.

Chapter 5 is an overview of Integrated Flood Analysis System (IFAS) studies in Insufficiently Gauged Catchments: Approaches, Challenges and Prospects, by Chow. The chapter develops Flood Analysis System (IFAS) to predict the flood event in insufficiently gauged catchments. IFAS can automatically collect the geographical data, soil type, land uses, and satellite rainfall data to set up the river basin model for flood simulations. They find that IFAS can better simulate the flood in large river basins compared to small river basins. Flood forecasting with calibrated satellite rainfall data in IFAS model performed higher reproducibility than satellite rainfall without calibration particularly for the beginning and peak of hydrograph.

Chapter 6 is titled Heat Vulnerability Index development and application in Medan City, Indonesia, by Setiawati et al. The chapter aims to assess heat vulnerability in Medan City, by developing and applying Heat Vulnerability Index (HVI) based on the commonly used health indicators and principal component analysis (PCA). They find that though the impacts of heat stress are alarming, the integration of heat-related risk map into the master plan for spatial planning of Medan City have not been taken into consideration. This information will be very useful for local authorities when deciding on targeted campaign of urban climate adaptation and assist the heat warning system in the future.

Chapter 7 is titled Identification of Landslide Prone Areas through multiple GIS-based Susceptibility Assessments: Case study of Chepe River, Nepal, by Kripa et al. Statistical Index Model and Logistic Regression Model were compared for performance through Geographical Information System (GIS), to derive landslide hazard map of the Chepe River corridor. Eleven factors (slope, aspect, geology, distance to road, land use, rainfall, elevation, relief, drainage density, plan curvature, and profile curvature) were considered as possible key factors for the landslide susceptibility assessment. They find that logistic regression model seems to have better applicability in the Chepe river corridor in comparison to statistical index method using the same 11 triggering factors.

Chapter 8 is titled Resilience Capacity Index and Migration Patterns after disaster among the Japanese and Foreign Residents at the Municipality Level in Japan, by Matsuura. The chapter aims to contribute to the literature of resilience measurement in two ways. First, the chapter develops a Japanese version of the Resilience Capacity Index (RCI), which was ranked at the highest sub-national level resilience index in terms of maturity according to the UNDP's systematic review (2014). Second, it examines how high- and low-community resilience measured by RCI affects inter-municipality migrations among Japanese nationals and foreign residents. The results of this paper suggest the need for a strategy to reduce the number of municipalities with low resilience and encourage more inclusive and participatory communities for foreigners.

Part 2 is related to the governance and management of disaster risks. There are three chapters in this part based on studies in Indonesia, Nepal, and South Africa. Chapter 9 is on Understanding Social-Mediated Disaster and Risk Communication with Topic Model, by Xianlin Jina. It reviews the literature on disaster and risk communication along with the introduction of the Crisis and Emergency Risk Communication model (CERC). This chapter highlights the importance of utilizing big data tools to unpack the conundrum of social-mediated disaster and risk communication.

Chapter 10 is titled Integrating Risk Sensitive Land Use Planning into Municipal Planning Process: Experience from Nepal, by Hada et al. This chapter focuses on how to effectively enhance collaborative, participatory and interactive approach for risk-sensitive land use planning, and fully integrate into a mandatory planning process. The research is to postulate a framework that enables collaborative, participatory, and iterative approach for the implementation of the RSLUP through integration in municipal development plan.

Chapter 11 is titled Counting Down to Day Zero: Exploring Community-based Water Management Strategies in Western Cape Province Drought, South Africa (2017/2018), by Vhumbunu. This research explored the various community-based water conservation and water management strategies employed in the face of the Western Cape Province Drought that occurred in the season period of 2017/2018. The chapter recommends the strengthening of community-based water management strategies through the involvement of Government and non-state actors as well as prioritization and alignment of community-based water management strategies with municipal, provincial, national and global frameworks, and plans on disaster management and climate change adaptation so as to facilitate more robust coordination drought risk mitigation, responses, and preparedness.

Part 3 discusses the emerging topics in DRR research such as post-disaster recovery and reconstruction, build-back-better approach, public health, role of young scientists, multi-stakeholder engagement, gender, and roles of indigenous knowledge. Chapter 12 is Climate Change Impacts on Health and the potential of Primary Health Care (PHC) for Adaptation: A Case Study of Kemijen Village, Semarang, Indonesia, by Dwinantoaji et al. The aim of the study was to explore the adaptation strategies the Kemijen communities take to reduce the negative effects of floods on the human health. It is found that Primary Health Care (PHC) implemented by health cadres plays an important role in preparing for extreme events, monitoring, and responding to infectious disease outbreaks due to changing patterns of vector- and water-borne diseases by providing extra support for communities.

Chapter 13 is Capacity building for Community resilience: Lessons learnt from risk-informed development planning in Sri Lanka, by Saja. The Sri Lanka Community Resilience Framework (CRF) has been the driving concept behind the risk-sensitive approach in development planning. The state officials at the sub-national (district) levels were trained as trainers on CRF and risk-sensitive development planning process. The community-level state staff officials and local government officers were then trained by district state officials through cascade training programs.

This chapter details the capacity building strategy, process, and tools for creating risk-sensitive development plans as a locally led sustainable approach.

Chapter 14 is Leveraging Youth Engagement in Disaster Risk Reduction through Science, Engineering Technology, and Innovation in Indonesia, by Shidiq et al. This chapter provides an overview of Indonesian youth and young professionals' (YYP) engagement in Disaster Risk Reduction (DRR) and explores how the current national framework enables the environment for the YYP to best contribute in reducing disaster risks. DRR activities involving YYP have been growing for years.

Chapter 15 is Understanding Gender dimensions of disaster impacts on agriculture in the Global South, by Andlib. This review chapter is centred on cases of marginalized third world countries in Southern Africa and the Asian Pacific with the specific objective of understanding the nexus between gender, climate change adaptation and disaster risk reduction with a specific focus on agriculture. With this review, we intend to provide insight into gender dimensions for recovery in developing countries that are often overlooked by national and international policymakers, as well as bridging the gap between the policymakers, researchers, and farmers.

Chapter 16 is Preserving knowledge of indigenous housing structure in Disaster Reconstruction: Case of Nayala Village in Fiji, by Aquino et al. Traditional structures are among the cultural treasures that stand as testament to the ingenuity of people who have come before us. Three key things were found to cause the predominance of contemporary housing structures over the bure in the reconstruction phase: perspectives of key stakeholders on resilient housing, recognition of indigenous materials and methods in the code, and the inflexibility of government programmes and policies as regards to informal practices. Ultimately, empowering indigenous institutions to construct their traditional housing enhances the ability to achieve a timely and efficient recovery through the complementation of modern and traditional knowledge and contributions.

Chapter 17 is Adapting to Climate Change by *Building Back Better* in Disaster Recovery: Case study of Rarotonga, Cook Islands, by Mannakkara. Case study looking at the climate change challenges of coastal businesses on the island of Rarotonga, Cook Islands was conducted using the "BBB Framework" developed by Mannakkara and Wilkinson to identify best-practices for climate adaptation. The results showcased that framed under BBB, a combination of approaches is necessary to prepare for and respond to climate change impacts such as innovating and re-designing the built environment, re-thinking land use, appropriate education, effective early warning and evacuation practices, community cohesion, social and business networks, adaptive business practices, collaboration and alignment between all stakeholders involved, appropriate legislation, and long-term monitoring and evaluation.

Chapter 18 is by Sarah Lindbergh on Critical Infrastructure (CI). Starting with an overview of the emergence of CI protection in the twenty-first century, the chapter touches on the challenges related to the historical legacy of governmental plans in the U.S. that were formulated as a security and defense issue. The chapter argues that the transportation and energy system vulnerability to climate change hazards in California reinforces the current need to go beyond placed-based exposure and CI physical interconnectivity through two different and complementary approaches: (1) considering energy services "external" criticality, potentially different from traditional energy supply system criticality, and (2) to understanding CI organizational topologies, which in the case of complex supply chain systems often characterized with multi-stakeholders, can help identify issues of institutional fragmentation and improve cross-boundary risk governance.

1.3 A Description of the Intended Readership/Users

The intended readers of this book are researchers and scholars in the field of governance of sustainability and environmental governance. Postgraduate students will benefit this book within courses on environmental governance, on climate change governance, and on transformation and social change processes. Societal actors in climate change adaptation and other environmental governance fields on local, national, and international levels can benefit from the focus on societally relevant findings in the past 10 years of research on adaptiveness.

Part I
Characterisation of Hazard, Vulnerability, and Risk

Chapter 2
Application of a Machine Learning Technique for Developing Short-Term Flood and Drought Forecasting Models in Tropical Mountainous Catchments

Paul Muñoz, Johanna Orellana-Alvear, and Rolando Célleri

Abstract Floods and droughts are among the most common natural hazards worldwide. They produce major impacts on society, economy, and ecosystems. Even worst, the frequency and severity of hydrological extremes are expected to increase with climate change and land-use alteration. As a countermeasure, during last decades, implementation of flood and drought forecasting models have globally become an emerging field of research for water management and risk assessment. In mountainous areas, hydrological extremes forecasting is unfortunately more challenging considering that information other than precipitation and runoff is not commonly available due to budget constraints, remoteness of the study areas and extreme spatio-temporal variability of additional driving forces. This is especially true for the tropical Andes in South America, which is the longest and widest cool region in the tropics. Recent advances in computational science coupled with long-term data availability have boosted Machine Learning (ML) applications. Among the variety of ML techniques, there is a potential to use the Random Forest (RF) algorithm due to its simplicity, robustness and capacity to deal with complex data structures. We used a step-wise methodology to developed short-term flood and drought forecasting models for several lead times (4, 8, 12 and 24 h) for two catchment representative of the Ecuadorian Andes. We found that derived models can reach maximum validation performances (Nash–Sutcliffe efficiency, NSE) from 0.860 (4-h) to 0.545 (24-h)

P. Muñoz (✉) · J. Orellana-Alvear · R. Célleri
Departamento de Recursos Hídricos Y Ciencias Ambientales, Universidad de Cuenca, 010150 Cuenca, Ecuador
e-mail: paul.munozp@ucuenca.edu.ec

J. Orellana-Alvear
e-mail: johanna.orellana@ucuenca.ed.ec

R. Célleri
e-mail: rolando.celleri@ucuenca.edu.ec

J. Orellana-Alvear
Laboratory for Climatology and Remote Sensing, Faculty of Geography, University of Marburg, 35032 Marburg, Germany

R. Célleri
Facultad de Ingeniería, Universidad de Cuenca, 010150 Cuenca, Ecuador

R. Djalante et al. (eds.), *Integrated Research on Disaster Risks*, Disaster Risk Reduction,
https://doi.org/10.1007/978-3-030-55563-4_2

for optimal inputs composed only by features accounting for 80% of the model's outcome variance. Moreover, we found that a set of RF hyper-parameters can be transferred to a comparable catchment with a maximum model performance reduction of 0.10 (NSE). Overall, the forecasting of hydrological extremes (especially floods) is challenging mainly due to lack of relevant data (driving forces) and sufficient extreme events from which RF models learn. The applicability of this study is to assist authorities in flood and drought management to evaluate hazard risks and to found the basis for developing integrated action plans from a local and regional perspective.

Keywords Natural hazards · Early warning system · Flood · Drought · Timeseries forecasting · Hydroinformatics · Machine learning · Random forest

2.1 Introduction

Floods and droughts are among the most common natural disasters worldwide. Both hydrological extremes have major impacts on society (e.g., human losses, increased health risks, interruption of water and sewer services), economy (e.g., losses of agricultural production, damage to infrastructure), and ecosystems (e.g., hydro-geomorphic conditions, alteration of river and floodplain habitats and biodiversity) (Vos et al. 1999). Even worst, recent studies concur that the frequency and severity of floods are expected to increase with climate change and land use changes (Min et al. 2011; Sofia et al. 2017). For instance, in the Andes of Ecuador, flood and drought events cause human losses and perturb the everyday life of people by interrupting the water supply service, damaging transportation networks, among others (Chang and Hwang 1999).

A report of the Andean community[1] for the period 1970–2007 revealed that in the Andes of Ecuador 263 floods, 30 droughts, and 357 landslides (as a side effect, mostly in the city of Cuenca) caused 429 human deaths as well as destruction of 2149 houses. Cuenca, which is the third largest city of Ecuador (almost 0.6 million inhabitants), is crossed by four rivers, the Tomebamba, the Yanuncay, the Tarqui, and the Machángara. Together, the Tomebamba and the Yanuncay upper catchments provide nearly the 60% of the water demand of the city. Additionally, around 85% of the total area of the upper catchments is covered by páramo and native forests ecosystems. However, the water regulation function of these ecosystems is increasingly amended by natural and human processes (Buytaert et al. 2011, 2006). The vulnerability of the catchments led to declare them legally as protected zones[2] in 1985.

According to the local water company, the Empresa Pública Municipal de Telecomunicaciones, Agua potable, Alcantarillado y Saneamiento de Cuenca (ETAPA EP), on an annual basis, the city of Cuenca is affected by flood and hydrological drought events. Local media has reported inundation events causing human losses

[1] https://www.comunidadandina.org

[2] https://www.etapa.net.ec

and destruction of infrastructure (bridges). In addition, hydrological droughts cause water shortage problems, which obligate ETAPA EP to restrict the water supply of the city and surrounding rural areas. In response to this necessity, the official flood early warning system of the city was launched on 2014. It merely consists on monitoring (real time) the main currents at specific locations with the purpose to inspect the hydrograph transit (Fernández de Córdova Webster and Javier Rodríguez López 2016). The limitation is the dependence on instrumentation which could be damaged during extreme events. Additionally, the time in advance in which an alert can be emitted is in the order of hours (maximum concentration time of less than 6 h for the Yanuncay catchment). To date, there is no implementation of extreme runoff forecasting models for operational and warning purposes.

The main flash-flood driving forces are precipitation, soil humidity (humid areas) and topography (Braud et al. 2016; Ruin et al. 2008). However, in mountainous areas, flood and drought forecasting is even more challenging considering that information other than precipitation and discharge is not commonly available due to budget constraints, remoteness of the study areas and more importantly due to extreme spatio-temporal variability of the aforementioned driving forces. This is especially true for the Tropical Andes in South America, the longest and widest cool region in the tropics. Consequently, a simple approach, although useful, is the development of precipitation-runoff forecasting models.

In 2015, the Sendai Framework for Disaster Risk Reduction (SFDRR) 2015–2030 (UNISDR 2015) was adopted at the Third UN World Conference in Sendai, Japan. Seven global targets were proposed by 2030, in short, they are aimed to reduce disaster mortality, affected people, disaster economic loss, disaster damage, and to increase risk reduction strategies, international cooperation to developing countries, and early warning systems to the people. Similarly, the International Council for Science (ICSU) proposed a program entitled Integrated Research on Disaster Risk (IRDR). It was launched to address the challenge of natural and human-induced environmental hazards. Its first objective is devoted to hazard characterization, vulnerability, and risk through development of forecasting capacities.

Moreover, the Science Plan of the IRDR program is related to geophysical, oceanographic and hydrometeorological trigger events, flooding, storms, droughts, climate change, etc. Both the SFDRR and the Science Plan of the IRDR claim the development of studies aimed to forecast hazards (e.g., floods and droughts) and to mitigate their associated impacts. Therefore, runoff forecasting-related studies fit well with the scope of the initiatives previously mentioned.

As countermeasure against floods and droughts, forecasting activities have globally become an emerging field of research for water management and risk analysis (Chang and Hwang 1999). The ultimate goal is to perform an integrated social, economic, and environmental impact assessment to support the decision-making regarding flood control policy (Brouwer and Van Ek 2004). Several different types of models can be used for hydrological forecasting. Traditionally, fully distributed models are enhanced to describe the physical processes of a catchment in a detailed way. Nevertheless, severe data scarcity and the assumptions of the process-based structure applied to specific catchments often limit the model operational value (overparameterization and intense computational requirements) and the predictive capability for extreme flows (Brath et al. 2004; Chang and Hwang 1999; Dawson and Wilby 2001; Galelli and Castelletti 2013; Gupta et al. 2008; Willems 2014). An arising alternative is the use of Machine Learning (ML) techniques for predictive modeling. ML popularity has recently and remarkably increased mainly due to its flexibility and scalability properties in pattern extraction (Bontempi et al. 2012). ML models are characterized by a more compact representation and high predictive potential, with considerably fewer parameters to calibrate when compared to fully distributed models (Galelli and Castelletti 2013; Mosavi et al. 2018).

Moreover, since there is no assumption on a global function describing the data, ML techniques are particular relevant for problems of non-stationarity, missing features, and measurement errors (Bontempi et al. 2012). Several ML methods can be used for flood forecasting: artificial neural networks (ANNs (Kim et al. 2016)), support vector machines (SVMs) (Martens et al. 2007), and decision trees (DTs) (Kubal et al. 2009; Wang et al. 2015) among others. These methods exhibit some weaknesses such as overfitting for ANNs (Jin et al. 2005), the complexity of mathematical functions for SVMs (Martens et al. 2007) and the considerable effort needed for pre-processing data for DTs (Kubal et al. 2009). Generally, there is a global tendency for using the Random Forest (RF) algorithm due to its simplicity, robustness and capacity to cope with complex data structures (Kühnlein et al. 2014).

The objective of this study is to construct flood and drought forecasting models of varying time duration (4, 8, 12 and 24 h). All modes are based on the Random Forest (RF) algorithm. We selected two catchments, the Tomebamba and the Yanuncay catchments, to be representative of the Tropical Andes in Ecuador.

2.2 Study Sites and Dataset

The first Tropical Andean catchment, the Tomebamba catchment, is delineated upstream the Matadero-Sayausí station, in the Tomebamba river. It is located in the south-eastern flank of the Andes, and the outlet is positioned 10 km away from the city of Cuenca. The Tomebamba catchment has a drainage area of 300 km^2, and its elevation ranges between 2800 and 4100 m above the sea level (m asl). The Tomebamba catchment is part of the Cajas National Park, declared by UNESCO as a World Biosphere Reserve in 2013. The second catchment is allocated right next to the first one, in the Yanuncay river at the Yanuncay A.J. Tarqui station. The Yanuncay catchment has an area of 420 km^2, spanning from 2480 to 4280 m asl. The outlet of the catchment is located at the entrance of the city of Cuenca. Both the Tomebamba and Yanuncay catchments drain to the Amazon river toward the Atlantic ocean.

Data comprise rainfall and discharge hourly timeseries for the periods Jan/2015 to Sep/2018 and Jan/2015 to Jul/2017 for the Tomebamba and Yanuncay catchments, respectively. From Jul/2017 on, measurements of precipitation and discharge at the Yanuncay catchment are not reliable, and thus limited the use of a similar data period for both catchments. However, this issue will not compromise the analyses below since we are not focused in comparing specific events but rather overall model performances.

For the Tomebamba catchment, we used the information of 3 rain gauges, Toreadora, Chirimachay, and Virgen, at elevations of 3955, 3626, and 3298 m asl, respectively. Similarly, for the Yanuncay catchment, we obtained measurements of 3 rain gauges, Ventanas, Izhcayrrumi, and Huizhil, at elevations of 3592, 3748, and 2773 m asl, respectively. For both catchments, rain gauges are installed within the catchment and along their altitudinal gradients (Fig. 2.1). For model developing purposes,

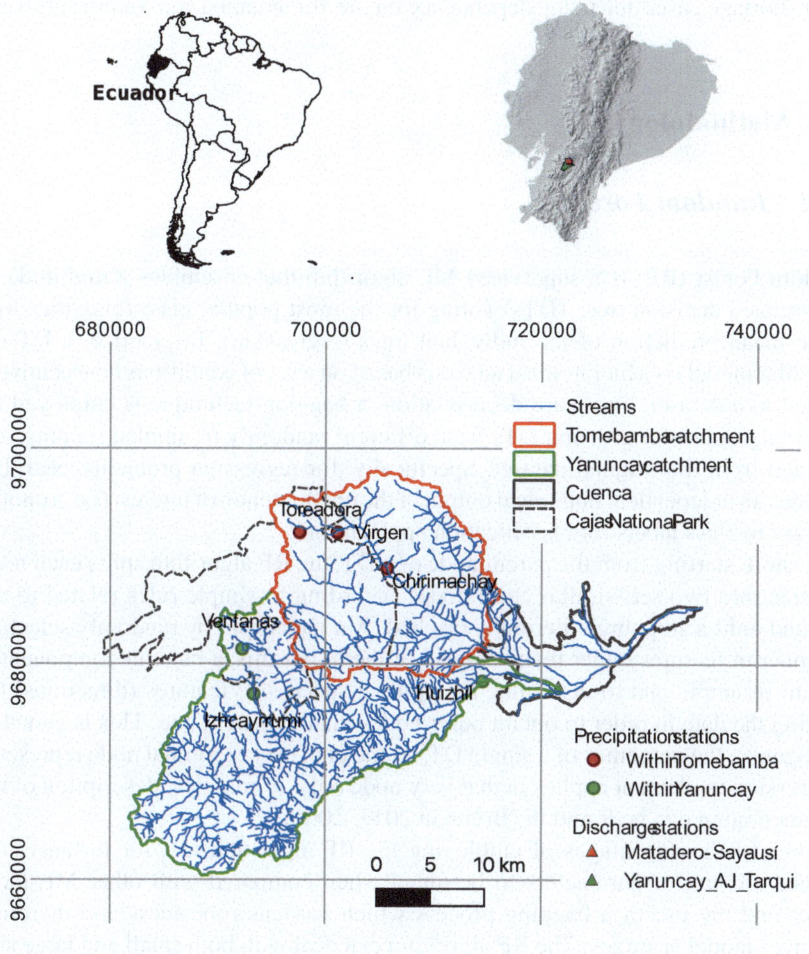

Fig. 2.1 Location of the tomebamba and yanuncay catchments in the andean cordillera of ecuador, South America (UTM coordinates) *Source* Author

we split the length of the data for training and test. For the Tomebamba catchment, training run from Jan/2015 to May/2017 and test from Feb/2017 to Jan/2019. Whereas, for the Yanuncay catchment, training run from Jan/2016 to July/2017 and test from Jan/2015 to Jan/2016 to capture more extreme events in the calibration period.

In both catchments, flash-floods events might occur when consecutive precipitation events saturate the upper parts, and a further event (not necessarily extreme) triggers a rapid response. It might also occur that a very intense event (above 100 mm hour^{-1}) saturates and produces a flash-flood event regardless of antecedent soil saturation conditions in the catchment. On the other hand, hydrological droughts cause water shortage cases due to the dependence on the Tomebamba and Yanuncay rivers.

2.3 Methodology

2.3.1 Random Forest

Random Forest (RF) is a supervised ML algorithm that ensembles a multitude of decorrelated decision trees (DTs) voting for the most popular class (classification) or the mean prediction of the individual trees (regression). In practice, a DT (or particular model) is a hierarchical analysis based on a set of conditions consecutively applied to a dataset. To assure decorrelation, a bagging technique is employed by the RF algorithm for growing DTs from different randomly resampled training sets obtained from the original dataset. Specifically, for regression problems, each DT provides an independent numerical output of the phenomenon of interest (i.e., runoff), contrary to class labels for classification applications.

In short, starting from the parent node of a DT, the RF algorithm splits each node of a tree into two self-similar child nodes according to simple rules related to the data and until a stopping criterion is reached. A node is split by randomly selecting a number of features rather than using all of them. For this, a random component is used to resample and to determine the optimal successive features (directions) for splitting the data in order to obtain purer nodes than its parent one. This is aimed to homogenize the outcomes of a single DT. At the end, every terminal node represents a regression model that applies in that very node only. A complete description of the RF functioning can be found in (Breiman 2017, 2001).

Some of the advantages of employing the RF algorithm are, for instance, the reduced amount of parameters to be tuned when compared with other ML techniques and the use of a bagging process which hardens robustness and therefore improves model accuracy. The RF algorithm can deal with both small and large size samples, high dimensionality, and complex data structures (Biau and Scornet 2016).

Breiman's original RF algorithm has already been implemented in programming languages such as R® and Python®; we selected the Python language through the scikit-learn package for ML (Pedregosa et al. 2011). Details of the package can be found online in https://scikit-learn.org.

2.3.2 *Algorithm*

The RF algorithm for regression can be summarized as follows:

1. Grow every DT based on a random selection of a number of bootstrap samples (n_estimators parameter) drawn with (or without) replacement from the training dataset. To compose each bootstrap, a different subset (roughly two-thirds) of the dataset is selected in a process known as out-of-bag (OOB).
2. Split each parent node into two descendant ones by using the best split criteria. The best split decision is taken by considering only a number of features (max_features parameter) of the total number of predictor variables of the dataset (n_features).
3. Construct n_trees as much as possible (largest extent) by repeating steps (1) and (2). This must be done until reaching a defined number of nodes in the tree. The depth of each tree is controlled by the max_depth and the min_samples_leaf parameters; where min_ samples_ leaf is the minimum number of samples required to be at a leaf node.
4. Determine the outcome of the RF model as the mean response from all DTs.

Some additional indications when using the RF algorithm:

- The OBB technique is intended to achieve unbiased estimates of the regression and to estimate the importance of the features used for the tree construction process (Probst et al. 2018).
- Conditioning max_ features to be lower than n_ features ensures the nonexistence of duplicated DTs in the forest. It aims to mitigate overfitting. For regression problems, Breiman (2001) recommends to set max_ features equal to the root square of n_ features.
- The depth of each tree is controlled to reduce the structural complexity of the trees (models). This is known as pruning criteria (Rodriguez-Galiano et al. 2014).
- Determination of the best splits are chosen based on the mean squared error (MSE) for regression problems. The minimum number of samples required to split a node is controlled by the min_ samples_ split parameter.
- The optimal number of trees is reached when the OOB error stops decreasing significantly (depends on the research objective).

2.3.3 *RF Hyper-Parameterization*

Model hyper-parameters determines the structure of the forest and its level of randomness (Probst et al. 2018). Although the algorithm can be run with default parameters,

Table 2.1 Random Forest most relevant model hyper-parameters

Hyper-parameter	Value
n_estimators*	50–700
max_features	'auto', 'sqrt', and 'log2'
min_samples_split	2, 5, and 10
min_samples_leaf	1, 2, and 4
max_depth*	10–700

*Increment of 10 units

higher accuracies have been obtained by tuning the most relevant hyper-parameters to the algorithm (Table 2.1) on the training dataset (Muñoz et al. 2018). This can be simply done by employing a Randomized Grid Search (RGS) procedure aimed to find the best combination (lower model residual) of hyper-parameters from a previously defined grid of parameter ranges. To avoid overfitting during the RGS process, a K-fold cross-validation scheme must be performed. We selected a three fold cross-validation scheme.

2.3.4 *Runoff Forecasting Model Construction*

Identification of the information required for the model to learn about the system (catchment) plays a key role on model performance. For our case studies, all models are to be merely trained with precipitation and discharge information. Consequently, the "model construction" phase consists on an autoregressive exogenous analysis to determine the necessary number of previous timesteps (lags) of precipitation and discharge that have a major influence when simulating a next step output variable.

Physically, the addition of precipitation lags to the model's input is aimed to mimic antecedent soil moisture conditions in the system. This is required since during dry periods, the soil is below field capacity and therefore, it needs additional rain water to get saturated and to generate streamflow (underestimation in forecasts). The opposite happens during wet periods, where the soil requires less rain water to generate streamflow (overestimation in forecasts) (Willems 2014).

To determine the number of precipitation and discharge lags to be used, we performed a qualitative analysis that relies on the statistical properties of the time series, such as cross, auto- and partial-auto-correlation functions. This approach relies on the linear relationship between the variables, however, the effect of an additional variable cannot be assessed.

The approach was proposed by (Sudheer et al. 2002) and it has been satisfactory tested by Muñoz et al. (2018), Wu and Chau (2011), and Wang et al. (2006). For discharge, the analyses focus on both the auto- and partial-auto-correlation functions with 95% confidence levels, which suggest the influencing antecedent flow patterns in the discharge at a given time. Whereas for precipitation, we selected a number of previous timesteps according to a Pearson cross-correlations applied to the timeseries.

2.3.5 Runoff Forecasting Model Assessment and Feature Reduction

It is commonly agreed that a direct comparison between model outputs and observations through goodness-of-fit statistics is not sufficient for correctly evaluating model performance. When goodness-of-fit statistics are used alone, model assessment is restricted to the mean model performance without taking into account the unbalanced influence of outliers and extreme high (floods) or low (droughts) values. Therefore, to complement model evaluation, we used graphical interpretation techniques to further identify model weaknesses and applicability.

2.3.6 Goodness-Of-Fit Statistics

Among the variety of performance metrics, model residual mean (ME) measures the average systematic difference between simulated and observed values. Whereas, the average random differences can be measured by the mean squared error (MSE) and the model residual variance (S_{EQ}^2). For a enough number of observations, $MSE = ME^2 + S_{EQ}^2$. Therefore, the MSE comprises both a systematic (bias in the model) and a random component (after bias correction) (Willems 2009). The systematic error, which is the objective function, can be minimized through calibration. On the contrary, the random component cannot be reduced since it is related to the inherent stochastic nature of the inputs.

Nevertheless, the major disadvantage of the ME, MSE and S_{EQ}^2 is their high dependence on the magnitude of the variable of interest (e.g., runoff). Thus, we selected the Nash–Sutcliffe efficiency (NSE) coefficient to measure the overall model accuracy. The NSE coefficient is less sensitive to high extreme values when compared to the previously mentioned performance metrics. In fact the NSE coefficient is an scaled version of the MSE; the NSE coefficient is the fraction of variability in the observations explained by the model. It can be calculated as follows:

$$NSE = \left[1 - \frac{\sum_{i=1}^{n} (Q_m(i) - Q_o(i))^2}{\sum_{i=1}^{n} \left(Q_o(i) - \bar{Q}_o\right)^2}\right] = \left[1 - \frac{MSE}{S_{Q_o}^2}\right],$$

where Q_o is the mean observations value. The NSE coefficient outputs values in the range from $-\infty$ to 1.0, being $NSE = 1$ the optimal one. NSE values between 0.0 and 1.0 are generally agreed as acceptable performance (depending on the application). Whereas, negative values indicate that the mean observed value is a better prediction than the simulated value (unacceptable performance) (Moriasi et al. 2007).

Ideally, when producing ML algorithms for extreme discharge forecasting, it is recommended to use target functions specifically designed for extreme flows (e.g., mean peak difference, see (Peleg and Gvirtzman 2010)). However, this demands extensive data for properly training and testing models (enough independent flood and drought events). In our case, less than 4 years of hourly information was not sufficient for using target functions. To overcome this issue, our approach consisted firstly in using the RF algorithm, which is capable to deal with small size samples and has been already tested in a previous flood forecasting study using 2.5 years of hourly information (Muñoz et al. 2018). Secondly, we proposed to train the model for all flows, and to enrich model's input with additional information specifically aimed to improve the prediction of extreme high and low flows.

2.4 Graphical Techniques

We encountered two main issues when performing graphical analyses to the observed and forecasted timeseries. The first one is known as homoscedasticity problem, where the MSE and S_{EQ}^2 increase with higher flows. To overcome this issue, we employed a Box-Cox (BC) transformation to the discharge timeseries, according to the recommendations of Willems (2009).

$$BC(q) = \frac{q^{\lambda} - 1}{\lambda},$$

where q is discharge and the parameter λ can be calibrated graphically until reaching homoscedasticity in the residuals (constant standard deviation). We set $\lambda = 0.25$, following the indications of Willems (2009) for runoff transformation.

The second problem is the serial dependence of flow magnitudes to the timescale selected. The graphical evaluation of flows occurring at all hourly timesteps will imply a higher representation of low flows. Moreover, the serial dependence for extreme high flows (floods) is stronger for shorter timesteps (hourly or smaller than the recession constant of the quickest subflow component). Here, the solution relies on selecting nearly independent observations obtained by splitting the discharge timeseries in events and using one value per event (peak-over-the-threshold approach, see Willems (2009)).

Once both problems have been solved, we complemented model assessment with a graphical inspection of the flow frequency distribution plots for extreme high and low flows (observations vs. forecasts). This is aimed to evaluate the overall model performance for both extreme flow conditions, floods, and droughts. It consists on analyzing the behavior of the distribution curve toward its tail (higher values for high frequency distribution and lower values for low flows frequency distribution). Underestimation or overestimation of the tail toward more extreme values unmask the weaknesses of a model for extreme flow forecasting.

2.4.1 Feature Reduction

In the pursuit of model parsimony, we reduced the input dimension of each model through a process known as feature selection. It consists on the calculation of the relative importance of each feature to the model´s output with the purpose to keep only the most important features (defined criterion) for model construction. Apart from shorten computation times, in some cases, feature selection even improves model accuracy (Tang et al. 2007).

There is a number of techniques for feature selection, based on a variance sensitivity analysis, based on univariate statistical tests, recursive elimination, etc. We employed the variance sensitivity analysis introduced by (Cortez 2010). This technique measures the output's variance produced by a single feature alone. With this approach, there is no consideration of the influence of features interaction. Thus, the impact of each feature can be isolated and attributed to the feature itself. A relevant feature to the model is to produce a higher output variance, therefore, the variance (V_k) and its relative importance (R_k) are as follows:

$$V_k = \frac{\sum_{j=1}^{L} \overline{(\widehat{y}_{t-k}(j) - \widehat{y}_{t-k}(j))}^2}{L-1}$$

$$R_k = \frac{V_k}{\sum_{i=1}^{m} V_i} x100$$

where, $\widehat{y}_{t-k}$ is the model output obtained by holding all m input features at their average values except $\widehat{y}_{t-k}$, which varies according to the sample or time step, along the interval $j \in \{1, \ldots, L\}$.

The criterion to obtain parsimonious models is to retain the most relevant features until reaching, at least, 80% of the total relative importance. The remaining features can be considered unimportant and therefore we trimmed them off from the model's input.

Figure 2.2 summarizes the methodology proposed for the construction and evaluation of flood and drought forecasting models. It is an extension of the methodology proposed by Muñoz et al. (2018) for flood forecasting. We applied this scheme for each lead time and for each catchment, separately. In short, we start by using an autoregressive (discharge) model as the base forecasting model. The input is then enriched with precipitation information to improve extreme flow forecasting. Finally, we reduced the dimension of the input through a feature selection process to retain only the most relevant information. Model hyper-parameterization is performed for each model (input data scenario).

Each model forecasts runoff for a defined timestep, and at the end, sequential forecasts results in a forecasted runoff timeseries. The forecasted and observed timeseries are ultimately used to perform a detailed model assessment specifically focused on floods and droughts.

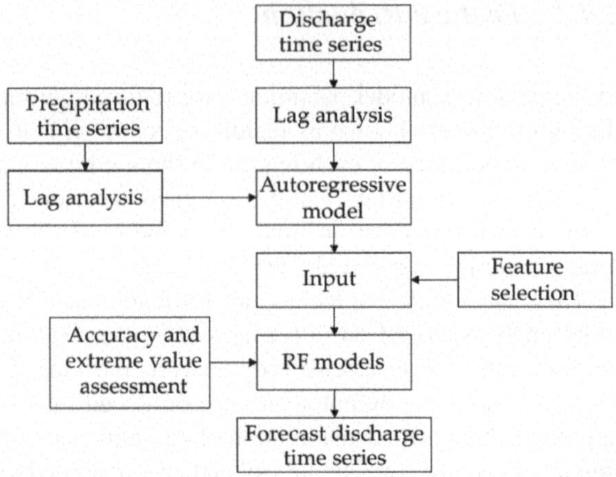

Fig. 2.2 Methodology scheme for parsimonious model development

2.5 Results and Discussions

As this study follows the methodology proposed by Muñoz et al. (2018) for constructing RF precipitation-runoff forecasting models, we firstly present in this section the results of the model construction stage for the case of the 4-h lead time model of the Tomebamba catchment. The procedure is then similar for all forecast horizons and for the Yanuncay catchment. At the end of this section, we present the results of the evaluation of flood and drought forecasts for all lead times and for both catchments. Notice that all RF models, which are specialized in extreme flows, serve to forecast both extreme high (floods) and low (droughts) flows.

For all lead times and both study catchments (cases), we determined the number of precipitation and discharge lags according to specific statistical analyses depending on the case. For instance, for the 4-h forecasting model of the Tomebamba catchment, the following summarizes the process for determining the number of discharge and precipitation required to be able to forecast extreme values.

For discharge, we calculated the autocorrelation function (ACF) and the corresponding 95% confidence interval from lag 1 up to 400 (hours), the highest autocorrelation occurring at the first lag (Fig. 2.3). We found a significant correlation up to lag 300, and thereafter, the correlation fell within the confidence band. The systematic ACF decay revealed a dominant autoregressive process. On the other hand, Fig. 2.4 presents the PACF and its 95% confidence band from lag 1 to 25. We found a significant correlation up to lag 8. The rapid decay of the PACF indicates a dominance of the autoregressive over the moving-average process. All in all, considering the ACF and PACF analyses, it seemed reasonable to include 8 discharge lags (hours) for the case of a 4-h forecasting model for the Tomebamba catchment.

Whereas, for precipitation, Fig. 2.5 plots the Pearson's cross-correlation between each precipitation station and discharge. For all stations, we found a maximum correlation at lag 4 (maximum 0.3323 for Toreadora). Nevertheless, we fixed a correlation

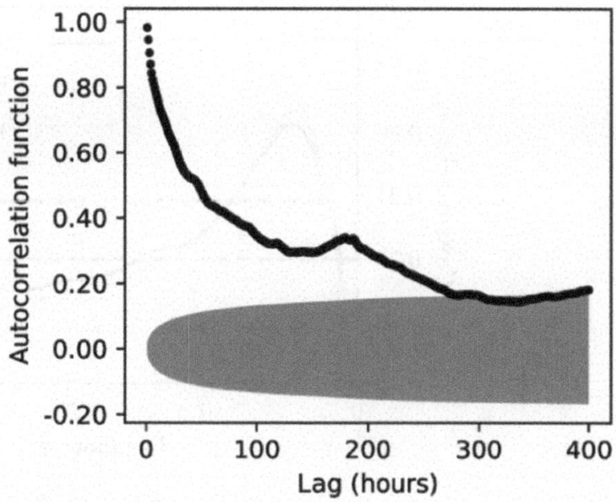

Fig. 2.3 Autocorrelation function (ACF) of the Matadero-Sayausí (Tomebamaba catchment) discharge series. The gray hatch indicates the 95% confidence interval

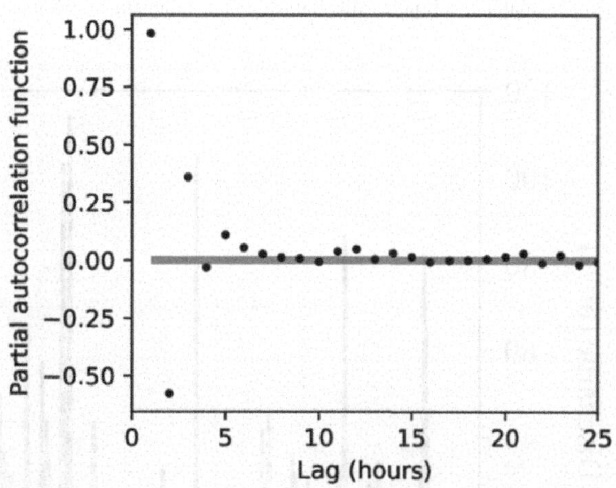

Fig. 2.4 Partial autocorrelation function (PACF) of the Matadero-Sayausí (Tomebamba catchment) discharge series

threshold of 0.20 to determine the number of lags (from each station) to be included as inputs to the 4-h forecasting model of the Tomebamba catchment. With this criterion, we included 24, 10, and 15 lags for Toreadora, Virgen, and Chirimachay stations, respectively.

Figures 2.6 and 2.7 show the 4-h RF forecasting model results for both the training and test periods of the Tomebamba catchment. Overall results reveal that flows up to approximately 50 m^3/s are well represented by the model developed with the aforementioned considerations.

The previous process to select lags provides a starting point for constructing RF forecasting models, however, in the pursuit of model parsimony, we applied a feature selection process based on the outputs' variance. The idea is to only include only the most relevant features, to the model, until achieving 80% of the total relative

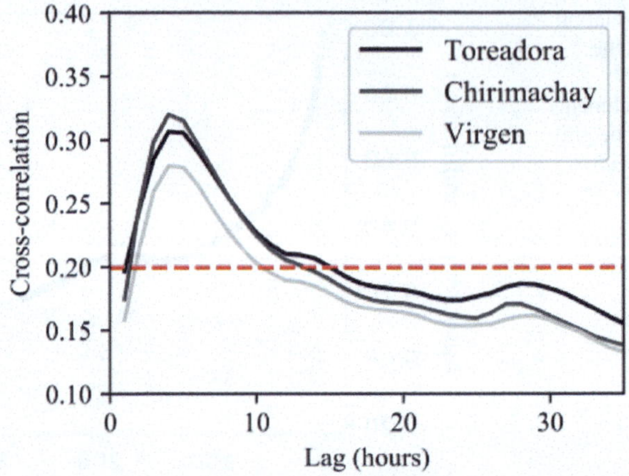

Fig. 2.5 Pearson's cross-correlation comparison between the Toreadora (3955 m.s.n.m.), Virgen (3626 m.s.n.m.), and Chirimachay (3298 m.s.n.m.) precipitation stations and the Matadero-Sayausí (Tomebamaba catchment) discharge series. Note the red horizontal line at a cross-correlation of 0.20

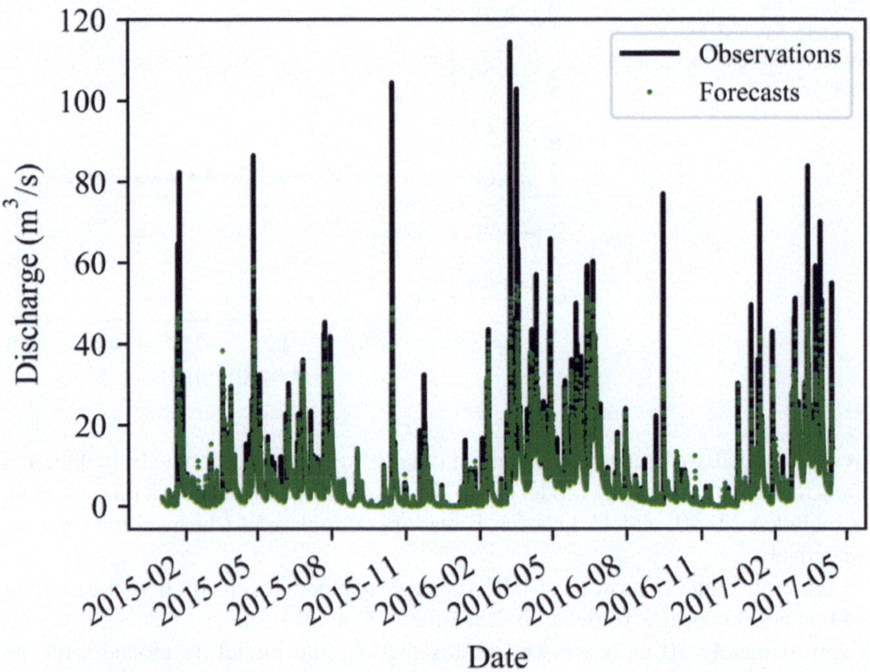

Fig. 2.6 Model results of the parsimonious 4-h discharge forecasting model of the Tomebamba catchment (training period)

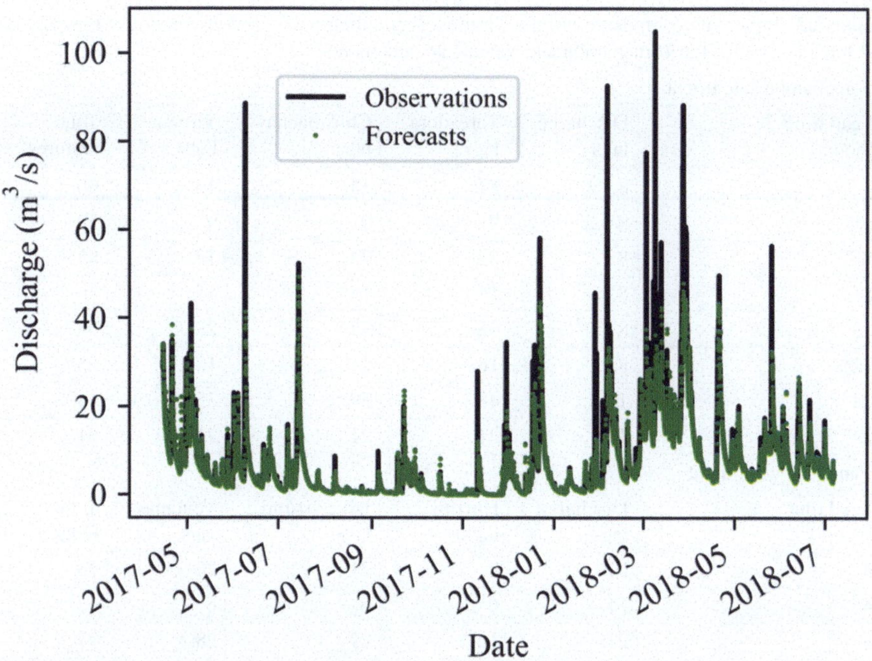

Fig. 2.7 Model results of the parsimonious 4-h discharge forecasting model of the Tomebamba catchment (test period)

importance of the model. For the 4-h forecasting model of the Tomebamba catchment, we found that including 9 lags from each precipitation station and 8 discharge lags would be enough to achieve 80.36% of the total relative importance. The percentage of reduction of model features was 58%. We did the same for each case; the percentage of reduction ranged from 40 to 70%, and from 40 to 60% for all forecasting models of the Tomebamba and Yanuncay catchments, respectively.

Table 2.2 summarizes the input data composition and total number of features utilized for the RF forecasting models for all cases. Whereas, Table 2.3 presents the obtained model performances in terms of the NSE coefficient. Notice that the NSE coefficients were calculated for the whole spectrum of flows. Table 2.3 also contrasts the NSE coefficients of the full-input and their parsimonious model obtained through a feature selection process. Results prove that NSE coefficients obtained from the parsimonious models do not differ significantly from the correspondent full-input version of the model (maximum differences in calibration and validation of 0.01 and 0.02 for the Tomebamba and Yanuncay catchments). In some cases, parsimonious models even outperformed their correspondent full-input models.

Regarding RF model overfitting, we found maximum differences between the NSE coefficients of the calibration versus the validation period of 0.20 and 0.64 for the Tomebamba and Yanuncay catchments, respectively. Significant differences for the Yanuncay RF models can be partly explained by the fact that a shorter calibration

Table 2.2 Input data composition of the RF models and their parsimonious versions (4, 8, 12 and 24-h lead time) for the Tomebamba and Yanuncay catchments

Tomebamba catchment					
Lead time [hours]	Discharge lags	Toreadora lags	Chirimachay lags	Virgen lags	Total features
4	8	24	15	10	60
4[a]	8	9	9	9	38
8	8	32	23	19	85
8[a]	8	15	15	15	56
12	8	36	27	23	98
12[a]	8	18	18	18	65
24	15	48	39	35	140
24[a]	15	21	21	21	81
Yanuncay catchment					
Lead time [hours]	Discharge lags	Huizhil lags	Izhcayrrumi lags	Ventanas lags	Total features
4	21	8	21	20	73
4[a]	21	6	6	6	42
8	21	16	29	28	97
8[a]	21	10	10	10	54
12	21	20	33	32	109
12[a]	21	12	12	12	60
24	21	32	45	44	145
24[a]	21	13	13	13	63

[a]Parsimonious version

Table 2.3 Model performance of the RF models and their parsimonious versions (4, 8, 12 and 24-h lead time) for the Tomebamba and Yanuncay catchments

Tomebamba				Yanuncay		
Lead time [hours]	Total features [#]	NSE		Total features [#]	NSE	
		Training	Test		Training	Test
4	60	0.9193	0.8604	73	0.9014	0.7572
4[a]	38	0.9211	0.8682	42	0.898	0.7675
8	85	0.8486	0.7494	97	0.8558	0.6260
8[a]	56	0.8441	0.7523	54	0.8572	0.6425
12	98	0.8131	0.6759	109	0.8131	0.4122
12[a]	65	0.8074	0.6799	60	0.8161	0.4049
24	140	0.7541	0.538	145	0.7577	0.1159
24[a]	81	0.7483	0.5454	63	0.7626	0.1189

[a]Parsimonious version

period (≈ 1.5 years) was available when compared to ≈ 2.5 years in the case of the Tomebamba catchment.

At first, we calibrated each RF model (specific catchment and lead time) with the RGS approach. However, another experiment consisted in calibrating hyper-parameters only for the RF models of the Tomebamba catchment (4, 8, 12 and 24 h), and simply transferring the optimal set of hyper-parameters to the correspondent RF models of the Yanuncay catchment. For this, we found a maximum difference in the NSE coefficient of 0.10. Therefore, we conclude that for comparable catchments, in terms of altitude, topography, and climate, the RF hyper-parameters are not sensitive and what determines model performance is the quantity, quality, and variety (discharge driving forces) of data available.

2.5.1 Evaluation of Flood Forecasts

Figures 2.8 and 2.9 show the empirical extreme high value distributions of all forecast horizons models (4, 8, 12 and 24 h) for both observation and simulations, and for the Tomebamba and Yanuncay catchments, respectively. For this, we employed the simulations obtained from the so-called parsimonious models. Overall results for both catchments, revealed that the underestimation of peak flows toward the upper tail of the distribution becomes stronger as the lead time increases. For the Tomebamba catchment, we found maximum underestimations of 48, 53, 57, and 66% for the 4 8, 12, and 24-h forecasting models, respectively. Whereas for the Yanuncay catchment,

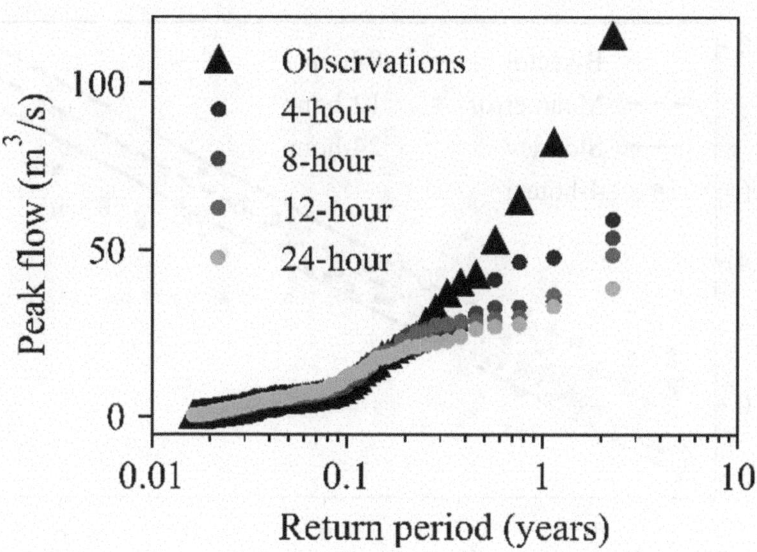

Fig. 2.8 Empirical extreme value distribution of peak flows (floods) for the Tomebamaba catchment

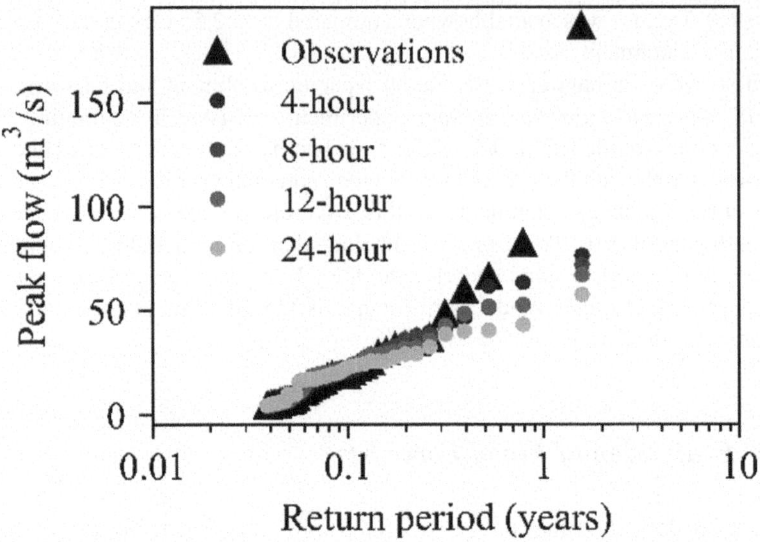

Fig. 2.9 Empirical extreme value distribution of peak flows (floods) for the Yanuncay catchment

the maximum underestimations obtained were 59, 61, 64, and 69% for the 4 8, 12, and 24-h forecasting models, respectively.

On the other hand, Figs. 2.10 and 2.11 present a scatter plot of forecasts (vertical axis) and observed discharge (horizontal axis) for extreme high flows. Here, model

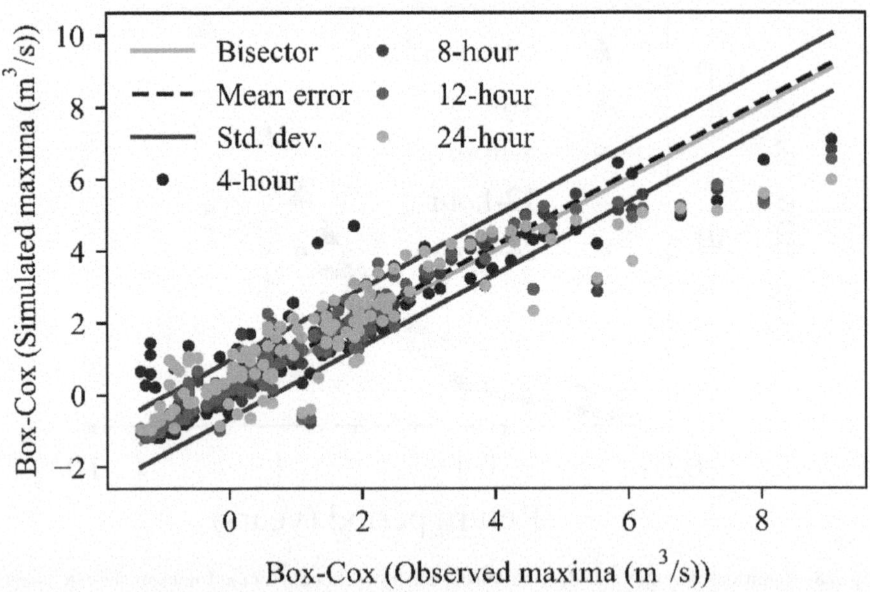

Fig. 2.10 Comparison of nearly independent peak flow maxima for the Tomebamba catchment

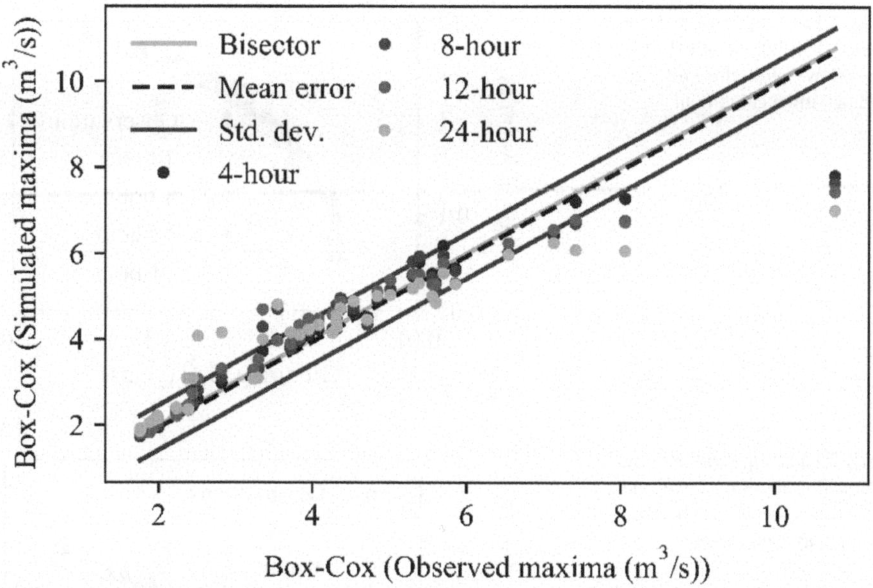

Fig. 2.11 Comparison of nearly independent peak flow maxima for the Yanuncay catchment

residuals are represented by the horizontal and vertical differences between each point and the bisector line. The dependence of the standard deviation on the flow magnitude was disrupted (constant standard deviation) with a λ-value of 0.25. Results confirm the observed in Figs. 2.8 and 2.9, where higher scatters and biases were found for longer forecast horizons.

The gross simplification of precipitation-runoff forecasting models (they do not include crucial discharge driving-forces) explain the difficulty to accurately forecast extreme high flows. In addition, for paramo ecosystems, it is well-known that soils govern flow processes, and therefore, lack of direct measurements limits the forecasting of extreme flows. Apart from missing data, extreme spatial and temporal variability of precipitation in mountainous regions, especially in the Andes of Ecuador, is hardly collected by a few rain gauge stations within the catchment. Precipitation events, might occur in regions not covered by rain gauges, and thus, the necessity to use spatial rainfall estimations (remote sensing imagery) arises.

2.5.2 *Evaluation of Drought Forecasts*

Contrary to the strong differences found between forecasts and observations for peak flows (Figs. 2.8 and 2.9), overall results for extreme low flows, for all forecast horizons and for both catchments, show good fits varying between underestimations and overestimations (see Figs. 2.12 and 2.13). For the Tomebamba catchment, we

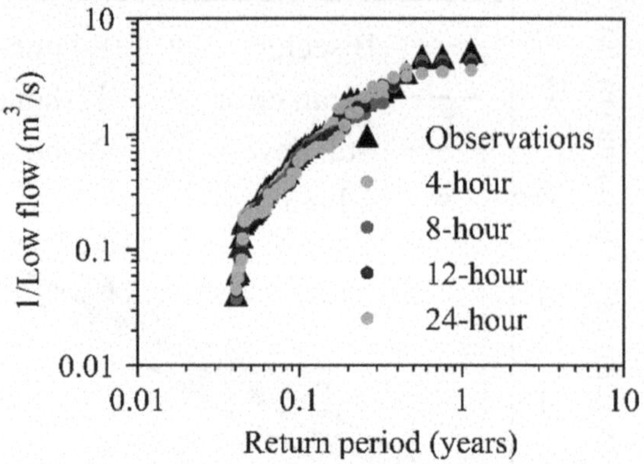

Fig. 2.12 Empirical extreme value distribution of extreme low flows (droughts) for the Tomebamba catchment

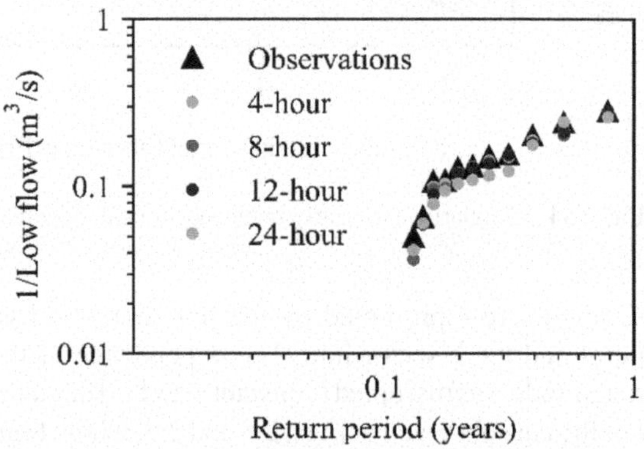

Fig. 2.13 Empirical extreme value distribution of extreme low flows (drouhgts) for the Yanuncay catchment

found maximum differences of 31, 37, 36, and 36% for the 4, 8, 12, and 24-h forecasting models, respectively. Whereas for the Yanuncay catchment, the maximum differences obtained were 16, 26, 16, and 16% for the 4, 8, 12, and 24-h forecasting models, respectively.

Similar to the extreme value findings, for extreme low flows, the dependence of the standard deviation on the flow magnitude was disrupted with a λ-value of 0.25. On the contrary, Figs. 2.14 and 2.15 demonstrate that scatters and biases are more or less constant regardless of the forecast horizon.

2.6 Conclusions

The present study commits with the efforts and claims of the SFDRR and the Science Plan of the IRDR program to develop capacities in forecasting hazards such as floods and droughts. The ultimate goal is to prevent and mitigate their impacts considering the susceptibility of lowland areas to catastrophic socio-economic impacts.

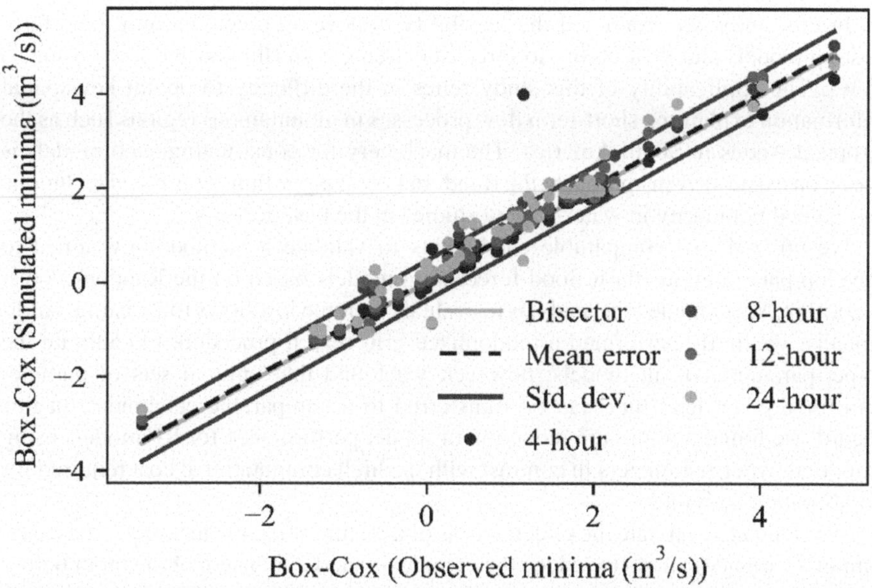

Fig. 2.14 Comparison of nearly independent extreme low flow minima for the Tomebamba catchment

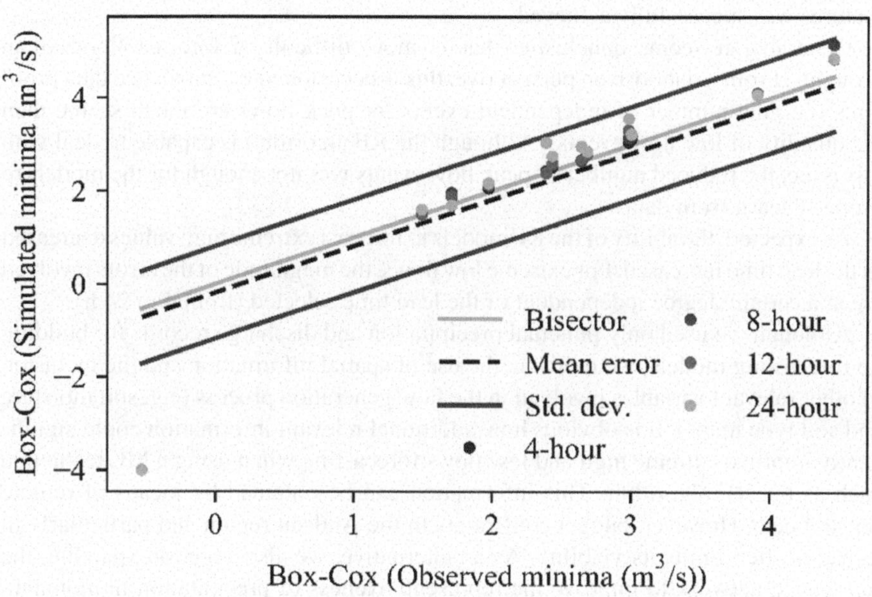

Fig. 2.15 Comparison of nearly independent extreme low flow minima for the Yanuncay catchment

In this study, we examined the feasibility to develop precipitation-runoff forecasting models and their ability to forecast extreme high (floods) and low (drought) flows. The applicability of this study relies in the difficulty to obtain key-spatial information explaining short-term flow processes in mountainous regions such as the Tropical Andes in South America. The machinery for constructing such models is based on a data-driven approach, the Random Forest algorithm. This novel technique has gained popularity in water-related studies in the past decades.

We utilized two comparable catchments to validate a methodology aimed to develop parsimonious flash-flood forecasting models based on the Random Forest algorithm. We extended the analysis to evaluate extreme low flows forecasting. Additionally, we firstly performed a randomized grid search procedure to calibrate the hyper-parameter of all models; however, we found that optimal sets of parameters for a given lead time can be transferred to a comparable catchment. In this regard, we found a minimum reduction of model performance for RF models using imported hyper-parameters in contrast with the high computational cost required for recalibration activities.

We could also validate the effectiveness of a feature selection technique (based on output's variance) aimed to reduce the complexity and dimension of the input before training a model. Retaining only the features accounting for 80% of the model's variance did not compromise forecasts but rather optimized computation times. Only slight differences in NSE coefficients proved that the selection of the most important features was successfully achieved.

Generally, it seems conclusive that is more difficult to forecast floods than droughts. From a data-driven perspective, this is occasioned by imbalance data problems (i.e., the number of independent events for peak flows are much scarce than the quantity of low flow events). Although the RF algorithm is capable to deal with this issue, the reduced number of peak flow events was not enough for the models to properly learn from data.

As expected, the ability of the RF models to forecast extreme high values decreased as the lead time increased. For extreme low flows, the magnitude of the errors involved are at a certain degree independent of the lead time selected (from 4 to 24 h).

Although we used only punctual precipitation and discharge records for building up forecasting models, we motivate the use of spatial information and the inclusion of other relevant variables involved in the flow generation process (e.g. soil moisture and soil type maps). It is obvious how additional relevant information could significantly improve extreme high and low flows forecasting when using a ML technique such as the RF algorithm. This information can be obtained by means of remote sensing data. However, budget constraints in the Andean region and particularly in Ecuador often limits its viability. As an alternative, we also suggest expanding the rain gauge network to improve the representativeness of precipitation in mountain catchments. This will improve, at a certain degree, model performances. However, it must be taken into account that the major shortcoming of the use of rain gauges in the Andean region is the occurrence of local rains due to complex topography. Thus, an adequate representation of the spatial variability of precipitation is rarely available for forecasting applications.

For the city of Cuenca, the current flood early warning system consists simply on a real-time monitoring of control points located upstream zones of interest (urban areas). When the flow at a control point exceeds a certain threshold, authorities activate warning alarms for prevention purposes. The main disadvantage is that the anticipation of an extreme event is limited by the transit time between the control station and the zone of interest. The methodology proposed in this study is therefore a further step for dealing with flood and hydrological drought events. We proposed forecasting extreme flows with lead times up to 24 h.

Although further exploration of the RF technique is still required for improving model performances, the models and the methodology followed by this study can be immediately used and the results interpreted by decision-makers and politicians. The next logic step is building up a platform for intelligently communicating results with the people, bearing in mind that they are not necessarily familiar to computer and engineering sciences.

References

Biau G, Scornet E (2016) A random forest guided tour. Test 25:197–227

Bontempi G., Taieb SB, Le Borgne Y-A (2012) Machine learning strategies for time series forecasting. EBISS 62–77

Brath A, Montanari A, Toth E (2004) Analysis of the effects of different scenarios of historical data availability on the calibration of a spatially-distributed hydrological model. J Hydrol 291:232–253. https://doi.org/10.1016/j.jhydrol.2003.12.044

Braud I, Ayral P-A, Bouvier C, Branger F, Delrieu G, Dramais G, Le J, Leblois E, Nord G, Vandervaere J.P (2016) Advances in flash floods understanding and modelling derived from the Flood-Scale project in South-East France. FLOODrisk 2016—3rd Eur. Conf. Flood Risk Manag. https://doi.org/10.1051/e3sconf/20160704005

Breiman L (2017). Classification and regression trees. Routledge.

Breiman L (2001) Random forests. Mach Learn 45:5–32. https://doi.org/10.1023/A:1010933404324

Brouwer R, Van Ek R (2004) Integrated ecological, economic and social impact assessment of alternative flood control policies in the Netherlands. Ecol Econ 50:1–21. https://doi.org/10.1016/j.ecolecon.2004.01.020

Buytaert W, Célleri R, De Bièvre B, Cisneros F, Wyseure G, Deckers J, Hofstede R (2006) Human impact on the hydrology of the Andean páramos. Earth-Science Rev 79:53–72. https://doi.org/10.1016/j.earscirev.2006.06.002

Buytaert W, Cuesta-Camacho F, Tobón C (2011) Potential impacts of climate change on the environmental services of humid tropical alpine regions. Glob Ecol Biogeogr 20:19–33. https://doi.org/10.1111/j.1466-8238.2010.00585.x

Chang FJ, Hwang YY (1999) A self-organization algorithm for real-time flood forecast. Hydrol Process 13:123–138. https://doi.org/10.1002/(SICI)1099-1085(19990215)13:2%3c123::AID-HYP701%3e3.0.CO;2-2

Cortez P (2010). Sensitivity analysis for time lag selection to forecast seasonal time series using neural networks and support vector machines. Int. Jt. Conf. Neural Netw. (IJCNN) 2010: 1–8. https://doi.org/10.1109/IJCNN.2010.5596890

Dawson CW, Wilby RL (2001) Hydrological modelling using artificial neural networks. Prog Phys Geogr 25:80–108

Fernández de Córdova Webster C, Javier Rodríguez López Y (2016) Primeros resultados de la red actual de monitoreohidrometeorológico de Cuenca, Ecuador. Ing. Hidráulica y Ambient 37: 44–56
Galelli S, Castelletti A (2013) Assessing the predictive capability of randomized tree-based ensembles in streamflow modelling. Hydrol Earth Syst Sci 17:2669–2684
Gupta HV, Wagener T, Liu Y, (2008). Reconciling theory with observations: elements of a diagnostic approach to model evaluation. Hydrol Process https://doi.org/10.1002/hyp.6989
Jin L, Kuang X, Huang H, Qin Z, Wang Y (2005) Study on the overfitting of the artificial neural network forecasting model. Acta Meteorol Sin 19:216–225
Kim S, Matsumi Y, Pan S, Mase H (2016) A real-time forecast model using artificial neural network for after-runner storm surges on the Tottori coast. Jpn Ocean Eng 122:44–53. https://doi.org/10.1016/j.oceaneng.2016.06.017
Kubal C, Haase D, Meyer V, Scheuer S (2009) Integrated urban flood risk assessment-adapting a multicriteria approach to a city. Nat Hazards Earth Syst Sci 9:1881
Kühnlein M, Appelhans T, Thies B, Nauss T (2014) Improving the accuracy of rainfall rates from optical satellite sensors with machine learning—A random forests-based approach applied to MSG SEVIRI. Remote Sens Environ 141:129–143. https://doi.org/10.1016/j.rse.2013.10.026
Martens D, De Backer M, Haesen R, Vanthienen J, Snoeck M, Baesens B (2007) Classification with ant colony optimization. IEEE Trans Evol Comput 11:651–665
Min SK, Zhang X, Zwiers FW, Hegerl GC (2011) Human contribution to more-intense precipitation extremes. Nature 470:378–381. https://doi.org/10.1038/nature09763
Moriasi DN, Arnold JG, Van Liew MW, Bingner RL, Harmel RD, Veith TL (2007) Model evaluation guidelines for systematic quantification of accuracy in watershed simulations. Trans ASABE 50:885–900
Mosavi A, Ozturk P, Chau KW (2018) Flood prediction using machine learning models: literature review. Water (Switzerland) 10:1–40. https://doi.org/10.3390/w10111536
Muñoz P, Orellana-Alvear J, Willems P, Célleri R (2018). Flash-flood forecasting in an andean mountain catchment-development of a step-wise methodology based on the random forest algorithm. Water (Switzerland) 10. https://doi.org/10.3390/w10111519
Pedregosa F, Varoquaux G, Gramfort A, Michel V, Thirion B, Grisel O, Blondel M, Prettenhofer P, Weiss R, Dubourg V, Vanderplas J, Passos A, Cournapeau D, Brucher M, Perrot M, Duchesnay E (2011) Scikit-learn: machine learning in {P}ython. J Mach Learn Res 12:2825–2830
Peleg N, Gvirtzman H (2010) Groundwater flow modeling of two-levels perched karstic leaking aquifers as a tool for estimating recharge and hydraulic parameters. J Hydrol 388:13–27
Probst P, Wright M, Boulesteix A-L (2018). Hyperparameters and tuning strategies for random forest. 1–18
Rodriguez-Galiano V, Mendes MP, Garcia-Soldado MJ, Chica-Olmo M, Ribeiro L (2014) Predictive modeling of groundwater nitrate pollution using random forest and multisource variables related to intrinsic and specific vulnerability: a case study in an agricultural setting (Southern Spain). Sci Total Environ 476:189–206
Ruin I, Creutin JD, Anquetin S, Lutoff C (2008) Human exposure to flash floods—relation between flood parameters and human vulnerability during a storm of September 2002 in Southern France. J Hydrol 361:199–213. https://doi.org/10.1016/j.jhydrol.2008.07.044
Sofia G, Roder G, Dalla Fontana G, Tarolli P (2017) Flood dynamics in urbanised landscapes: 100 years of climate and humans' interaction. Sci Rep 7:1–12. https://doi.org/10.1038/srep40527
Sudheer KP, Gosain AK, Ramasastri KS (2002) A data-driven algorithm for constructing artificial neural network rainfall-runoff models. Hydrol Process 16:1325–1330. https://doi.org/10.1002/hyp.554
Tang Y, Reed P, Van Werkhoven K, Wagener T (2007) Advancing the identification and evaluation of distributed rainfall & hyphen runoff models using global sensitivity analysis. 43: 1–14. https://doi.org/10.1029/2006WR005813
United Nations Office for Disaster Risk Reduction (UNISDR) (2015) Sendai framework for disaster risk reduction 2015–2030. United Nations Off. Disaster Risk Reduct. 32.

Vos R, Velasco M, Labastida E (1999) Economic and social effects of "El Nino" in Ecuador, 1997–8. ISS Work Pap Ser Ser 292:1–55

Wang W, Gelder PHAJMV, Vrijling JK, Ma J (2006) Forecasting daily streamflow using hybrid ANN models. J Hydrol 324:383–399. https://doi.org/10.1016/j.jhydrol.2005.09.032

Wang Z, Lai C, Chen X, Yang B, Zhao S, Bai X (2015) Flood hazard risk assessment model based on random forest. J Hydrol 527:1130–1141. https://doi.org/10.1016/j.jhydrol.2015.06.008

Willems P (2014) Parsimonious rainfall-runoff model construction supported by time series processing and validation of hydrological extremes–Part 1: Step-wise model-structure identification and calibration approach. J Hydrol 510:578–590. https://doi.org/10.1016/j.jhydrol.2014.01.017

Willems P (2009) A time series tool to support the multi-criteria performance evaluation of rainfall-runoff models. Environ Model Softw 24:311–321

Wu CL, Chau KW (2011) Rainfall-runoff modeling using artificial neural network coupled with singular spectrum analysis. J Hydrol 399:394–409. https://doi.org/10.1016/j.jhydrol.2011.01.017

Chapter 3
Increasing Trends in Tropical Cyclone Induced Surge Impacts Over North Indian Ocean

Md. Abdus Sattar and Kevin K. W. Cheung

Abstract Tropical cyclone (TC) is a well-known natural disaster that can devastate much of a society, environment, economy and result in people's deaths. The North Indian Ocean (NIO) is one ocean basin that is very prone to TC. TCs often cause huge human casualties in densely populated communities like Bangladesh, India and Myanmar around the Bay of Bengal (BoB) region of the NIO. Therefore, it is urgent to analyse the impacts of TC-induced storm surge activity over NIO under different climatic scenarios for better preparation. This study has been conducted aiming to analyse potential impacts of TCs from storm surges both for the past (1990–2010) and future (2075–2099). To fulfil research objective, future TCs scenarios were obtained from the U.S. Geophysical Fluid Dynamics Laboratory climate model under the IPCC RCP warming scenarios. Then, JMA-MRI storm surge model for NIO were used for estimating TCs induced surge heights. While almost all intense TCs had historically made landfall around the BoB region, climate model projection simulated more intense TCs over the AS, which indicates larger and more destructive TCs especially around the coastal areas of the AS. It is also found the probability of TC track shift from the eastern part to the western part of both basins at the end of this century. The storm surge model estimates twice the size of the surge in the future and especially over the AS. Furthermore, spatial variation of TCs activities and their associated surge heights are found which largely depends on TC intensity and topography. In conclusion, it is expected that higher maximum surge level over the NIO especially over the AS basin will occur if the projected TC activity is robust. This study provides crucial information that can be used for short- and long-term disaster preparation.

Keywords Tropical cyclone · Storm surge · Climate projection · Arabian Sea · Bay of Bengal

---

Md. A. Sattar (✉) · K. K. W. Cheung
Department of Environmental Sciences, Macquarie University, Sydney, Australia
e-mail: abdus.sattar@pstu.ac.bd

Md. A. Sattar
Department of Disaster Risk Management, Patuakhali Science and Technology University, Dumki, Patuakhali, Bangladesh

R. Djalante et al. (eds.), *Integrated Research on Disaster Risks*, Disaster Risk Reduction,
https://doi.org/10.1007/978-3-030-55563-4_3

3.1 Introduction

Tropical cyclone (TC) is a well-known natural disaster that can devastate much of a society, environment, economy and result in people's deaths. The North Indian Ocean (NIO) is one ocean basin that is very prone to TC. TCs often cause huge human casualties in densely populated communities like Bangladesh, India and Myanmar around the Bay of Bengal (BoB) region of the NIO. Most of the population resides in low-lying coastal areas and most settlement areas are formed by riverine sedimentation. These areas suddenly and excessively inundate due to TC landfall along the Bangladeshi coast of the BoB, which causes massive flooding and immense damage and losses in material goods and people's lives. Jisan et al. (2018) applied the hydrodynamic model Delft3D in their study and reported that if TC activity over the Bangladeshi coast remains the same as what is currently happening, there will still be an increase in surge height and inundation areas due to sea-level rise (SLR). In an another study, Woodruff et al. (2013) reported that for the densely populated coastal areas, global impacts from flooding due to TC activity may even be larger than that from SLR. The issue is complicated by the fact that TC impacts are not only concentrated within inundated areas but can extend beyond those areas that vary with wind speed.

It is known that several factors are responsible for coastal inundation that could make the coast more vulnerable (Gayathri et al. 2017). For example, sea-level rise (SLR), coastal erosion, changes of geomorphology and man-made disturbances will cause permanent flooding, while TC, tsunami and tide strikes will trigger short-term flooding. When a TC approaches the coast, it brings an enormous amount of water with it and subsequently increases the sea level. In addition, landfall of TC brings excessive rainfall together with windy to wild weather conditions. Numerical models play a vital role in predicting storm surges and their associated flooding outcomes. Storm surge modelling research began in the 1970s for the BoB basin since most TCs formed there. Das (1972) took the first attempt to develop a numerical storm surge model for the BoB region, especially the east coast of India and Bangladesh. In general, this model simulated surge height as being too high compared with observed values by the tide gauges. Since then, several initiatives have been taken by other modellers. The study by Gayathri et al. (2017) argued that although remarkable achievements have been reported in terms of storm surge and inundation forecasting, it is still critical to investigate the coastal risk associated with TC landfall because SLR is expected to increase in the future.

Past studies mainly focused on predicting peak surge height. For instance, the SPLASH (Special Programme to List Amplitudes of Surges from Hurricanes) model (Jelesnianski 1972), the FVCOM (Finite-volume coastal ocean) model (Chen et al. 2006), the SLOSH (NOAA) and ADCIRC (Luettich et al. 1992) were developed for storm surge forecasting. Recently, several attempts have been made to address coastal area inundation caused by TC and a few other studies developed tools and models for the assessment of coastal inundation and their associated impacts worldwide (Cheung et al. 2003; Graeme and Kathleen 1999; Lian et al. 2004).

Both vulnerability and model performance vary from one region to another. Thus, location-specific validation of the storm surge inundation model is very important. In line with this view, Dube et al. (2000) argued for location-specific storm surge inundation model development and they devised models for the Andhra and Orissa coasts of India. This study found a reasonable agreement between model simulated inland flooding and what was observed. An integrated forecast system has been developed by Madsen and Jakobsen (2004) for Bangladeshi coast by using the MIKE 21 together with other models' components. This model served to analyse the storm surge height and inland flooding caused by Super Cyclone in April 1991. In addition, two more TC scenarios were analysed, and the authors found a reasonable agreement of model results with the observed surge height.

Changes in the track characteristics of TCs, especially in some ocean basins, have been observed. For instance, the TC tracks have been identified to move more poleward (Kossin et al. 2014) and their translation speeds have slowed down (Kossin 2018). Until recently, there have been very few studies that analysed the impacts from TC-induced storm surge under past and future climate change scenarios exclusively for NIO region. Therefore, it is vitally important to analyse the impacts of both past and future TC activities over NIO to reduce the carnage. For the projected TC tracks and intensities, uncertainties always exist and thus it is essential to examine the model bias for the climatology period before evaluating the future storm surge height and their associated impacts. Current study has been conducted aiming to: (a) examine both simulated and observed TC climatology for the periods 1990–2010 and 2075–2099; and (b) analyse TC- induced storm surge height for the same period. This study found crucial information for disaster managers regarding disaster preparedness and responses.

3.2 Data and Methodology

3.2.1 Data Sources

3.2.1.1 Observed Data

TC information for running the Japan Meteorological Agency-Meteorological Research Institute (JMA-MRI) storm surge model was obtained from the best track data of the India Meteorological Department (IMD) because the IMD best tracks include the mean-sea-level pressure, which is necessary for inclusion in the storm surge model. Radius of maximum wind and that for 30-kt wind are not available for all storms. Recently, the U.S. Navy Joint Typhoon Warning Center (JTWC) reported this data for some TC cases over the NIO. When the radius of wind is not available from both operational centres, estimate of TC size is done based on TC reports and satellite images. This information is part of the TC track input to the storm surge model.

3.2.1.2 Simulated Data

In this study, future projection of TC genesis and track in the NIO are considered. These projection data were provided by Murakami et al. (2017), in which the authors simulated both the past (1970–2010) and future (2075–2099) period. In an earlier study, Murakami et al. (2013) generated TC formation scenarios under the International Panel on Climate Change (IPCC) A1B scenario and fifteen ensemble experiments were performed using the JMA-MRI general circulation model (GCM). It was reported that the ensemble mean simulated reasonable TC frequency for the NIO basin in comparison to the observed TC frequency. Furthermore, this study projected a significant increase of TC frequency (46%) for AS, but a substantial decrease of TC frequency (31%) for the BoB basin. Murakami et al. (2017) came to a similar conclusion based on the US Geophysical Fluid Dynamics Laboratory model.

3.3 The JMA-MRI Storm Surge Model

The JMA-MRI storm surge model has been applied for estimating surge height against each TC landfall. This storm surge model is a two-dimensional ocean model and vertically integrated, which can be run by using either observed TC data or numerical weather prediction (NWP) model data. Two main equations governing the model are the momentum flux equation and water continuity equation, which have been illustrated in Eqs. 3.1 and 3.2, respectively.

$$\begin{cases} \frac{\partial Du}{\partial t} + \frac{\partial Du^2}{\partial x} + \frac{\partial Duv}{\partial y} = -\frac{1}{\rho_w g} D \frac{\delta(\varsigma - \varsigma_0)}{\delta x} - \frac{1}{\rho_w}(\tau_{ax} - \tau_{bx}) + fDv \\ \frac{\partial Dv}{\partial t} + \frac{\partial Duv}{\partial x} + \frac{\partial Dv^2}{\partial y} = -\frac{1}{\rho_w g} D \frac{\partial(\varsigma - \varsigma_0)}{\partial y} - \frac{1}{\rho_w}(\tau_{ay} - \tau_{by}) - fDu \end{cases} \quad (3.1)$$

$$\frac{\partial \varsigma}{\partial t} + \frac{\partial Du}{\partial x} + \frac{\partial Dv}{\partial y} = 0 \quad (3.2)$$

where,

(x, y) = horizontal direction
U = (u, v) current components
ς = height deviation
ς_0 = balance level with surface pressure
ρ_w = sea water density
f = Coriolis parameter
g = gravitational acceleration
D = the local water depth
$\tau_a = (\tau_{ax} - \tau_{ay})$ = the surface stress by winds and
$\tau_b = (\tau_{bx} - \tau_{by})$ = the bottom stress by winds.

After successfully installing the JMA-MRI storm surge model, past and future TCs are simulated. The model requires seven inputs data which are (a) date: time (month, day, hour; mmddhh); (b) Pcenter: central pressure (hPa); (c) lon, lat: point of TC center (in degree); (d) Rref: radius of referenced pressure contour (in degree); (e) R0: radius of maximum wind (in km); (f) C_f: coefficients $c1$ (=0.70 in usual case); and (g) P∞: environmental pressure (hPa). The Rref usually set as 1000 hPa or it can set from R0.

3.4 Model Validation

The performance of the JMA storm surge model was reported in several studies (Choudhury 2014; Tsuboki et al. 2015; Mio and Tetsuo 2015). Currently, the Bangladesh Meteorological Department (BMD) has adopted this model into the operational forecasting systems and reported reasonable results for the demonstration areas (Kohno et al. 2018). One of the salient features of the model is that it considers astronomical tide during surge simulation. To examine model performance for the entire NIO, we compare the reported TC-induced maximum surge heights (m) with modelled values. Reported surge levels were available only for 16 TCs out of 28 from different sources (Bangladesh Meteorological Department; Dube et al. 2009; Needham et al. 2013; Pakistan Meteorological Department 2010; Roy et al. 1999; Unisys 1999, 2007, 2008).

The results show that the modelled surge levels are in reasonable agreement with reported surge levels (Fig. 3.1). The correlation coefficient between reported and model estimated surge level is 0.69, which is highly significant ($p < 0.05$ at 99% confidence level). This correlation value clearly indicates that the model performs well. The model underestimates surge levels for all cases except one TC. Both the model error and the quality of the reported data cause deviations between modelled and reported surge levels. Various factors, for instance, radius of maximum wind, wind speed, storm track, storm's central pressure, landfall location, coastal elevation and morphology of the coast largely affect the surge level. Furthermore, we only estimated maximum surge levels within 48 h of TC landfall that may also contribute to the lower surge levels.

In conclusion, the JMA-MRI storm surge model can be applied for simulating surge levels for the NIO as the simulated values are comparable with the reported values. This conclusion is based on our model validation and on the other validations reported in other studies.

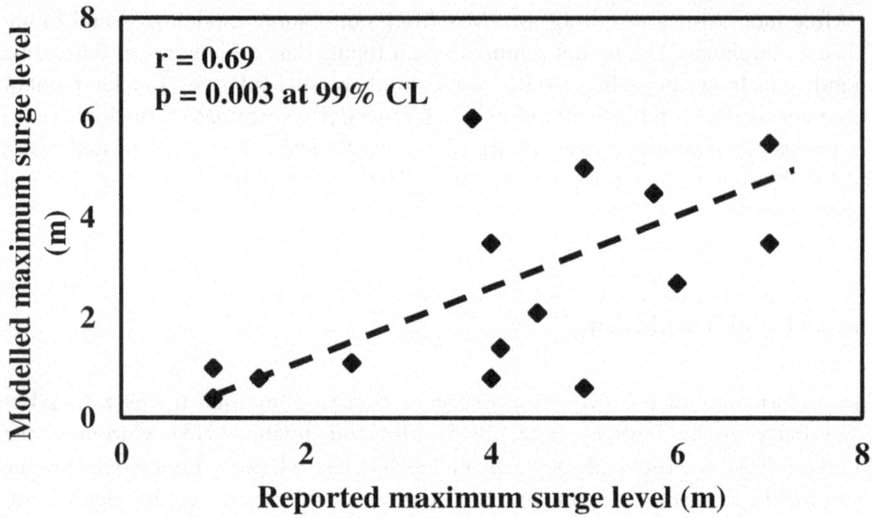

Fig. 3.1 Modelled versus reported tropical cyclone induced maximum surge height (m) over the North Indian Ocean during 1990–2010. The dashed line indicates trendline and r represents for correlation coefficient

3.5 Results

3.5.1 Past and Future Trends of TC Genesis and Tracks

Murakami et al. (2013), who utilised the Japan Meteorological Research Institute general circulation model (MRI-AGCM) under the IPCC A1B warming scenario, did one of the early studies that simulated future TC activity in the NIO. More recently, Murakami et al. (2017) conducted experiments using the U.S. Geophysical Fluid Dynamics Laboratory climate model under the IPCC Representative Concentration Pathway (RCP) warming scenarios also with future TC activity over the NIO in mind. As will be seen later, although there are model biases, both studies demonstrated there will be reduced (enhanced) TC frequency in the BoB (AS). In order to estimate future impacts in terms of storm surge, inundation and even social/economic aspects given such trends of TC frequency, the simulated tracks from Murakami et al. (2017) for both the historical and future periods were acquired from the authors for the following study.

Both past and future tracks were analysed for the entire NIO region. To validate the model, simulated TC tracks for the past (1990–2010) were compared with the observed TC best tracks of IMD (Fig. 3.2). During 1990–2000, IMD recorded 86 TCs for NIO basin and among these 69 and 17 formed over the BoB and AS, respectively (Fig. 3.2a). It is well demonstrated that most of the TC formed and made landfall over the BoB region. About 24% of TCs were intense having an intensity ≥64 knots (equivalent to hurricane category-1), which caused large numbers of human deaths

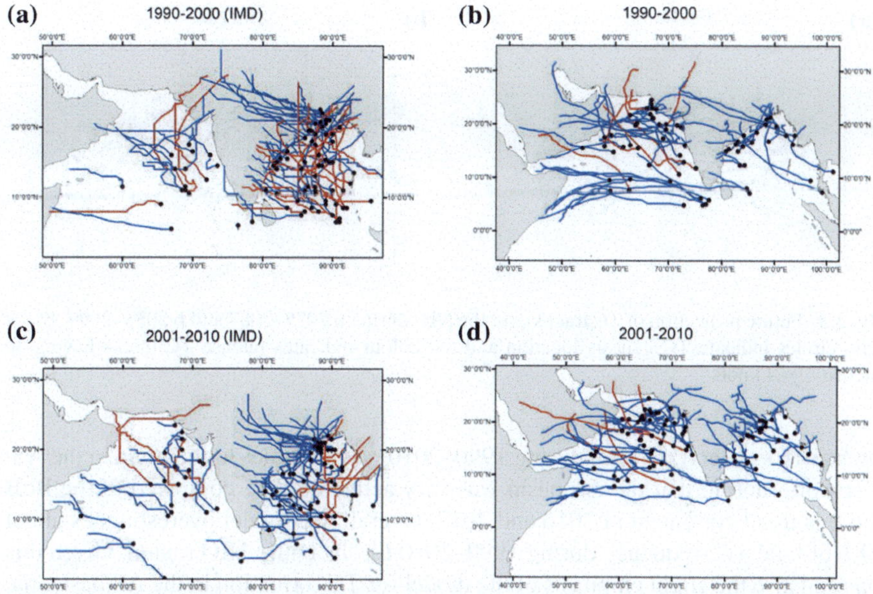

Fig. 3.2 **a** Observed versus **b** simulated TC tracks over NIO during 1990–2000. **c** and **d** are for the period 2001–2010. Black solid circles indicate TC formation locations and red tracks indicate intense TCs having intensity ≥64 knots

and property damage. As for TC frequency, there was also contrast between AS and BoB in terms of intense TC activity. For instance, only 4 out of 20 hurricane category-1 TCs were recorded for AS and the remaining ones in the BoB region. It was also observed that the entire BoB basin mostly had northwestward and northeastward tracks. In contrast to the BoB, only the AS's eastern part close to the BoB basin had TC activity (Fig. 3.2c). During the decade 2001–2010, 58 (25) TCs were observed over the BoB (AS). Of these TCs, 9 were recorded as hurricane category-1 and only 3 were observed over the AS region. Nevertheless, TC activity over AS has increased in recent years. For example, two intense TCs formed and made landfall over the AS coast during 2015. Furthermore, TCs tracks seem to have shifted from the eastern to western region of the NIO.

Compared to the observed TC activity, overall the model simulated lower TC frequency (67) during 1990–2000, whereas, 86 TCs were observed during this time (Fig. 3.2b). Furthermore, the model overestimated (underestimated) TC frequency as well as intensity for the AS (BoB) basin. During this time, no intense TC was simulated for the BoB, although 26 intense TCs were recorded by IMD. Similarly, for the period 2001–2010, the model estimated TC frequency of 35 (52) for BoB (AS) basins in the NIO (Fig. 3.2d). Similar to the observed tracks, the model simulated mostly westward TC tracks. The model significantly underestimates TC frequency as well as intensity for the BoB but not for the AS. During 1990–2010, 11 intense TCs were hindcasted for AS and none of them for the BoB, although the number of

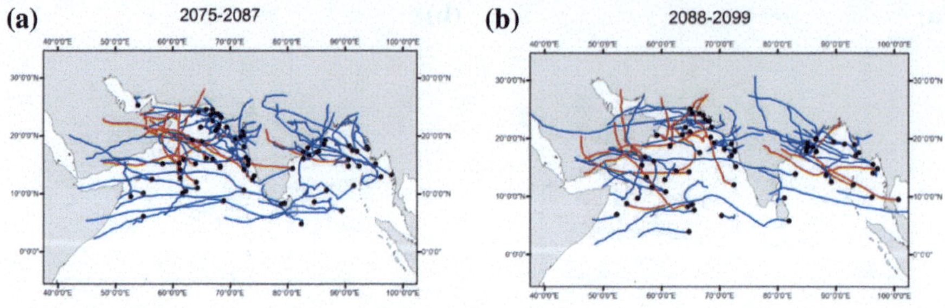

Fig. 3.3 Future projection of TC tracks over the NIO during **a** 2075–2087 and **b** 2088–2099. Black solid circles indicate TC genesis location and red colour indicates intense TC tracks having an intensity ≥64 knots

intense TCs observed over it during 1990–2010 was six times higher than in the AS. It is worth noting that the AS basin was very active in 2015 compared to the BoB and this trend continued in 2016 and 2017. Overall, the model overestimates about 30% of total TC frequency during 1990–2010 for the entire NIO region. Given this bias and also the slight emphasis of the model for TC formation in the AS basin, the model's performance is still considered reasonable and can be used for estimating future storm activity in the NIO, as has been applied in Murakami et al. (2017).

Future projections of TC formations and tracks for NIO showed enlarged contrast between the AS and BoB in terms of TC frequency and intensity (Fig. 3.3). There will be higher TC activity over the AS compared to the BoB by the end of this century. The model projected TC frequecy for the AS basin almost double that in the BoB for both 2075–2087 and 2088–2099. The model also predicted 5 (19) intense TCs (intensity ≥64 kt) out of 143 TCs for BoB (AS). The results suggest that future TCs tracks will be towards the western part of the BoB and AS basin and there will be many TC formations in the coastal regions of the northeast AS. These results clearly demonstrated the potential shifting of TC activity from the eastern to western part of BoB and AS. It is expected that the AS basin will be more active in terms of TC formation, development and landfall. Therefore, coastal areas around AS basin will have larger impacts and disaster managers must get ready to tackle this future calamities.

3.6 Past and Future Trends of Maximum Surge Height

The simulated TC tracks reported in Murakami et al. (2017) were applied as input to the JMA/MRI storm surge model. No tidal cycle was activated in the storm surge model, and thus only waves generated by the storms have been simulated. The climatology of maximum surge height associated with TCs that reach intensity at least 64 kt both for the past (1990–2010) and future (2075–2099) period have been analysed. Maximum surge has been computed for each intense TC for 48 h while the TC approaches the coastal areas and makes landfall afterwards. During this period, the TC will cause major damage and there is a great risk of surface inundation. The box plot of all the simulated maximum surge levels for the hindcast period (1990–2010) is illustrated in Fig. 3.4a. The average maximum surge was estimated at around 2 and 1 m for the BoB and AS, respectively. During this period, two extreme cases were identified: one for each basin where the estimated surge height was 7 m for the BoB

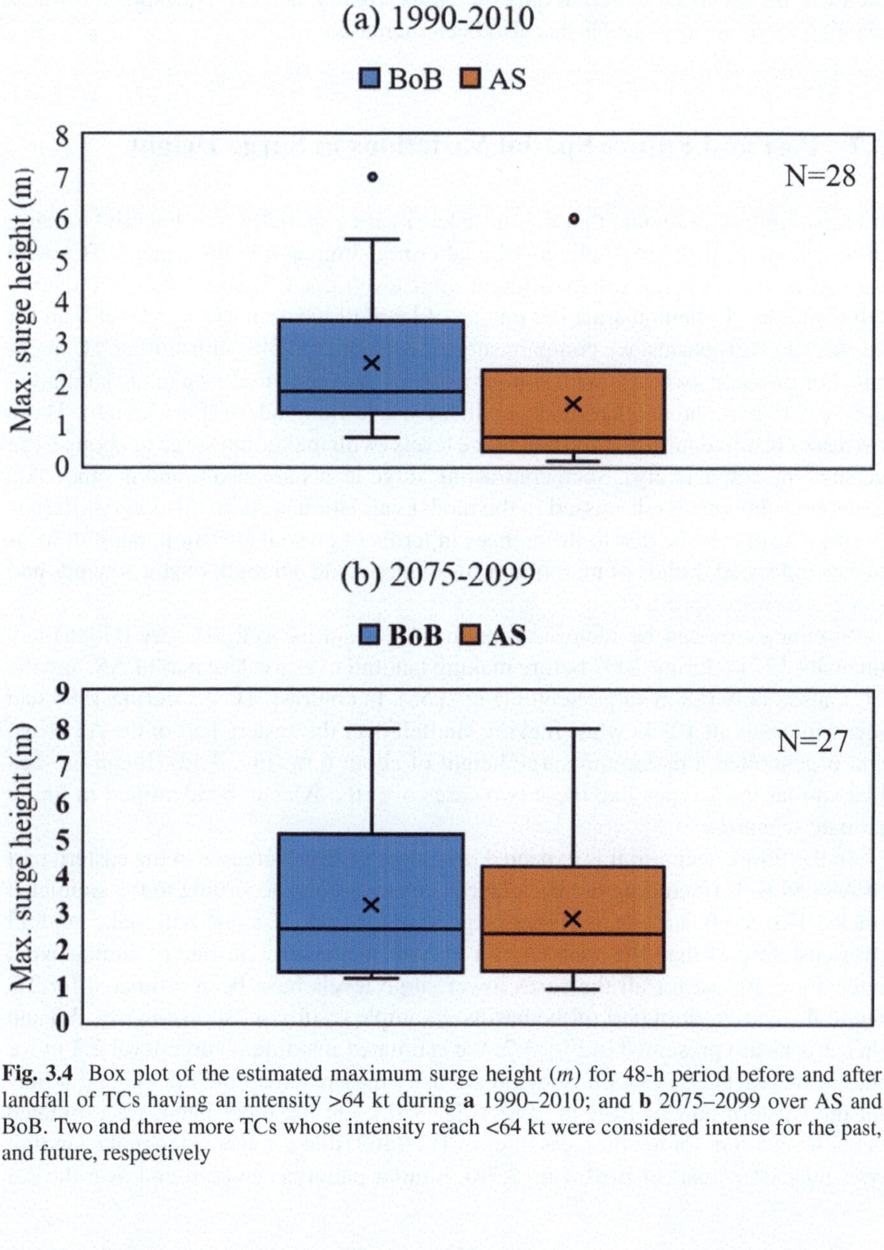

Fig. 3.4 Box plot of the estimated maximum surge height (*m*) for 48-h period before and after landfall of TCs having an intensity >64 kt during **a** 1990–2010; and **b** 2075–2099 over AS and BoB. Two and three more TCs whose intensity reach <64 kt were considered intense for the past, and future, respectively

and 6 m for the AS. When all cases are considered, the mean value of the maximum surge (middle line in Fig. 3.4a) for the BoB was estimated to be almost double of that for the AS.

For the future TC scenario, there is a significant increase in estimated surge height especially for the AS basin (Fig. 3.4b). Mean surge level is about 2.5 m for both basins, while the two extreme cases are both about 8 m in surge height. There is consequently a clear increase in future surge level for AS compared to the climatology period. For the BoB, the mean surge height only increases slightly, however, the spread towards the high extreme surge levels has also been increased.

3.7 Past and Future Spatial Variations in Surge Height

It is also important to note that the surge level varies spatially over the NIO because of its topography and tidal effects (the latter not simulated in this study). TCs with similar intensities will result in different surge levels due to their formation and landfall locations. To demonstrate the impact of landfall location to surge level both for the AS and BoB basins, we compare surge levels for landfalls with similar TC intensity. For instance, two TCs with similar intensity are compared: one made landfall to the western part and another made landfall close to Bangladesh (Fig. 3.5a, b). These two cases resulted in very differrent surge levels (with maximum surge of about 2.1 m versus 7 m, respectively). Such contrasting surge levels are also found in other TCs cases (not shown). As discussed in the model validation section, this large diffrence in surge level may be due to differences in terms of coastal elevation, landfall location, wind speed, radius of mamimum sustanined wind, strength of storm winds and storm's forward speed, etc.

Similar cases can be identified for the AS basin as well. TC 3A (Gonu) had intensity 127 kt during 2007 before making landfall over western part of AS, and the TC caused only 0.8 m surge height (Fig. 3.5c). In contrast, TC 2A during 1999 had lower intensity of 105 kt while making landfall over the eastern part of the AS coast, and it generated a maximum surge height of about 6 m (Fig. 3.5d). It can be seen that similar track types like these two cases over the AS can be identified in future climate scenarios.

In the future scenario it is expected that surge height increases in the eastern part of AS and BoB (including the Bangladesh coast of BoB) according to the simulated tracks. This study simulated higher surge levels for the TCs that will make landfall over eastern part than the western part of both the basins. Despite of similar wind intensities, almost for all the cases lower surge levels have been estimated for TC landfalls over western part of the basins. Example results are shown in Fig. 3.6 and their tracks are presented in Fig. 3.7. We estimated maximum surge level 2.2 m for the simulated TC B1930 whose wind intensity may reach up to 78.73 kt and it may hit the western part of BoB in 2081 (Fig. 3.7a). On the other hand, we estimated surge level 5.5 m for another less intense TC B4631that is expected to make landfall over the eastern part of BoB (Fig. 3.7b). Similar patterns can be found over the AS

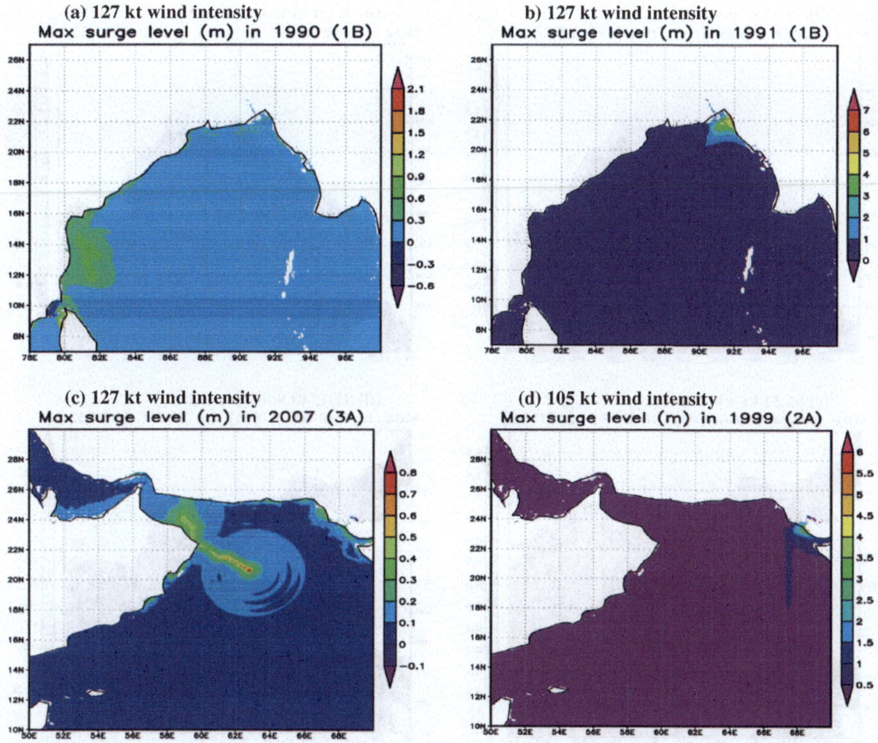

Fig. 3.5 Spatial differences of surge levels for the AS and BoB basins for selected cases during 1990–2010. The best tracks of these cases can be found in Fig. 3.7

basin as well. For instance, the westward track of TC A7040 generated 1.2 m surge level (Fig. 3.7c), while even a less intense TC A6455, which is expected to landfall over eastern part of AS, generated 5-m surge level. Such differences between the east and west in landfall location exists in the entire NIO. Further studies are necessary for elucidating the factors behind the spatial differences of storm surge height, such as topography, landfall direction and astronomical effects.

3.8 Discussion and Conclusion

Climatology of the simulated intense TCs over the NIO has been analysed for the period 1990–2010 and 2075–2099. Results indicate more intense TCs over the BoB in the past, however, simulated results show a rising number of intense TC over the AS in the future climate scenario. The mean maximum surge will be twice the size in the future and especially over the AS. The general TC track will also shift from the eastern part to the western part of both basins. Furthermore, spatial variation of surge

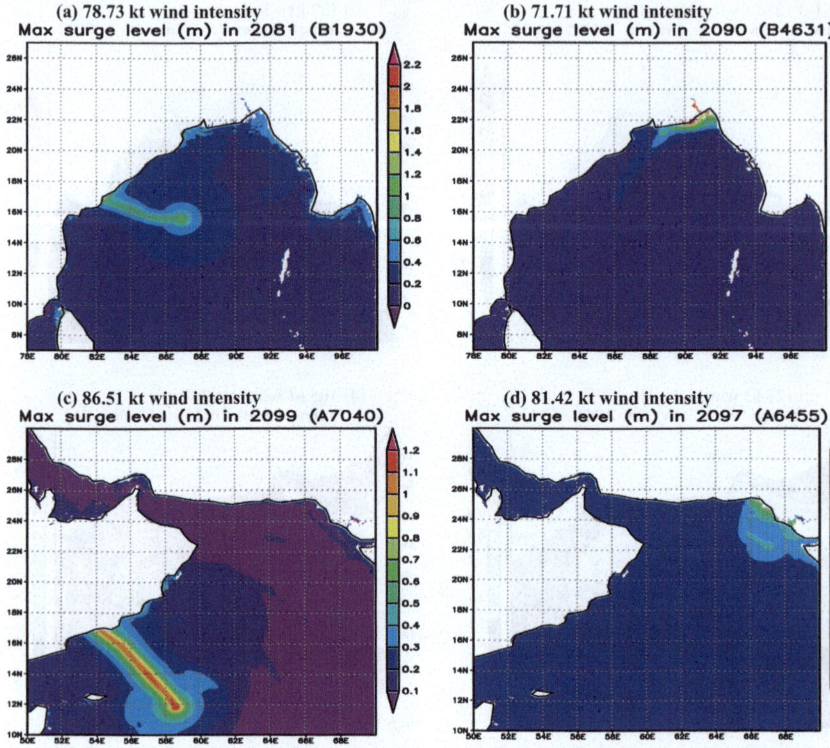

Fig. 3.6 Spatial differences of surge levels for the AS and BoB basins during 2075–2099

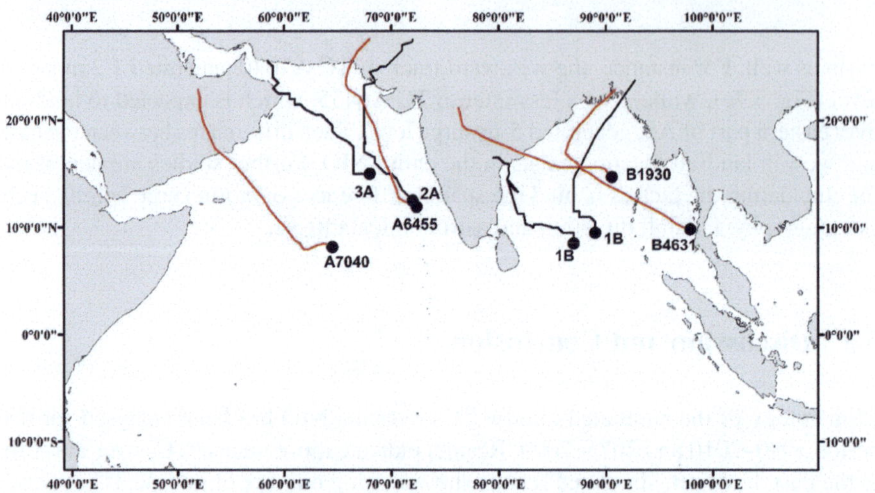

Fig. 3.7 Identified past (black) and future (red) TCs tracks for representing east and west differences over NIO

height has been demonstrated through case studies. A large difference in surge level is found for TC cases making landfall over the eastern part of the NIO versus the western part. However, further study is necessary to explain such spatial variation in the surge level. In conclusion, it is expected that higher maximum surge level over the NIO especially over the AS basin will occur if the projected TC activity is robust.

In this study, intense TCs (hurricane category-1 with intensity >64 kt) in the climatology and future periods simulated by a climate model have been applied as inputs to a storm surge model. Results clearly demonstrated that maximum surge height will be higher over the NIO in the future. Compared to the BoB, the increasing surge level over the AS is more remarkable, which is more than double compared to the past. These results are consistent with the study by Murakami et al. (2013, 2017) who predicted the occurrence of more intense TCs over the NIO and especially over the AS basin. The uniqueness of this study is that model simulated TC information has been applied for generating inundation depth while some other studies used synthetic TC tracks for estimating inundation depth. For instance, Sahoo and Bhaskaran (2018) used synthetic TC tracks for analysing inundation due to TC activity for India's east coast. Another study by Dasgupta et al. (2013) used an estimated inundation risk map for Bangladesh for the year 2050. They took climate change into account and reported an inundation depth of more than 6 m for some regions including Bhola and Hatiya island. This study found similar results for the year 2090.

References

Bangladesh Meteorological Department (BMD). Historical cyclones. http://bmd.gov.bd/p/Historical-Cyclones/

Chen C, Beardsley RC, Cowles G (2006) An unstructured grid, finite-volume coastal ocean model (FVCOM) system. Oceanography 19:78–89

Cheung KF, Phadke AC, Wei Y, Rojas R, Douyere YM, Martino CD, Houston SH, Liu PF, Lynett PJ, Dodd N, Liao S (2003) Modeling of storm-induced coastal flooding for emergency management. Ocean Eng 30:1353–1386

Choudhury SA (2014) Effective tropical cyclone warning in Bangladesh. Country report

Das PK (1972) Prediction model for storm surges in the Bay of Bengal. Nature 239:211

Dasgupta S, Huq M, Khan ZH, Ahmed MMZ, Mukherjee N, Khan MF, Pandey K (2013) Cyclones in a changing climate: the case of Bangladesh. Climate Dev 6:96–110

Dube S, Jain I, Rao A, Murty T (2009) Storm surge modelling for the Bay of Bengal and Arabian Sea. J Int Soc Prev Mitig Nat Hazards 51:3–27

Dube SK, Chittibabu P, Rao AD, Sinh PC, Murty TS (2000) Sea levels and coastal inundation due to tropical cyclones in Indian coastal regions of Andhra and Orissa. Mar Geodesy 23:65–73

Gayathri R, Bhaskaran PK, Jose F (2017) Coastal inundation research: an overview of the process. Curr Sci 112:267–278

Graeme DH, Kathleen LM (1999) A storm surge inundation model for coastal planning and impact studies. J Coastal Res 15:168–185

Jelesnianski CP (1972) SPLASH (Special Programme to List Amplitudes of Surges from Hurricanes) and landfall stormsTech. Memo, NWS TDL-46

Jisan MA, Bao S, Pietrafesa LJ (2018) Ensemble projection of the sea level rise impact on storm surge and inundation at the coast of Bangladesh. Nat Hazards Earth Syst Sci 18:351–364

Kohno N, Dube SK, Entel M, Fakhruddin SHM (2018) Recent progress in storm surge forecasting. Trop Cyclone Res Rev 7:128–139

Kossin JP (2018) A global slowdown of tropical-cyclone translation speed. Nature 558:104–107

Kossin JP, Emanuel KA, Vecchi GA (2014) The poleward migration of the location of tropical cyclone maximum intensity. Nature 509:349

Lian X, Pietrafesa LJ, Machuan P (2004) Incorporation of a mass-conserving inundation scheme into a three dimensional storm surge model. J Coastal Res 20:1209–1223

Luettich RA Westerink JJ, Scheffner NW (1992) ADCIRC: an advanced three-dimensional circulation model for shelves, coasts, and estuaries, report 1: theory and methodology of ADCIRC-2DDI and ADCIRC-3DL Dredging Research Program Technical Report DRP-92–6

Madsen H, Jakobsen F (2004) Cyclone induced storm surge and flood forecasting in the northern Bay of Bengal. Coast Eng 51:277–296

Mio M, Tetsuo N (2015) Early warning products for severe weather events derived from operational medium-range ensemble forecasts. Meteorol Appl 22:213–222

Murakami H, Sugi M, Kitoh A (2013) Future changes in tropical cyclone activity in the North Indian Ocean projected by high-resolution MRI-AGCMs. Clim Dyn 40:1949–1968

Murakami H, Vecchi GA, Underwood S (2017) Increasing frequency of extremely severe cyclonic storms over the Arabian Sea. Nat Clim Change 7:885–889

Needham HF, Keim BD, Sathiaraj D, Shafer M (2013) A global database of tropical storm surges. Eos, Trans Am Geophys Union 94:213–214

Pakistan Meteorological Department (PMD) (2010) Tropical Cyclone "PHET" in North Arabian Sea

Roy GD, Kabir ABMH, Mandal MM, Haque MZ (1999) Polar coordinates shallow water storm surge model for the coast of Bangladesh. Dyn Atmos Oceans 29:397–413

Sahoo B, Bhaskaran PK (2018) Multi-hazard risk assessment of coastal vulnerability from tropical cyclones—a GIS based approach for the Odisha coast. J Environ Manage 206:1166–1178

Tsuboki K, Yoshioka MK, Shinoda T, Kato M, Kanada S, Kitoh A (2015) Future increase of supertyphoon intensity associated with climate change. Geophys Res Lett 42:646–652

Woodruff J, Irish J, Camargo S (2013) Coastal flooding by tropical cyclones and sea-level rise. Nature 504:44–52

Unisys Weather Information Services (1999) http://weather.unisys.com/hurricane/indian_oc/1999/index.html

Unisys Weather Information Services (2007) http://weather.unisys.com/hurricane/indian_oc/2007/index.html

Unisys Hurricane Database (2008) Unisys Corporation: http://weather.unisys.com/hurricane/n_indian/index.html

Chapter 4
Classifying the Forest Surfaces in Metropolitan Areas by Their Wildfire Ignition Probability and Spreading Capacity in Support of Forest Fire Risk Reduction

Artan Hysa

Abstract The main objective of this work is to develop a cost-free and rapid method for categorizing the forest surfaces in metropolitan areas based on their Wildfire Ignition Probability Index (WIPI) and Wildfire Spreading Capacity Index (WSCI). The original method applies a multi-criterion (social, environmental, and physical) framework and utilizes commercial software for spatial analysis and collecting environmental data. Instead, this study utilizes QGIS as an open-source software during all geospatial analytical phases. At this stage, the method is tested on the metropolitan area of Tirana (Albania), relying on a variety of open-source geospatial databases. First, the forest surfaces are identified based on Urban Atlas land cover and are translated into regular point grid (100 m). Each point is loaded with unique values regarding each criterion. The diversity among values of different criteria is normalized by redistributing them into 10 classes based on Jenks natural break method. Class values of each criterion are introduced into the indexing equation multiplied by their respective impact factor as weighted via Analytical Hierarchy Processing. Finally, each representative point is calculated as a final WIPI and WSCI value. The locations possessing relatively highest values indicate areas of significant wildfire ignition and spreading likelihood. The results of the study validate a rapid and cost-free method for forest fire risk assessment being applicable and reproducible on similar study areas at metropolitan scale. The method presented in this chapter is aimed to support of forest fire risk reduction agendas at local level in the developing countries.

Keywords Forest fire risk assessment · Wildfire ignition probability · QGIS · Urban atlas · Analytical hierarchy process · Albania

---

A. Hysa (✉)
Faculty of Architecture and Engineering, Epoka University, Tirana, Albania
e-mail: ahysa@epoka.edu.al

R. Djalante et al. (eds.), *Integrated Research on Disaster Risks*, Disaster Risk Reduction,
https://doi.org/10.1007/978-3-030-55563-4_4

4.1 Introduction

The main objective of this chapter is to present a rapid and cost-free method based on Geographic Information System (GIS) for indexing the forest surfaces in metropolitan areas by their Wildfire Ignition Probability (WIPI) and Wildfire Spreading Capacity Index (WSCI). This is achieved by utilizing open-source software and open-access spatial data during all phases of the analytical work.

The metropolitan area is accepted among the first priority territories to be considered in respect to forest fire regimes. This is because it includes the intermediate zone between natural and artificial surfaces, which an impact of wildfire hazards on socio-economic and landscape systems is the highest. This transitory zone known as well as the Wildland–Urban Interface (WUI) includes the human settling activities which are located close or within the natural lands (Davis 1990; Radeloff et al. 2005; Theobald and Romme 2007).

Within this work, the wildfire phenomena are considered as processes rather than as events. In other words, they are accepted as socio-natural hazards implicating several social, environmental, and geomorphological factors of the place. In literature, forest fires are defined as events that under certain circumstances are increased in occurrence much beyond their natural probabilities (ISDR 2009). The dominant presence of anthropogenic factors in affecting wildfire regimes is well reported by several scholars (Martínez et al. 2009; San-Miguel-Ayanz et al. 2013; Zambon et al. 2019).

This approach leads to the definition of a set of diverse factors having measurable implications with wildfire systems. As an integrated disaster risk assessment, this study covers three main categories of causing factors; (i) social/anthropogenic, (ii) geophysical, and (iii) hydrometeorological. Distance from human activity, such as distance to urban centers, distance to settlements, distance to main roads, distance to any road, and distance to agricultural lands are among the social criteria defined in this study. While the geophysical ones include altitude, aspect, slope, and distance to surface water. In addition, there are certain environmental factors such as solar radiation, maximum temperature, precipitation, and wind speed, which are accepted crucial in wildfire spreading phases (de Souza et al. 2015).

At this stage, the study is focusing on a specific study area located in the central Albania named "Durana Metropolitan Area". It includes Tirana as the capital city, the city of Durres as the main port of the country, and the area between them. The total surface of the study area is 1,670 km^2, with a total of 614 km^2 forest surfaces. These surfaces are the core focus of this study being indexed by their WIPI and WSCI values. Yet, the method presented in this work is flexible enough to be applied to other metropolitan areas, given that its geospatial data are available.

This study is a direct contribution to the wildfire disaster risk reduction at the local level at which Albania is majorly urging for. According to the Official Statement made by the Albanian ex-President H.E. Bujar Nishani, at the fifth session of the Global Platform for Disaster Risk Reduction held in May 2017 in Mexico, Albania is declared to be proactive in the implementation of 2030 agenda and is fully committed to contribute according to all its capacities and capabilities in reducing the risk of natural disasters (Nishani 2017). But at the local level the case is not promising, and will be discussed further in this chapter during the interpretation of the results.

Finally, this chapter is aiming to contribute to the *understanding of disaster risk* which is being strongly marked as the Priority 1 under Sendai Framework for Disaster Risk Reduction 2015–2030 (UNISDR 2015). Similarly, it is aligned with the Objective 1 of the Science Plan for Integrated Research on Disaster Risk (IRDR) by contributing to the *characterization of wildfire hazard*. More specifically, this work presents a framework for identifying the hazards, and vulnerabilities leading to forest fire risk (Objective 1.1), contributing to the forecasting of forest fire hazards and assessing their risks (Objective 1.2), and developing a dynamic modeling of wildfire risk (Objective 1.3) (ICSU 2008).

4.2 Methods

This study puts forward a cost-free and rapid method for categorizing the forest surfaces being closely related to metropolitan areas based on their Wildfire Ignition Probability Index (WIPI) and Wildfire Spreading Capacity Index (WSCI). The original method applies a multi-criteria (social, environmental, and geophysical) framework and utilizes commercial software such as ArcGIS for spatial analysis and Meteonorm for collecting environmental data (Hysa and Başkaya 2019).

Instead of relying on commercial software, which may not be accessible to a wider DRR stakeholder, this study utilizes QGIS as an open-source software during all geospatial analytical phases. Furthermore, data used are mainly gathered from various free and open-sourced geospatial databases; e.g., E-OBS for environmental raster data; Copernicus Portal for Urban Atlas-UA, DEM, and Vegetation density; Open Street Map for transportation network.

Furthermore, in previous studies the method was tested on rural forested lands, quite remote from the urbanized territories. But in this work, the context of the metropolitan zone as the selected study area is specific and very sensitive in relation to the wildfire risk and forest fires regimes as mentioned in the introduction of this chapter.

4.2.1 Study Area

At this stage, the method is tested on "Durana Metropolitan Area", covering Tirana as the capital of Albania and Durres as the largest national port city (Extent coordinates; 19.39–20.24 E, 41.12–41.59 N). Physically the area can be defined as a valley opening toward the west-facing Adriatic Sea, while its landscape is characterized as canvas of disconnected ecologies: the terraced hills, foothill villages, a linear urban agglomeration, industrial connecting strip, the inhabited agricultural plain, and the embracing forest surfaces (Fabi and Piovene 2016).

Tirana-Durres area is defined as metropolitan zone based on being a large urban area, composed of several interconnected agglomerations, having at least 500,000 inhabitants, being distinct for their economic potential, and possessing considerable national and international level impact (Ahrend et al. 2014; INSTAT 2014). Durana is considered a metropolitan "region" having the features of a vague space, defined by strong geographical and environmental conditions, common production patterns, transportation networks, labor markets, social interaction, and complex social-ecological interdependencies (Keating 1998).

The total area of the case study is 166,885 ha based on the Urban Atlas data boundary as shown in Fig. 4.1. While the forest surfaces cover an area of 61,412 ha, marking slightly more than 1/3rd of the total region. The area is located in the center of the territory of Albanian Republic aligned to the Adriatic coastline. The dominant layers are the urbanized lands shown in red in Fig. 4.1, and forest surfaces represented in green.

Yet, the method is potentially applicable in other metropolitan areas, particularly if relevant geospatial data are available and accessible, such as from an open source. Similar metropolitan areas from developing countries, especially in the Western

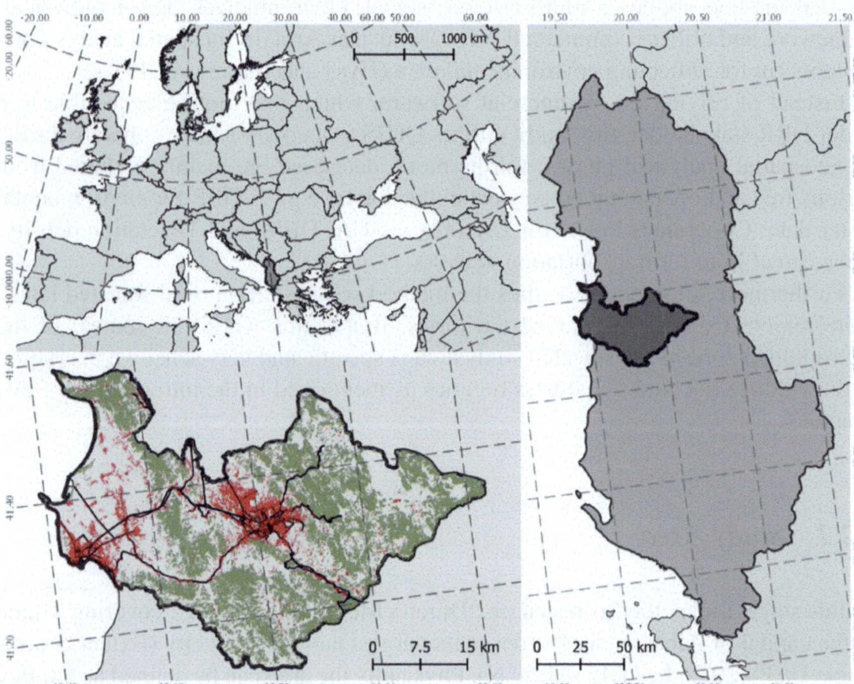

Fig. 4.1 The study area within the European continent and the Republic of Albania

Balkans can apply the method to rapidly prepare maps of indexing the forest surfaces by their wildfire ignition probability and wildfire spreading capacity.

4.2.2 *Procedure of Data Analysis*

The workflow of the applied method consists of ten consecutive steps organized into four main phases as shown in Table 4.1. The preliminary works phase initiates with the study area selection and extracting the forest surfaces as the main target. Then the reference points that overlap with the targeted surfaces are extracted from regular point's grid which is generated by covering the full extent of the study area. The final reference points will be loaded unique values for each criterion via the point sampling tool plugin in QGIS.

Table 4.1 Workflow of the method utilizing QGIS software

Phase		Goal	Method	
Preliminary	1	Defining the study area	Extracting forest surfaces from Urban Atlas data	
	2	Generating regular points grid	Vector/Research tools/Regular points (100 m)	
	3	Generate reference points	Extracting the points overlapping with forests	
Inventory	4	Calculating values for each criterion	Calculating the values of all criteria for each point via QGIS plugins such as NNJoin for measuring distances	
	5	Multi-criteria inventory	Projecting separate values into an aggregate reference points layer via QGIS plugin "point sampling tool"	
Analysis	6	Data clustering	Clustering the values of each criterion into 10 classes according to Jenks natural breaks reclassification method	
	7	Ignition vs. Spread	Sub-dividing factors by their relevancy either to ignition or to spreading phase of wildfire through literature review	
	8	Defining weighting factors	Assign weight to each criterion based on Analytical Hierarchy Process (AHP, pairwise comparison)	
Indexing	9	Calculating WIPI	Field calculator	
			$\text{WIPI} = \sum_{i=1}^{n} \alpha_i C_i$	Equation (1)
	10	Calculating WSCI	Field calculator	
			$\text{WSCI} = \sum_{j=1}^{m} \beta_j C_j$	Equation (2)

The second phase consists of a multi-criteria inventory procedure (Hysa et al. 2017), through which there are measured absolute values of each criterion joining to a cumulative multi-criteria inventory. As it is expected, the range of values of different criteria are diverse among them. For example, the range of values for criterion "distance to urban centers" (ranging from 0 m to several km) is much different than the values range of "maximum temperature" (ranging from 0 °C to at most 50 °C).

The variety among absolute values of different criteria is normalized during the Analysis phase of the workflow. This is achieved by redistributing them into 10 classes based on Jenks natural break method as provided under QGIS layer properties. The Jenks natural breaks classification method, also known as the Jenks optimization (Jenks 1967), is a data classification method aimed to define the best reorganization of values into sub-classes by reducing of the variance within classes and maximizing the variance among them (Stefanidis and Stathis 2013). This is done by minimizing the standard deviation within each class and make the most of the standard deviation between classes (McMaster and McMaster 2002).

Since the analysis phase is the last one before the final double Indexing phase, the criteria are clustered into two main groups based on their relation with the ignition and spreading phases of the wildfire. Some criteria may have no effect on either ignition or spreading phase of the forest fire. As a result, there are two separate sets of criteria for WIPI and WSCI indexing.

Furthermore, based on the assumption that not all criteria have the same impact on the wildfire events, criteria of each set are introduced into a weighting process based on Analytical Hierarchy Processing (AHP) (Mu and Pereyra-Rojas 2017).

At the final stage, Jenks classification values of each criterion are introduced into the equation (Eqs. 1 and 2 from Table 4.1) multiplied by their respective impact factor as weighted via AHP. Finally, each representative point is loaded a final value of WIPI and WSCI. In principle, the locations possessing relatively highest values indicate for areas of high wildfire ignition and spreading risk.

4.3 Results and Discussions

The ten steps introduced above have been applied in the case of Durana metropolitan area, with each of four main phases lead to distinct results. First, the preliminary work and inventory phases both deliver a final shapefile layer of reference points, which consist of all inventory measured values for 13 criteria. As a result, raw statistical data on each criterion is produced.

Then, during the phase of the analysis, some inferential statistical data about the value distribution of each criterion are generated. During this phase, a set of maps that show the spatial distribution of Jenks classification of values per each criterion are produced. Besides, during this phase the weighted impact factor values via AHP are produced by being calculated as the average values as evaluated by eight professionals.

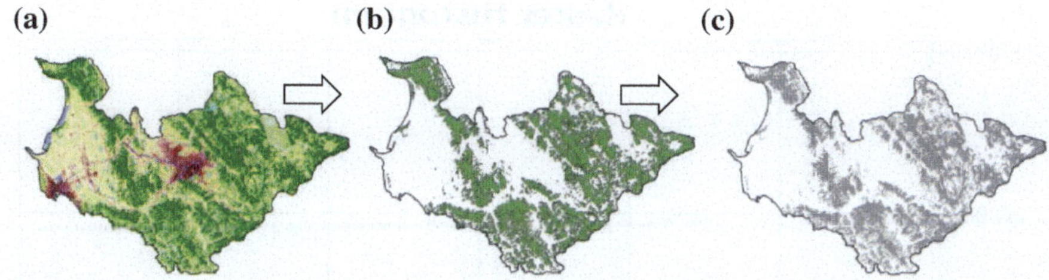

Fig. 4.2 Data transferring from UA land cover (**a**) to extracted forest surfaces (**b**), and to Reference points cloud (**c**)

At the final phase, two main maps representing the spatial distribution of WIPI and WSCI values are produced. Beyond the visual spatial distribution, the WIPI and WSCI values are represented and discussed via their histogram charts showing the frequency distribution of indexing values.

4.3.1 Multi-criteria Inventory for Measuring Exposure and Vulnerability of Forest Surfaces Toward Wildfires

First, the forest surfaces are identified based on UA evidence and are translated into regular points grid (100 m), where each point represents an area of 1 ha (Fig. 4.2c). Initially, there are 61,478 points generated from the regular grid that overlap with the forest surfaces within Durana region. The points that could not be loaded the value of at least one criterion are excluded from further processing.

The lack of some values is related to the non-total spatial coverage of environmental data which have been acquired for free in raster format.[1] Thus, according to the attribute table of the final layer of reference points, there are 58,294 points that have values for all criteria and are finally introduced into the process of WIPI and WSCI indexing. While 3,184 points (about 5% of the total) are excluded from further processing.

The remaining points are loaded unique values regarding each criterion in separate point shapefile layers via *NNjoin plugin* in QGIS (distance to urban centers, settlement, main roads, any road, agricultural areas, and water sources). The shapefiles are converted into raster data. The remaining criteria (altitude, aspect, slope from DEM file, solar radiation, maximum temperature, precipitation, and wind speed) are already acquired as raster completing the full set of 13 criteria in raster format.

Then, their values are collected into a single reference point via *point sampling tool* plugin. According to the attribute table of the final reference point layer, the absolute values of different criteria are diverse in range due to the variety in units as represented by the histogram in Fig. 4.3. In Fig. 4.3a, the full histogram map of all

[1]Source: http://surfobs.climate.copernicus.eu.

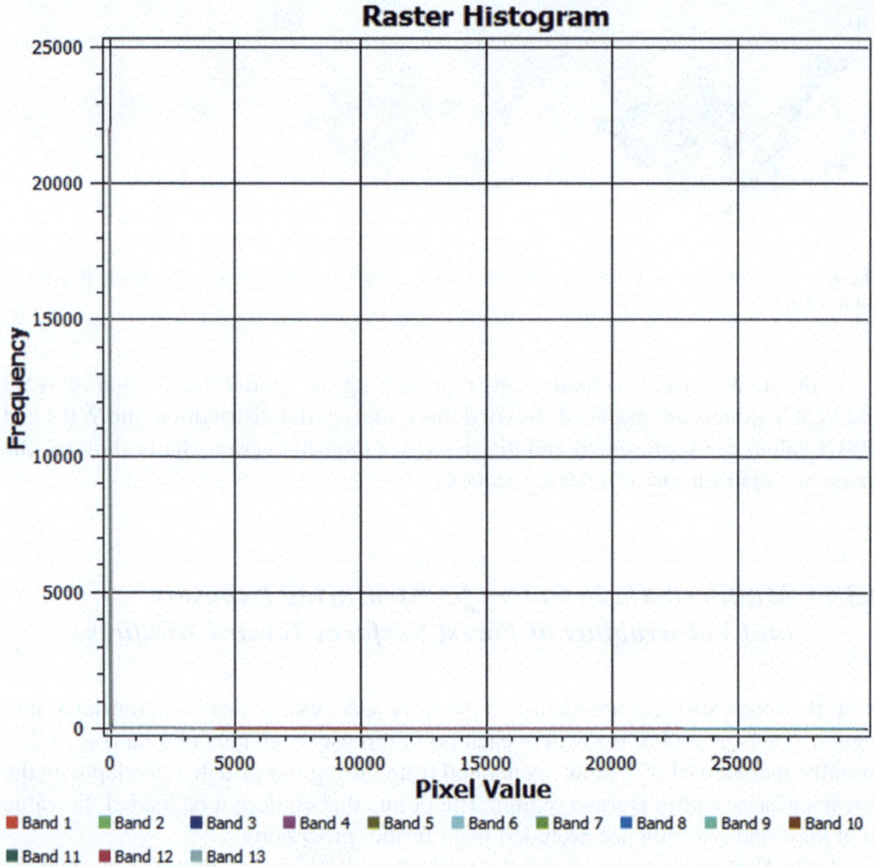

Fig. 4.3 Histogram chart showing absolute values of all criteria (**a**) and a zoom-in section showing their diverse distribution (**b**)

13 bands (13 criteria) is shown, and the high diversity within each criterion range of absolute values makes it hard to get visual information from the chart. The 13 bands that are shown closer in Fig. 4.3b represent partially the 13 criteria measurement values' distribution by their frequency. Moreover, the diversity is understandable from the ranges of values of each criterion referring to "min" and "max" columns of Table 4.2.

Beyond the range of values, the diversity among the absolute values is inferable from the Coefficient of Variation (CoV) in the final column of Table 4.2. In principle, the smaller CoV the smaller the variance among the values and the larger the variance among CoV of different criteria the greater the inconsistency between them. For example, the values of criterion E1 (solar radiation) mark the lowest CoV values since the standard deviation value is 0.15. On the other hand, the highest CoV values

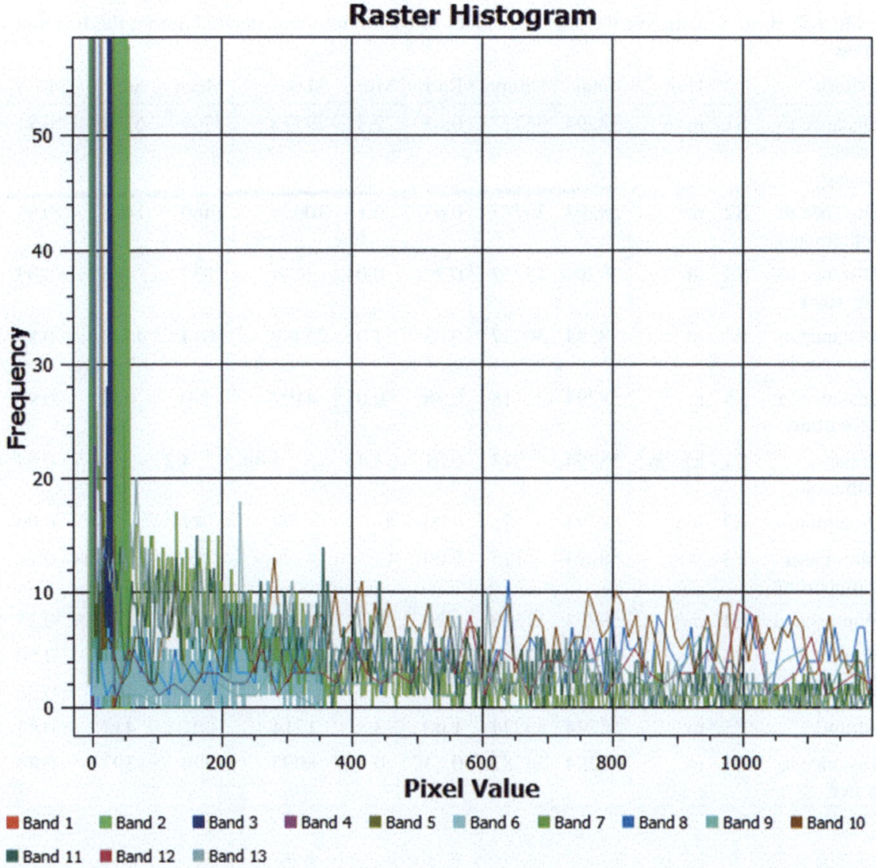

Fig. 4.3 (continued)

belong to criterion S2 (distance to settlements) scoring 0.98 relying on a standard deviation value of 1428.

Even though the highest standard deviation value is scored by criterion S1 (distance to urban centers) as 6271, due to the higher mean value (7769) it remains at 0.81 of CoV value. As the results in Table 4.2 show, CoV represents a measure independent from the unit of measurement and it represents another dimension of variance among criteria values (Brown 1998). Thus it further supports the irrelevancy of the direct inclusion of the absolute values into the indexing formula.

As a consequence, the absolute values of 13 criteria per each reference point (58294 in total) are introduced into the normalizing data clustering procedure based on *Jenks natural break* method. In this study, it is decided to reclassify into 10 classes via "symbology" procedure under layer properties as shown in Fig. 4.4. The automatically generated range of values (Fig. 4.4a) are used as breaking points for the reclassification step via field calculator (Fig. 4.4b).

Table 4.2 Basic statistics of the absolute values of each criterion generated during the Inventory phase

Criteria		Unit	Total	Unique	Ratio	Min	Max	Mean	Std	CoV
Distance to urban centers	S1	m	58294	36777	0.63	2.3	29778	7769	6271	0.81
Distance to settlements	S2	m	58294	36777	0.63	0.16	10034	1460	1428	0.98
Distance to any road	S3	m	58294	15253	0.26	0.04	3684	523	485	0.93
Distance to main roads	S4	m	58294	36937	0.63	4.05	22989	5034	4428	0.88
Distance to agriculture	S5	m	58294	22118	0.38	0.03	4192	450	440	0.98
Solar radiation	E1	kW/m^2	58294	514	0.01	3.43	4.46	4.2	0.15	0.04
Precipitation	E2	mm	58294	25	0.00	33	57	47	4.1	0.09
Maximum temperature	E3	°C	58294	125	0.00	17.3	29.9	26	3.08	0.12
Wind speed	E4	m/s	58294	11556	0.20	1	11.7	4.6	1.15	0.25
Slope	P1	°	58294	57537	0.99	0.04	77	16	8	0.50
Aspect	P2	°	58294	58175	1.00	0	360	193	108	0.56
Altitude	P3	m	58294	58224	1.00	-3	1714	501	418	0.83
Distance to water	P4	m	58294	36782	0.63	0	8995	1636	1393	0.85

The example in Fig. 4.4 presents the case of distance to urban centers criteria (S1, for the first social criteria) in relation to forest fire ignition probability. Considering the high impact of anthropogenic factors to wildfire ignition phase as reported by the current literature, it is decided that the closer to urban centers (the smallest the distance number), the higher the chances of a forest fire ignition event.

As a result, there are 13 new fields produced with relative values between 1 and 10. The spatial representation of these values is delivered in Fig. 4.5. It includes the maps of all criteria reclassified into 10 classes and represented in red-yellow-blue color gradient. According to the legend, the reddish zones indicate the highest values. In other words, it indicates the areas within the forest surfaces that have the highest chances of a wildfire ignition event caused by each criterion individually.

According to Fig. 4.5, the higher the distance of reference points to the elements of social criteria the more bluish their color indicating the less likely areas of forest surfaces which may face a wildfire ignition. For example, the color ramp distribution in the cases of the distance to urban centers (S1) and settlements (S2) are different even though both are targeting the human settling activity. While the former is referring to the core urban areas (major cities), the latter is targeting any settlement including small villages as well.

Similarly, certain points could be highlighted discussing the maps belonging to meteorological and geomorphological criteria. For example, there exist remarkable similarity between the map of maximum temperature (E3) and altitude (P3). This is explainable based on the assumption that the higher the altitude the lower the temperature. Another resemblance occurs between the maps of wind speed (E4) and slope (P1).

The next challenge is to merge the individual values into a single cumulative multi-criterion indexing map by considering their relative impact to either ignition or spreading phase of the wildfire event. As previously introduced, the weighted values are generated via *AHP pairwise comparison* method. Since AHP is based on personal evaluation and a certain subjectivity is generally accepted, in this study the final values are calculated as the average values among eight respondents (Fig. 4.3) being professionals in the field of Disaster Risk Management and Fire Safety (DRM&FS).

Beyond the subjectivity of the responses, the study is more a methodical proposal than a weighted values set. Yet, the weighted factors can be further improved at least by enhancing the number of the respondents as well as their professional level and profile diversity. Moreover, AHP weighting procedure can become further inclusive and holistic, by integrating respondents from other stakeholder profiles such as inhabitants, fire-fighting teams, farmers, governing bodies. The individual evaluations could be easily checked about their Consistency Ratio (CR) by semi-automated AHP excel template as developed by Goepel (2013).

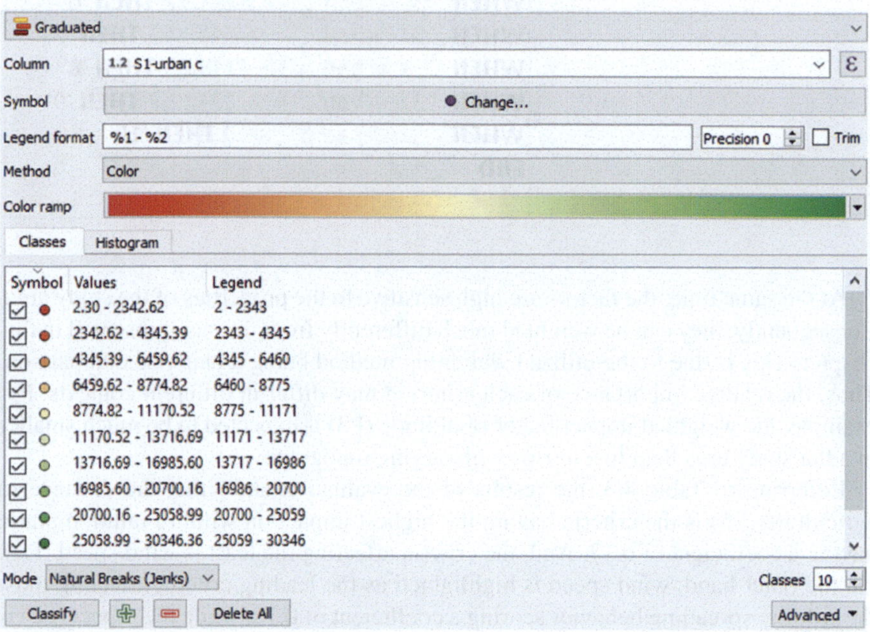

Fig. 4.4 Jenks natural break classification process via (**a**) "symbology" under layer properties and (**b**) field calculator under layer attribute table

Fig. 4.4 (continued)

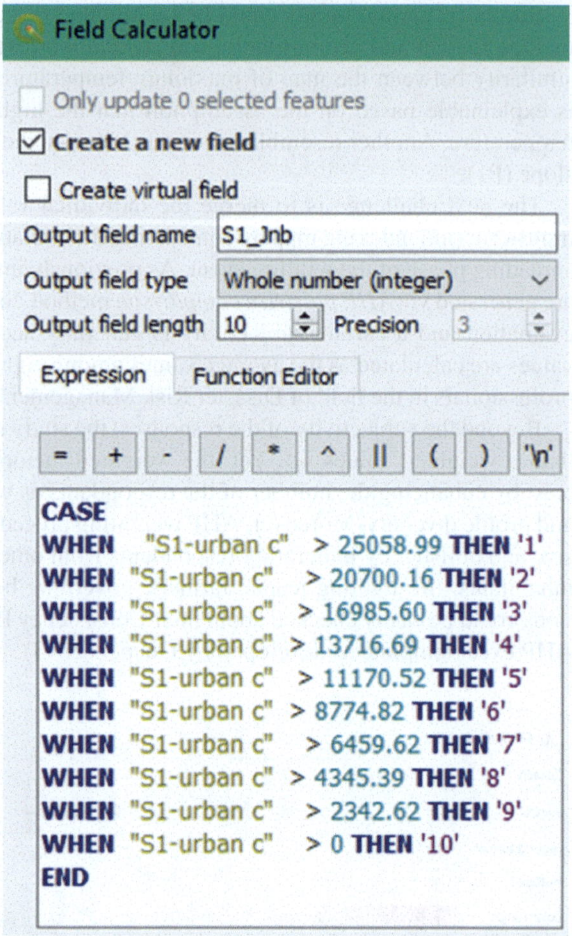

At the same time, the factors are high sensitive to the properties of the study area. Consequently, they can be weighted much differently from the case presented in this chapter. This is due to the utilized weighting method being a pairwise comparison. Thus, the relative importance of each criterion may differ in different contexts. For example, the weighted impact factor of altitude (P3) is expected to be much smaller in a flat study area than in a territory of varying topography.

Referring to Table 4.3, the results of the evaluation show that the distance to agriculture (S5) is the criteria having the highest impact on wildfire ignition phase having a coefficient of 0.17. And, the criteria affecting the least is wind speed (E4). On the other hand, wind speed is highlighted as the leading criteria affecting most the wildfire spreading behavior scoring a coefficient of 0.18, while the aspect (P2) is considered to have the lowest impact on the spreading phase recording at 0.05.

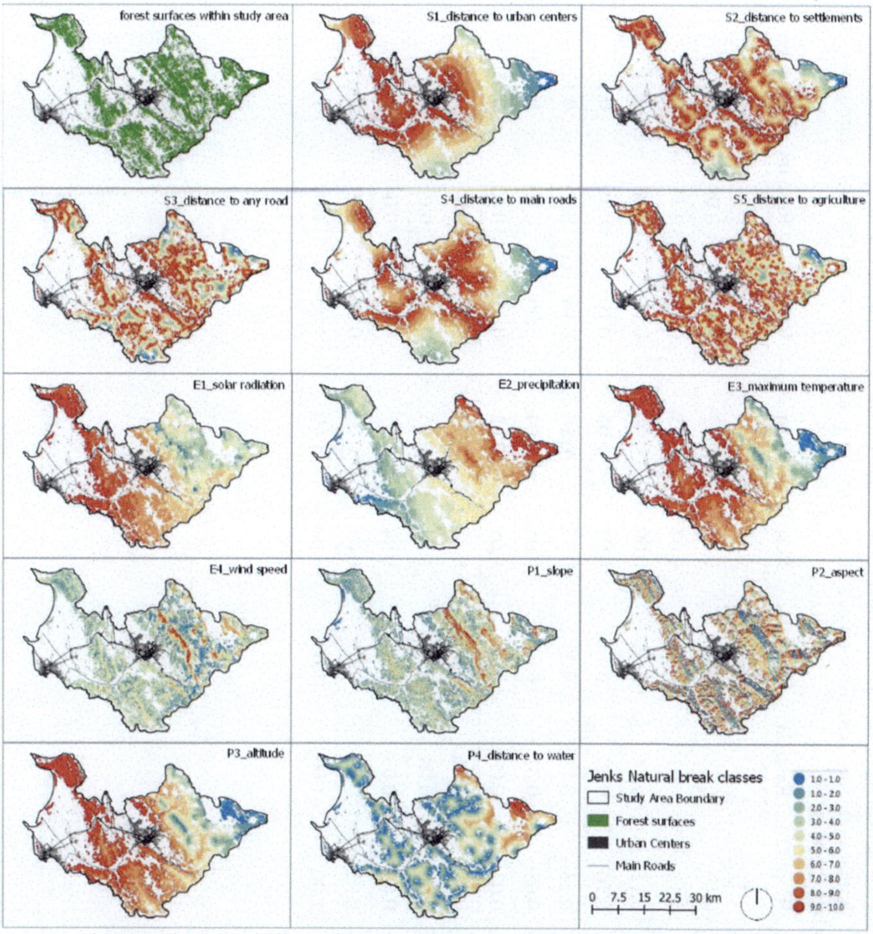

Fig. 4.5 Jenks natural break classification maps for each criteria

4.3.2 Calculation and Mapping of WIPI and WSCI Values

The fourth and final phase of Indexing consists of assignment of WIPI and WSCI values to the reference points. It relies on the field calculator operation under attribute table of the reference points' layer. According to Eqs. 1 and 2 in Table 4.1, the WIPI and WSCI index values consist of the sum of the products of the Jenks classification values (1–10) of each criterion multiplied by the respective weighted values as generated via AHP in the previous step. Thus, the final values range from 1 to 10, and the higher the values the higher the chances of either ignition or spreading of wildfire at the specific location within the forest surface.

The results of the study gives floor to a comparison between WIPI and WSCI values. According to the frequency distribution via the histogram charts in Fig. 4.6,

Table 4.3 Weighted impact factors of each criterion via AHP method based on 8 professional respondents (The group consist of practicing professionals with a background in civil engineering and spatial planning. Furthermore, they have been trained via a variety of graduate courses in DRM&FS during one year professional master program) for Ignition (a) and Spreading (b) phase

AHP ignition	Criteria		Resp 01	Resp 02	Resp 03	Resp 04	Resp 05	Resp 06	Resp 07	Resp 08	Average
(a)											
Dist. to urban areas	S1	−1	0.15	0.19	0.18	0.19	0.15	0.22	0.13	0.08	0.16
Dist. to settlements	S2	−1	0.14	0.15	0.07	0.13	0.07	0.15	0.11	0.11	0.12
Dist. to any road	S3	−1	0.12	0.13	0.11	0.06	0.09	0.06	0.15	0.16	0.11
Dist. to main roads	S4	−1	0.17	0.10	0.20	0.13	0.07	0.05	0.10	0.10	0.11
Dist. to agriculture	S5	−1	0.14	0.13	0.11	0.20	0.17	0.18	0.16	0.25	0.17
Solar radiation	E1	1	0.04	0.08	0.02	0.02	0.05	0.04	0.06	0.04	0.04
Precipitation	E2	−1	0.05	0.02	0.03	0.06	0.05	0.02	0.05	0.12	0.05
Max. temperature	E3	1	0.06	0.09	0.12	0.09	0.12	0.14	0.09	0.05	0.09
Wind speed	E4	1	0.04	0.06	0.08	0.04	0.14	0.05	0.07	0.01	0.06
Aspect	P2	1	0.04	0.03	0.03	0.04	0.04	0.03	0.04	0.04	0.03
Altitude	P3	−1	0.06	0.03	0.05	0.05	0.05	0.07	0.04	0.03	0.05
			1.00	1.00	1.00	1.00	1.00	1.00	1.00	1.00	1.00

(continued)

Table 4.3 (continued)

AHP spreading	Criteria		Resp 01	Resp 02	Resp 03	Resp 04	Resp 05	Resp 06	Resp 07	Resp 08	Average
(b)											
Dist. to urban areas	S1	−1	0.08	0.06	0.13	0.03	0.08	0.11	0.17	0.04	0.09
Dist. to settlements	S2	−1	0.05	0.06	0.07	0.05	0.09	0.10	0.11	0.01	0.07
Dist. to any Road	S3	−1	0.05	0.06	0.10	0.05	0.08	0.03	0.10	0.03	0.06
Dist. to main roads	S4	−1	0.04	0.05	0.11	0.04	0.09	0.02	0.07	0.05	0.06
Dist. to agriculture	S5	−1	0.12	0.22	0.16	0.13	0.09	0.17	0.16	0.02	0.13
Solar radiation	E1	1	0.07	0.14	0.01	0.07	0.08	0.08	0.05	0.05	0.07
Precipitation	E2	−1	0.02	0.03	0.03	0.02	0.09	0.02	0.05	0.18	0.06
Max. temperature	E3	1	0.15	0.19	0.10	0.16	0.09	0.14	0.06	0.10	0.12
Wind speed	E4	1	0.30	0.09	0.19	0.32	0.11	0.13	0.07	0.21	0.18
Slope	P1	1	0.05	0.03	0.03	0.05	0.08	0.12	0.06	0.12	0.07
Aspect	P2	1	0.05	0.04	0.04	0.05	0.06	0.06	0.04	0.06	0.05
Dist. to water sources	P4	1	0.03	0.02	0.03	0.03	0.05	0.02	0.06	0.12	0.05
			1.00	1.00	1.00	1.00	1.00	1.00	1.00	1.00	1.00

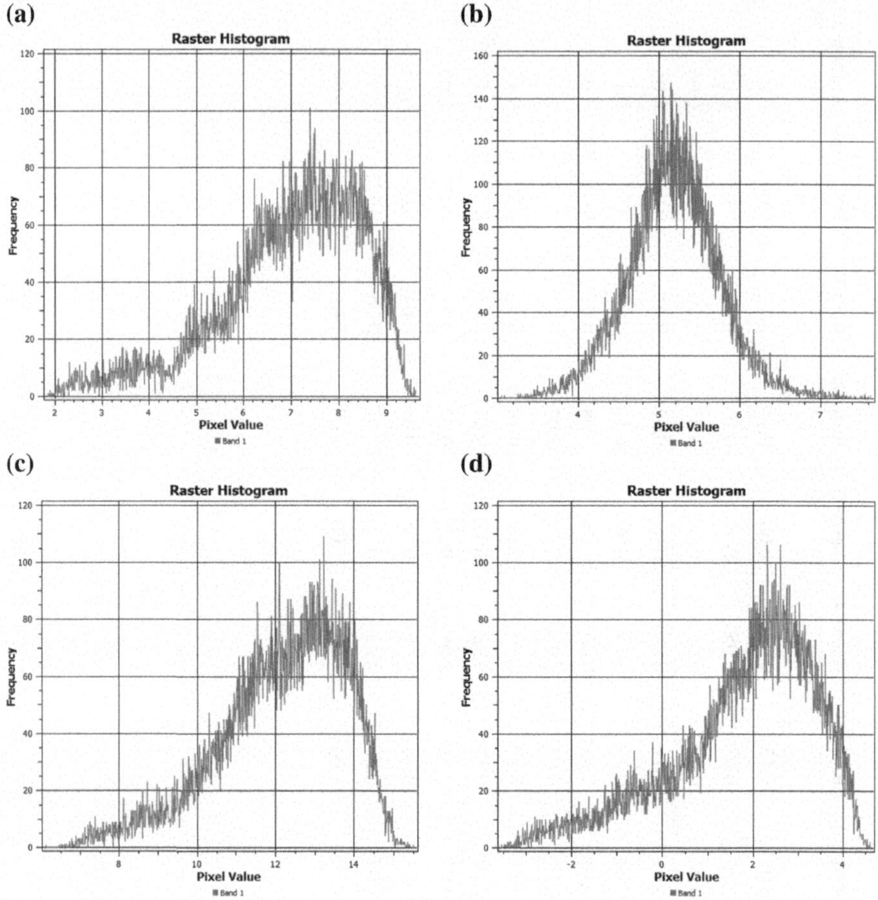

Fig. 4.6 Frequency distributions histogram of (**a**) WIPI, (**b**) WSCI, (**c**) WIPI + WSCI, and (**d**) WIPI − WSCI values

the WIPI values are concentrated relatively at higher range of values than WSCI. This fact is supported by the difference values (Fig. 4.6d) in which the results above 0 are the most dominant.

As a result, the WIPI and WSCI columns are added to the attribute table of reference points. Besides, the sum and the difference between WIPI and WSCI values is calculated. The cumulative sum values (WIPI + WSCI) is helpful in identifying the locations within the forest surfaces that can be considered as hotspots having higher chances of both wildfire ignition probability and spreading capacity. Furthermore, within the set of locations that have the highest sum values, those that have also the highest difference values (WIPI − WSCI), are the most critical ones. It indicates cases of the highest chances of forest fire ignition and relatively high spreading capacity of wildfire.

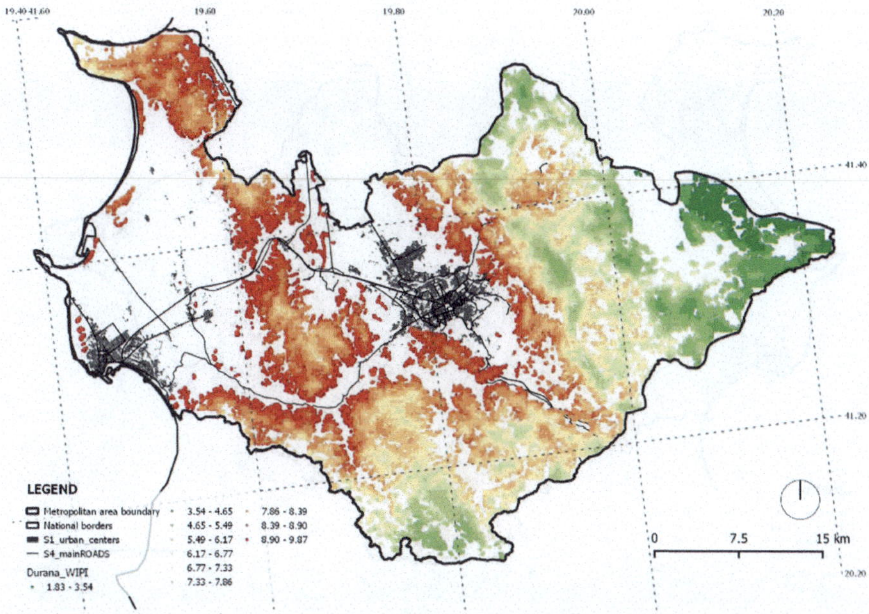

Fig. 4.7 Map of WIPI values of forest surfaces within Durana Metropolitan area

The spatial distribution of the WIPI and WSCI values are represented via indexing maps in Figs. 4.7 and 4.8. The maps show the set of reference points' locations covering the forest surfaces within the metropolitan area of Durana, being weighted according to the final WIPI and WSCI values. The green-yellow-red color ramp is selected to present in red the hotspots of endangered areas while in green the safer ones.

In general, the concept of risk is associated with social and urban systems vulnerability being exposed to certain hazards. But, in this study, the natural system's vulnerability is kept equally important to the former ones. In other words, the forest surfaces as precious assets of natural capital are considered vulnerable as well as elements at risk considering wildfire hazards. The forest fund vulnerability toward wildfire is accepted to be higher during the forest fire ignition phase since the majority of wildfire ignition events are advocated to be caused by human activity.

According to Fig. 4.7, the highest WIPI values belong to locations close to urbanized areas such as settlements and transportation network. At the same time, this can be considered a risk map showing the vulnerable forest surfaces under the wildfire ignition risk as the majority of wildfire ignitions are related to anthropogenic factors according to current literature. This remains on the same line with the weighted values reported in Table 4.3a where 67% of the ignition risk is related to human activity (non-natural).

On the other hand, Fig. 4.8 can be considered a risk and hazard map at the same time. It is presenting the spatial distribution of wildfire hazard potentially happening

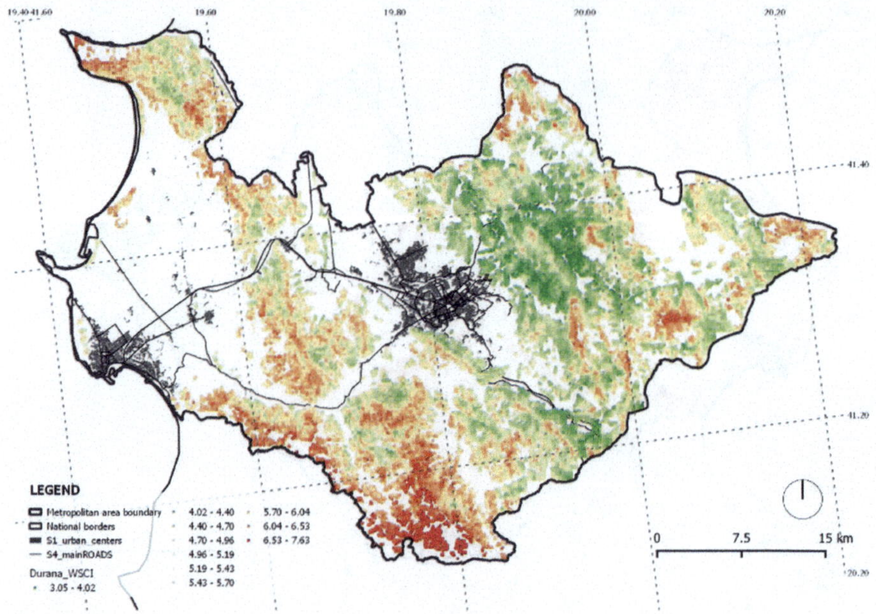

Fig. 4.8 Map of WSCI values of forest surfaces within Durana Metropolitan area

within the forest surfaces putting under risk not only the valuable forest fund but also the urban systems within the WUI. Even though, the highest WSCI values belong to remote locations, yet, there exist certain zones closer to urban areas with relatively high WSCI values. These locations can be considered as critical areas to be further studied at finer spatial scale in order to better assess elements at risk and vulnerability of social systems, especially within the WUI.

According to the new local administrative law within the framework of territorial and administrative reform in Albania, the local government is the main body managing the forest fund at municipal level (Hysa and Türer Başkaya 2018). Thus, both maps as well as the proposed methodology can serve as reference material for the local administration bodies of Tirana and Durres in preparing local strategic plans for disaster risk reduction in compliance with the target E of Sendai Framework (UNISDR 2015).

4.4 Conclusions and Recommendations

The results of the study validate a rapid and cost-free method for forest fire risk assessment in support of forest fire risk reduction agendas. In this respect, it can be considered a contribution to the discourse of open access science and research, which has been a vibrant topic in recent years (National Research Council 2004).

At this stage, it is successfully tested in the case of Durana metropolitan area, but potentially it is applicable and reproducible on similar study areas at metropolitan scale. Besides, its contribution to the open access science and research, this work is expanding the spatial scale scope of WIPI/WSCI method by presenting its application in metropolitan scale in which the sensitivity about wildfire events is highest. The wildland–urban interface is the most vulnerable zone, consisting of thc majority of elements at risk. Thus, the method presented here contributes to the development of useful methods in support of local level (metropolitan) DRR strategies being targeted to be widely prepared by a considerable amount of countries by 2020 (UNISDR 2015).

As earlier introduced in this chapter, Albania is persistently showing significant interest in participating and being involved within UNDRR agendas, but it remains weak in implementation. For example, in 2017 both Mr. Nishani in his speech (Nishani 2017) as well as the Sendai Framework Data Readiness Review Report about the Albanian case (Albania 2017), are targeting the approval of a new law which integrates the priorities, goals, and objectives of the Sendai Framework by 2019. Unfortunately, up to this date (July 30, 2019) the current law on civil emergencies being published on the official website of the government dates back to 2001. Similarly, the current "National Plan for Civil Emergencies" was prepared 15 years before in 2004 (Plani Kombetar per Emergjencat Civile 2004). Consequently, this study is crucial in developing a method which may provide reliable results to support the preparation of local strategic plans for DRR at metropolitan level.

In this study the method is applied via QGIS software, nevertheless it has also potential application by using Google Earth Engine (Gorelick et al. 2017) in order to generate global scale indexing maps of forest surfaces by their WIPI and WSCI values. The web-based application could make the process of wildfire risk assessment more inclusive as well as enhance the consciousness in the wider society about wildfire hazard at a global scale. As it is reported in other Earth Engine applications (Hansen et al. 2013; Pekel et al. 2016), such a work could be possible only through a cross-disciplinary research group. In this respect, this chapter can be considered an open call for collaboration in applying the method presented here, into the Google Earth Engine.

References

Ahrend R, Gamper C, Schumann A (2014) The OECD metropolitan governance survey: a quantitative description of governance structures in large urban agglomerations. OECD Reg Dev Work Pap 2014(4)

Albania (2017) Policy, plans & statements: national progress reports: Albania-Sendai framework data readiness review report. https://www.preventionweb.net/files/53183_albaniaalb.pdf. Accessed 30 July 2019

Brown CE (1998) Coefficient of variation. In: Applied multivariate statistics in geohydrology and related sciences. Springer, Berlin, Heidelberg, pp 155–157

Davis JB (1990) The wildland–urban interface: paradise or battleground? J For 88(1):26–31 de Souza FT, Koerner TC, Chlad R (2015) A data-based model for predicting wildfires in Chapada das Mesas National Park in the State of Maranhão. Environ Earth Sci 74(4):3603–3611

Fabi A, Piovene G (2016) Durana, Albania: a field of possibilities. Architect Des 86(4):114–117

Goepel KD (2013) Implementing the analytic hierarchy process as a standard method for multi-criteria decision making in corporate enterprises—a new AHP excel template with multiple inputs. In: Proceedings of the international symposium on the analytic hierarchy process, Kuala Lumpur

Gorelick N, Hancher M, Dixon M, Ilyushchenko S, Thau D, Moore R (2017) Google earth engine: planetary-scale geospatial analysis for everyone. Remote Sens Environ 202:18–27

Hansen MC, Potapov PV, Moore R, Hancher M, Turubanova SAA, Tyukavina A, Kommareddy A (2013) High-resolution global maps of 21st-century forest cover change. Science 342(6160):850–853

Hysa A, Başkaya FAT (2019) A GIS based method for indexing the broad-leaved forest surfaces by their wildfire ignition probability and wildfire spreading capacity. Model Earth Syst Environ 5(1):71–84

Hysa A, Türer Başkaya FA (2018) Land cover data as environmentally sensitive decision-making mediator in territorial and administrative reform. Cogent Environ Sci 4(1):1505326

Hysa A, Zeka E, Dervishi S (2017) Multi-criteria Inventory of Burned Areas in Landscape Scale; Case of Albania. In: K-FORCE first symposium, Novi Sad

ICSU (2008) A science plan for integrated research on disaster risk: addressing the challenge of natural and human-induced environmental hazards. International Council for Science, Paris

INSTAT (2014) A new classification scheme for the albanian urban-rural population: geographical typology of eu based on network of cells. INSTAT Publishing, Tirana

ISDR (2009) Terminology on disaster risk reduction. UNISDR, Geneva

Jenks GF (1967) The data model concept in statistical mapping. Int Yearb Cartogr 7:186–190

Keating M (1998) The new regionalism in western europe, territorial restructuring and political change. Edward Elgar Publishing Limited, Cheltenham

Martínez J, Vega-Garcia C, Chuvieco E (2009) Human-caused wildfire risk rating for prevention planning in Spain. J Environ Manage 90:1241–1252

McMaster R, McMaster S (2002) A history of twentieth-century American academic cartography. Cartography and Geographic Information Science 29(3):305–321

Mu E, Pereyra-Rojas M (2017) Understanding the analytic hierarchy process. In: Practical decision making. Springer, Cham, pp 7–2

National Research Council (2004) Open access and the public domain in digital data and information for science. In: Proceedings of an international symposium. National Academies Press, Washington

Nishani B (2017) Albania: statement made at the global platform for disaster risk reduction. UNDRR, Cancun

Pekel JF, Cottam A, Gorelick N, Belward AS (2016) High-resolution mapping of global surface water and its long-term changes. Nature 540(7633):418

MPVD, Plani Kombetar per Emergjencat Civile, Tirana: Ministria e Pushtetit Vendor dhe Decentralizimit, 2004

Radeloff VC, Hammer RB, Stewart SI, Fried JF, Holcomb SS, Mckeefry JF (2005) The wildland–urban interface in the United States. Ecol Appl 15(3):799–805

San-Miguel-Ayanz J, Moreno J, Camia A (2013) Analysis of large fires in European Mediterranean landscapes: lessons learned and perspectives. For Ecol Manage 294:11–22

Stefanidis S, Stathis D (2013) Assessment of flood hazard based on natural and anthropogenic factors using analytic hierarchy process (AHP). Nat Hazards 68(2):569–585

Theobald DM, Romme WH (2007) Expansion of the US wildland–urban interface. Landscape Urban Plann 88:340–354

UNISDR (2015) Sendai framework for disaster risk reduction 2015–2030. UN, Geneva

Zambon I, Cerdà A, Cudlin P, Serra P, Pili S, Salvati L (2019) Road network and the spatial distribution of wildfires in the Valencian community (1993–2015). Agriculture 9(5):100

Chapter 5
An Overview of the Integrated Flood Analysis System (IFAS) Studies in Insufficiently Gauged Catchments: Approaches, Challenges, and Prospects

M. F. Chow

Abstract Flooding problem is becoming a great challenge for developing countries because of climate change deforestation and urbanization processes. Lack of hydrological data and catchment information has hindered the local government to set up the flood early warning system where can predict the lead time of flood and thus reduce the vulnerability to flood disaster. Therefore, the Integrated Flood Analysis System (IFAS) is developed to predict the flood event in insufficiently gauged catchments. IFAS can automatically collect the geographical data, soil type, land uses, and satellite rainfall data to set up the river basin model for flood simulations. This paper provides an overview of approaches, challenges, and prospects of using IFAS model for flood prediction at several river basins with different catchment characteristics in Asian countries. The results of previous studies suggest that IFAS can better simulate the flood in large river basins compared to small river basins. Flood forecasting with calibrated satellite rainfall data in IFAS model performed higher reproducibility than satellite rainfall without calibration particularly for the beginning and peak of hydrograph.

Keywords Flood forecasting · Flood early warning system · Insufficiently gauged catchment · Satellite rainfall

5.1 Introduction

According to the IPCC report, frequent flood and drought events are occurring worldwide due to the intensification of hydrological cycle driven by climate change which results in the spatiotemporal fluctuation of rainfall (IPCC 2014). The damages due to the past 24 major flood events in Pakistan have caused an aggregated financial loss

M. F. Chow (✉)
Institute of Sustainable Energy (ISE), Universiti Tenaga Nasional, Jalan IKRAM-UNITEN, 43000 Kajang, Selangor, Malaysia
e-mail: mingfaichow12345@gmail.com

R. Djalante et al. (eds.), *Integrated Research on Disaster Risks*, Disaster Risk Reduction,
https://doi.org/10.1007/978-3-030-55563-4_5

of more than US$ 38.171 billion, the human death toll of 12,330, affecting approximately 197,275 villages with over 616,598 km^2 of area inundated by the floodwaters during the past 69 years (FFC 2017). Especially the 2010 Megaflood event in the Indus River which was driven by exceptional monsoon storms has affected almost all areas in Pakistan (NDMA 2010). Flooding has become a big challenge for many Asian countries in recent decades due to climate change and uncontrolled urban developments (Bormudoi et al. 2011; Sugiura et al. 2014a, b). Severe flood events had occurred in the past in countries like Myanmar (2008), Philippines (2009), Pakistan (2010), Thailand (2011), the Philippines (2013), and Malaysia (2014). Huge economic losses due to the damage of infrastructures and even human life losses are burdening the responsibility of the country's government. Since it is almost impossible to prevent from flood hazards completely with sufficient structural measures due to its high cost, implementation of flood forecasting and early warning system becomes the main strategy for the authorities to reduce the vulnerability against flood risk (Sugiura et al. 2014a, b; Chinh et al. 2014; Chow and Jamil 2017). The Sendai Framework for Disaster Risk Reduction 2015–2030 was adopted during the Third UN World Conference to achieve the substantial reduction of disaster risk and losses in lives, livelihoods, and health and in the economic, physical, social, cultural, and environmental assets of persons, businesses, communities and countries over the next 15 years. The Sendai Framework for Disaster Risk Reduction 2015–2030 outlines four priorities for action to prevent new and reduce existing disaster risks: (i) Understanding disaster risk; (ii) Strengthening disaster risk governance to manage disaster risk; (iii) Investing in disaster reduction for resilience; and, (iv) Enhancing disaster preparedness for effective response, and to "Build Back Better" in recovery, rehabilitation, and reconstruction.

Forecasting the flood event requires input data such as topography map, rainfall, streamflow, river cross section, land use, and soil type data. These data are required for the setup of hydrological and hydraulic models and undergo calibration and validation processes before implemented for forecasting the flood events. However, the problems of flood forecasting system installation in poorly gauged river basins are including (i) difficulty to get real time hydrological data in the upstream of a transboundary river basin; (ii) insufficient of implementation and maintenance of ground-based real-time hydrological observation stations, such as rain gauge and river discharge gauging station with data transmission system; (iii) lack of the data required for the creation of a flood forecasting model such as altitude, land use, and river channel network, etc.; (iv) lack of budget for flood forecasting system installation and; (v) insufficient framework to enhance the technical capabilities (Fukami et al. 2006; Miyamoto et al. 2014; Kimura et al. 2014; Sugiura et al. 2014b). Therefore, there is a great need for developing a flood forecasting system that able to acquire the necessary information in an ungauged river basin for efficient and accurate flood prediction. As such, Integrated Flood Analysis System (IFAS) is developed by ICHARM (International Centre for Water Hazard and Risk Management) and Public Works Research Institute (PWRI) of Japan as advanced technology to forecast the flood event using satellite rainfall data and Geographic Information System (GIS). The objective of this paper is to review the applications and challenges of IFAS for flood simulation at different river basins with different catchment characteristics in Asian countries. Some future studies are recommended for the applications of IFAS in forecasting the flood event.

5.2 Integrated Flood Analysis System (IFAS)

The Integrated Flood Analysis System (IFAS) is a flood forecasting system that used to perform rainfall–runoff simulation for river basins with insufficient hydrological data to reduce its vulnerability to flooding disasters. IFAS is developed by the International Centre for Water Hazard and Risk Management (ICHARM). The IFAS software is capable to import satellite-based rainfall data such as GSMaP and 3B42RT for insufficient ground observation river basin and other data includes geographic and land use data for almost the entire world (ICHARM 2011). Using satellite-based rainfall information, flood prediction can be performed even for river basins where no ground-based observation network exists, or one exists but is inadequate. If and when a ground-based observation network is put in place, it can be incorporated into the flood prediction system. The Public Works Research Institute Distributed Hydrological Model (PDHM) is employed in IFAS software as its runoff simulation model. The structure of PDHM is consists of a surface tank model, subsurface tank model, aquifer tank model, and river tank model. IFAS is equipped with numerous features, including automatic creation of river channel networks in a river basin, graphic display flow rate changes over time, and automatic alerts. The system is able to conduct a series of operations required for runoff analysis.

5.2.1 Model Structures

The structures of IFAS can be classified into four main components: input rainfall, basin model, runoff calculation engine, and results display interfaces as shown in Fig. 5.1. The input rainfall data for IFAS can be obtained either from ground-based rain gauges or satellite rainfall data. The global satellite rainfall datasets provided by National Aeronautics and Space Administration (NASA) and National Oceanic and Atmospheric Administration (NOAA), USA, and Japan Aerospace Exploration Agency (JAXA) offer different spatial and temporal resolutions and time of delay.

The IFAS runoff basin model can be developed by utilizing global GIS data. The GIS tool kits were used to obtain the geophysical data and construct the river basin model. The imported global GIS data includes catchment elevation, land use/land cover, soil type, and geology. The topography map and river channel network for the modeled catchment can be constructed by using the digital elevation model (DEM) data provided by the United States Geological Survey (USGS) as shown in Fig. 5.2. The land use data can be downloaded from the Global Land Cover Characterization (GLCC) database provided by USGS. Meanwhile, the soil type and

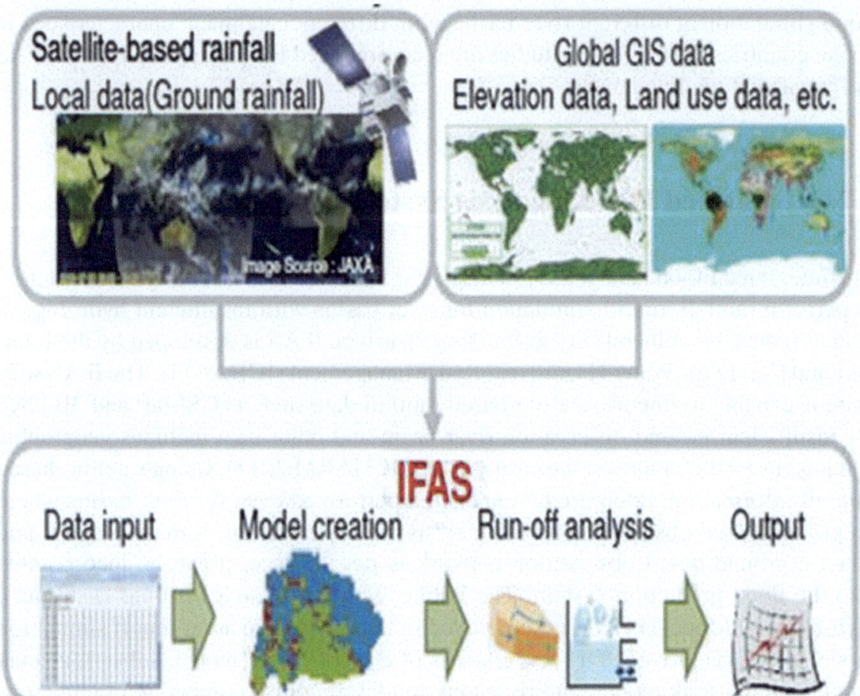

Fig. 5.1 Model structures of Integrated Flood Analysis System (IFAS)

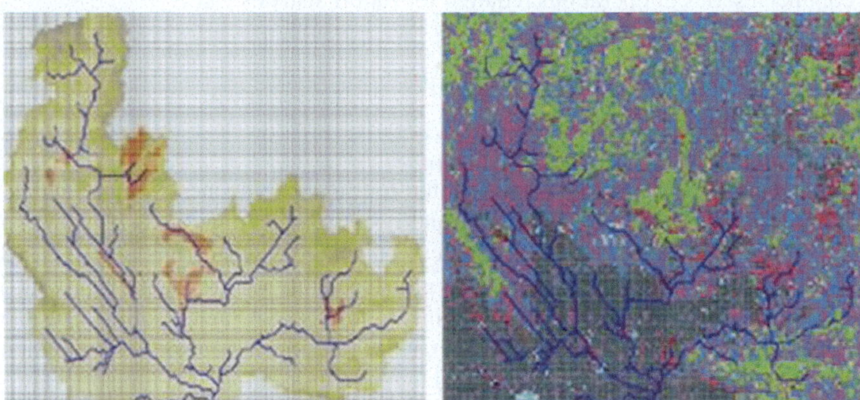

Fig. 5.2 River channel network and land use map in the IFAS basin model

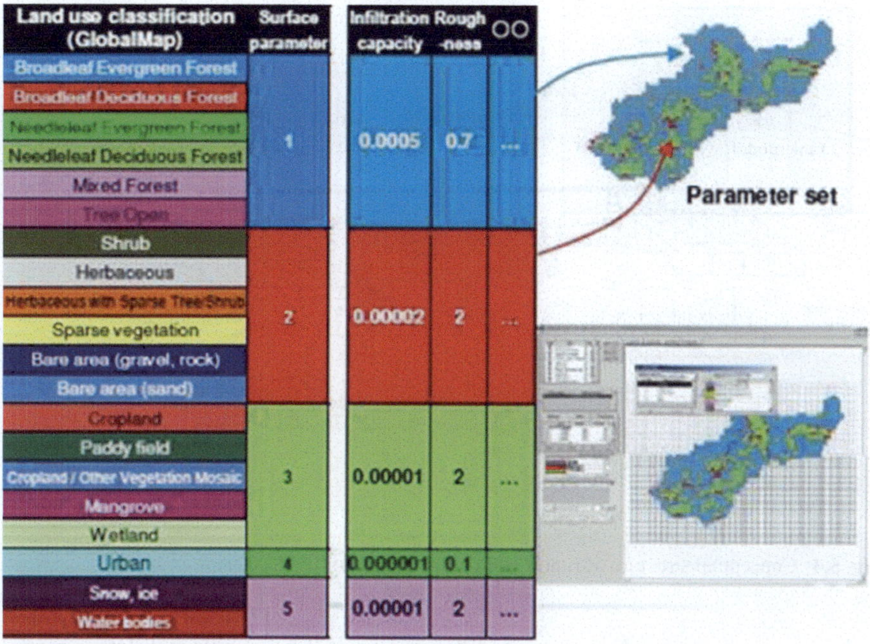

Fig. 5.3 Parameter setting function in IFAS basin model

geology data can be acquired through the database provided by the United Nations Environment Programme (UNEP) and Commission for the Geological Map of the World (CGMW), respectively. All these data are used to estimate the hydrological parameters for runoff simulation model in IFAS. Since all data can be obtained from global GIS databases and satellites, it is possible for IFAS to simulate the rainfall–runoff at an ungauged catchment where no or limited hydrological information. The coefficient and parameter values for each cell in the runoff model can be set up automatically based on the GIS data. Default values have been recommended for all surface, subsurface, aquifer, and river course tanks in the IFAS model (as shown in Fig. 5.3). This standard guideline allows the IFAS user to set the parameter values without the past hydrological information. The calibration for IFAS model simulation can be carried out by tuning these parameters in order to match the observed results. The parameters are tuned and set by using the trial and error method.

5.2.2 *Runoff Simulation*

The runoff simulation engine for IFAS is Public Works Research Institute (PWRI) distributed hydrological model. The PWRI distributed hydrological model is configured as three- or four-tank models that are surface, unsaturated if three layers, aquifer

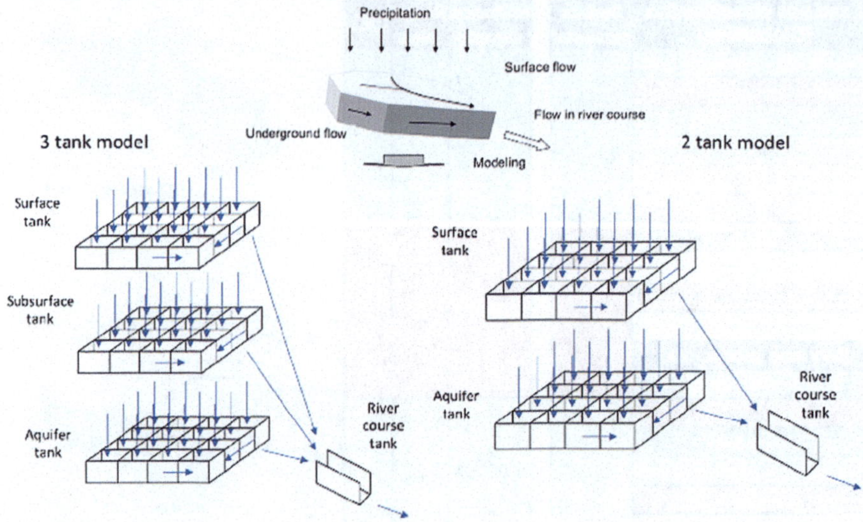

Fig. 5.4 Conceptual structure of runoff simulation engine for IFAS

and river course tanks. The conceptual structures of IFAS runoff simulation model are represented in Fig. 5.4. The times taken for executing the runoff simulation may vary from several minutes to a few hours which depends on the number of calculation cell sizes, simulation period, and computer capability.

The surface tank model separates the rainfall into surface, rapid intermediate, and ground infiltration flows as shown in Fig. 5.5a. The input parameters for surface tank model includes vertical hydraulic conductivity (f0), maximum water height (Sf2), height where rapid unsaturated subsurface flow occurs (Sf1), minimum height for infiltration to start (Sf0), surface roughness coefficient (N), mesh length (L), rapid unsaturated subsurface flow regulation coefficient (αn), and initial water height. The subsurface tank model is used to simulate the low flow conditions as well as long-term periods. The input parameters for subsurface tank model includes tank height (D), horizontal saturated hydraulic conductivity (Ksx), vertical saturated hydraulic conductivity (Ksz), saturated moisture content (θs), moisture content at field capacity (θFC), vertical hydraulic conductivity at field capacity (KFC_z), and initial water height as shown in Fig. 5.5b. The configuration of aquifer model is shown in Fig. 5.5c. The input parameters for aquifer tank are slow unsaturated subsurface flow regulation coefficient (Au), base flow coefficient (Ag), water height where the slow unsaturated subsurface flow occurs (Sg), and initial water height and the coefficient for unaccountable aquifer loss (αGW). For river discharge calculation, the equations used differ according to the cell type. The input parameters for river course tank model includes breadth of the river course, coefficient of the resume law: c, coefficient of the resume law: s, Manning's roughness coefficient, initial water level in the river course, infiltration from river tank to the aquifer tank, coefficient for cross section shape, and meander coefficient (as shown in Fig. 5.5d).

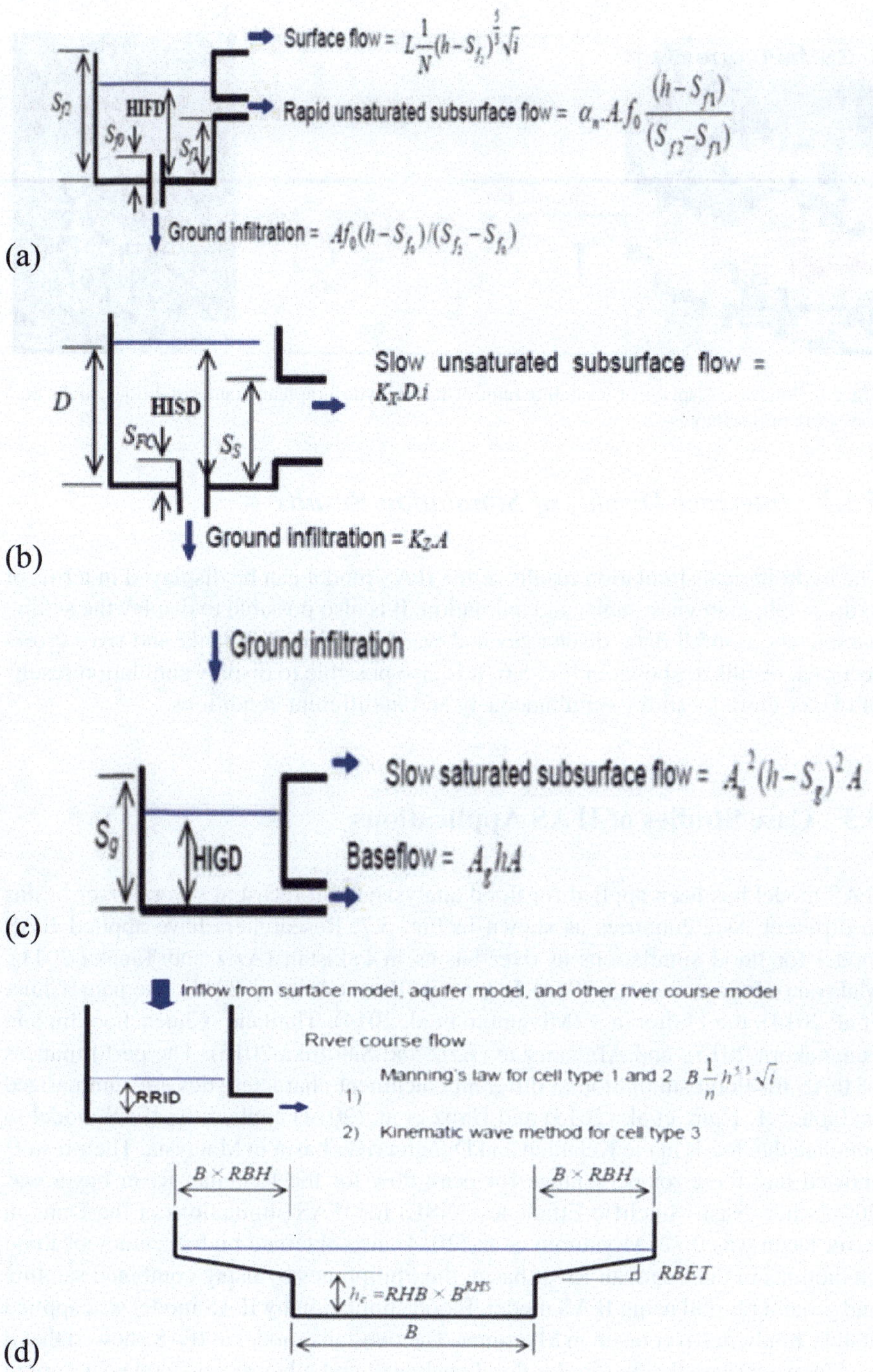

Fig. 5.5 Schematic representations of **a** surface, **b** subsurface, **c** aquifer, and **d** river course tanks

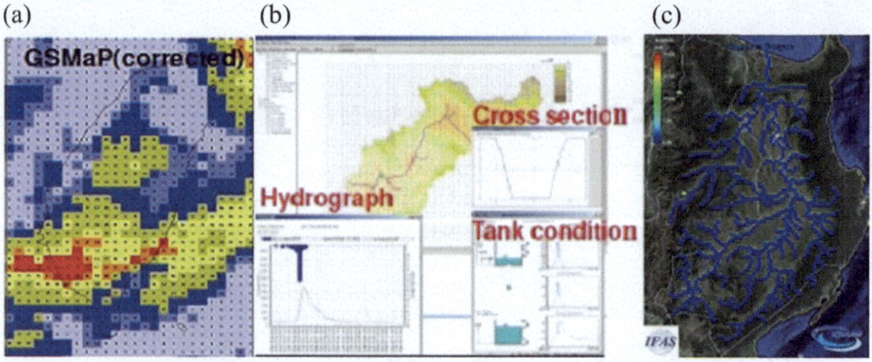

Fig. 5.6 Interface displays of **a** satellite rainfall data, **b** hydrograph and tank conditions, and **c** plan view of river discharges

5.2.3 *Interface Display of Simulation Results*

The hydrological simulation results of the IFAS model can be displayed in terms of hydrograph, plan view, table, and animation. It is also possible to display the setting parameters, rainfall data, discharges and water level in each tank, and river cross-sectional results as shown in Fig. 5.6. It is also possible to display simulation results of two or more locations simultaneously and in different conditions.

5.3 Case Studies of IFAS Applications

IFAS model has been applied for flood analysis and forecast at several river basins in different Asia countries as shown in Fig. 5.7. Researchers have applied IFAS model for flood simulations at river basins in Pakistan (Aziz and Tanaka 2011), Malaysia (Hafiz et al. 2013, 2014), Indonesia (Bormudoi et al. 2011), Vietnam (Chinh et al. 2014), the Philippines (Miyamoto et al. 2014), Thailand (Chuenchooklin and Pangnakorn 2015), and Afghanistan (Aziz and Satofuka 2015). The performances of IFAS for flood simulation at different catchment characteristics are summarized in Table 5.1. Hafiz et al. (2013) and Hafiz et al. (2014) applied the IFAS model to simulate the floods in the Kelantan and Dungun river basin in Malaysia. Their results showed that the error percentage for peak flow for the Kelantan River basin was 26%, while Nash–Sutcliffe Efficiency (NSE) for IFAS simulations at the Dungun River basin was 0.72. Miyamoto et al. (2014) also obtained high accuracy of flood predictions in the Cagayan River basin, the Philippines by using combined satellite and ground rainfall using IFAS model. Flood simulation by IFAS model was applied at the Chindwin River basin in Myanmar. The two-tank model in IFAS showed that it is able to simulate the first major flood peak well and other major flood peak timing well but underestimated the low flows. Meanwhile, the three-tank model reproduced

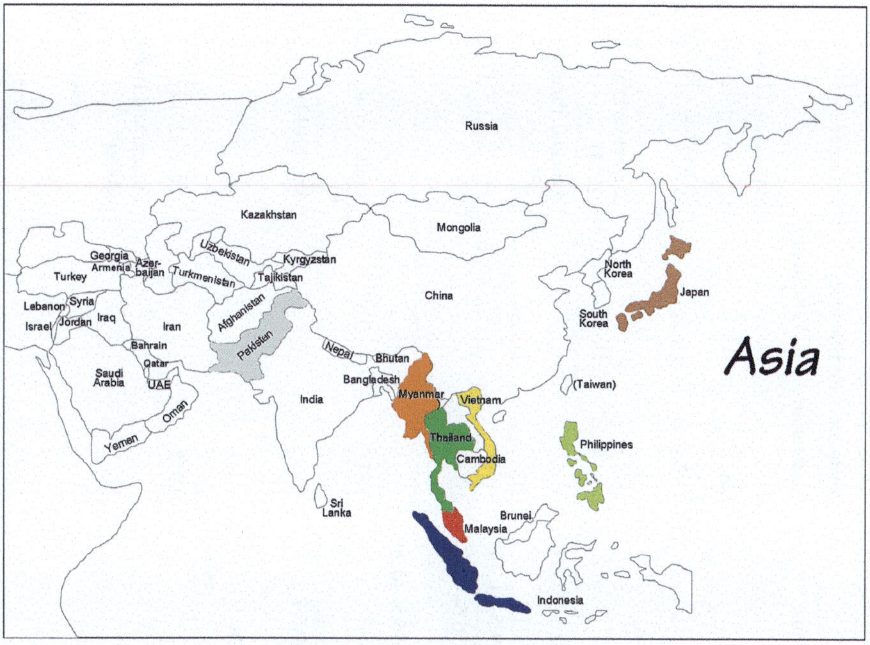

Fig. 5.7 Case studies of IFAS application for flood simulations in Asian countries

both major flood peaks (timing and level) and the river discharges better as shown in Fig. 5.8. Chinh et al. (2014) assessed the accuracy of the IFAS simulation model by a relative error of the flood peak (ΔQ) and Nash–Sutcliffe coefficient (R). Their results showed that the flow simulation model achieves high accuracy in large rivers ($\Delta Q = -13.63\%$, $R = 0.93$), while low accuracy in small river branches ($\Delta Q = +57.64\%$, $R = 0.44$). This may suggest that IFAS can better simulate large river basins as compared to small river basins. Similarly, a basin-wide flood prediction system based on IFAS model using open source geospatial and meteorological data is being used in the upper and mid reaches of the Indus and Kabul river basins (Sugiura et al. 2016). IFAS was also calibrated and validated for the upper Indus catchment with an average NSE of 0.8 (Sugiura et al. 2014a, b). IFAS model performance improved, as local soil data was used in the upper Indus catchment (Sugiura et al. 2016). Aziz and Tanaka (2011) compared the results of rainfall from multiple sources to model the upper middle Indus River. The trans-boundary Kabul River basin was successfully modeled for floods using IFAS (Aziz 2014). These studies showed the capacity of IFAS to accurately simulate the flood peaks in large-scale, data-scarce basins.

Table 5.1 Performances of IFAS for flood simulation at different catchment characteristics

No	Catchment	Area (km^2)	Rainfall type	Calibration (%)		Nash–Sutcliffe Efficiency		Reference
				Peak flow	Volume	Calibration	Validation	
1	Kelantan	11,099	Ground	26	–			Hafiz et al. (2014)
			Ground			0.8		Chow et al. (2019)
2	Dungun	1858	Ground	1.68		0.72		Hafiz et al. (2013)
			Satellite GSMAP	41.96%				
3	Cagayan	27,280	Ground	< 10%				Miyamoto et al. (2014)
4	Upper-Middle Indus	467,136	Ground	−85.8–44.9%	−62.7–27.3%			Aziz and Tanaka (2011)
			Satellite GSMAP	53.2–82.7%	−69.5–71.7%			
			Satellite GSMaP (corrected)	−21.7–48.1%	−156.5–48.8%			
			Satellite 3B42RT	−48.1–12.1%	−155.5–20.2%			
5	Upper Indus	133,300	Ground			0.61	0.67	Sugiura et al. (2014a, b)
6	Kabul	92,605	Satellite GSMaP (original)	87%	−69.5%			Aziz (2014)
			Satellite GSMaP (corrected)	−6.8%	−156.5%			
7	Bengawan Solo	16,100	Ground			0.955	1.106	Astuti (2014)
8	Thai Binh River, Thailand	17,580	Ground					Fukami and Herath (2009)

(continued)

Table 5.1 (continued)

No	Catchment	Area (km^2)	Rainfall type	Calibration (%)		Nash–Sutcliffe Efficiency		Reference
				Peak flow	Volume	Calibration	Validation	
9	Bang Giang river basin	4,200	Ground	−13.6–57.6%		0.44–0.93		Chinh et al. (2014)
10	Citarum Hulu River basin, Indonesia	1,771	Ground			0.51		Bormudoi et al. (2011)
11	Indragiri, Riau	7467	Ground		10.01%	1.3	1.83	Hendra et al. (2015)
			Satellite (corrected)		0.1%	0.52	2.15	
12	Pua, sub-basin of Nan River basin, Thailand	404	Ground	−5.2%				Chuenchooklin and Pangnakorn (2015)
13	Chindwin river basin	27,420	Ground	62%				Fukami and Herath (2009)
			Satellite	−9.65%				
14	Shijyushida Dam river basin	1196	Ground	< 10%				Sugiura et al. (2008)
	Kakogawa River basin (Oshima)	1291	Ground	< 10%				

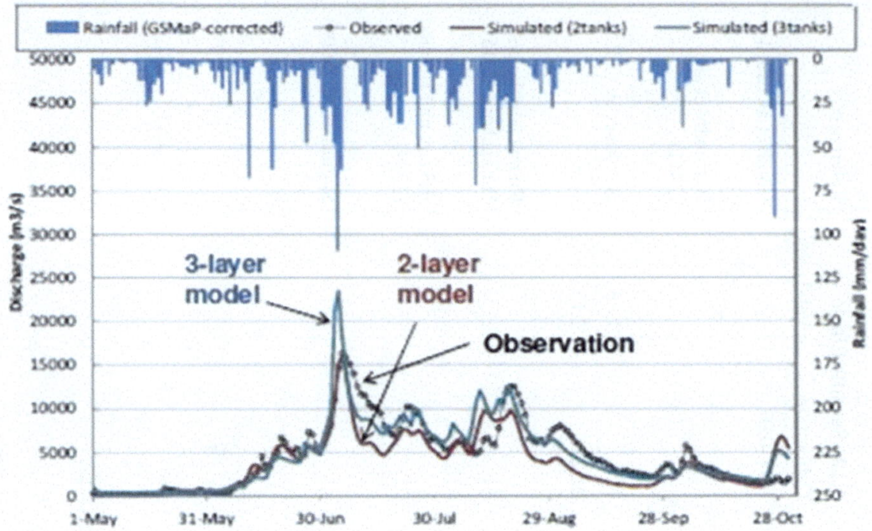

Fig. 5.8 Comparison of simulation results by using two-tank and three-tank models in IFAS for the Chindwin River basin, Myanmar

5.4 Challenges and Prospects

The IFAS model has the advantage function to download the satellite rainfall data as input data for flood simulation at all river basins worldwide. However, the accuracy using satellite data is normally lower than from ground-based observation data as shown in Fig. 5.9. For example, the peak discharge error percentage using satellite rainfall data is 41.96% compared to 1.68% using ground-based rainfall data for IFAS study in Dungun catchment, Terengganu, Malaysia (Hafiz et al. 2013). Correction is needed to adjust the satellite rainfall data for flood prediction in IFAS model. Shahzad et al. (2018) concluded that the satellite rainfall estimates must be corrected to improve the IFAS simulation results. Aziz and Tanaka (2011) obtained better flood simulation results using corrected GSMaP satellite rainfall data compared to the Satellite 3B42RT and GSMaP (original) for Pakistan flood event in 2010. The discharge calculated by the Satellite GSMaP_NRT (corrected) is well synchronized with the measured discharge. The flood duration and flood peak calculated by the Satellite GSMaP_NRT (corrected) also have the best agreement with the observed one. Meanwhile, the calculation results of the Satellite GSMaP_NRT (original) have low signals. The Satellite GSMaP-NRT (original) neither captured the flood duration nor did the flood peak. Since the satellite rainfall data is inferior to ground observation data in terms of mesh size and accuracy, these features can only be applied for major river basins of a certain size where ground-based observation is poor.

Another problem with satellite rainfall data is time intervals of these data are not in real time. Because satellite-based rainfall data is not in real-time, this system is not

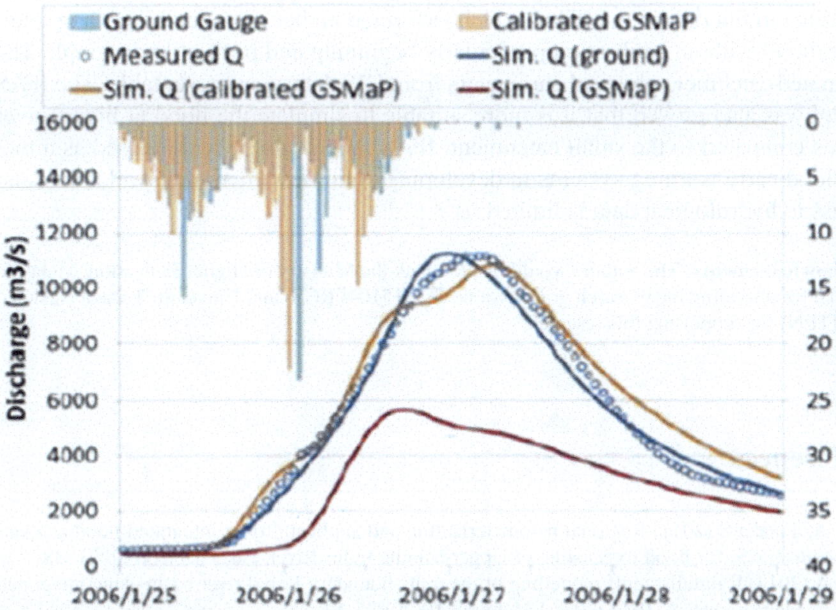

Fig. 5.9 Flood simulation results using observed, satellite, and corrected satellite rainfall data in a poorly gauged river basin

well suited for small rivers where floods can arrive quickly. Therefore, it is desirable to make use of ground-based data as soon as ground observation equipment has been installed in the future. Meanwhile, it is difficult to establish appropriate parameters and obtain good results without precise flow measurement in order to convert water levels to flow volumes. Thus, it is also important to conduct precise measurement of river cross sections and observation of flow volumes in order to accurately determine flow volumes from water levels. The IFAS model also manages to analyze the flow changes and flooding in the event if the river flow volumes increase to the point that the river breaches its banks. Nonetheless, this IFAS system can be used as supplementary information for rivers where local ground-based observation equipment is insufficient and analyze downstream river flow in international rivers that straddle more than one country.

5.5 Conclusions

This paper has reviewed the capability and limitation of the Integrated Flood Analysis System (IFAS) in flood forecasting. IFAS is able to integrate the satellite-based rainfall data and global GIS data for insufficiently gauged catchments at different river basins with various catchment characteristics. Flood forecasting with calibrated satellite rainfall data in the IFAS model performed higher reproducibility than satellite rainfall without calibration particularly beginning and peak of hydrograph. The estimated catchment physical parameters from GIS data are representable. The IFAS model was also proved that it is more suitable to simulate the flood at bigger river basins compared to the small catchment. IFAS model can be widely used as a tool for flood early warning systems in developing countries where the availability and access to hydrological data is limited.

Acknowledgements The authors would like to thank the Ministry of Higher Education, Malaysia (MOE) for providing the research grant (Vot no: F2015104FRGS) and Universiti Tenaga Nasional (UNITEN) for supporting this research.

References

Aziz A, Tanaka S (2011) Regional parameterization and applicability of integrated flood analysis system (IFAS) for flood forecasting of Upper-Middle Indus River. Pak J Meteorol 8:21–38

Aziz A (2014) Rainfall-runoff modelling of the trans-boundary Kabul river basin using integrated flood analysis system (IFAS). Pak J Meteorol 10(20):75–81

Bormudoi A, Hazarika MK, Budi Kartiwa Ir, Murniati E, Samarakoon L (2011) Hydrograph simulation and flood mapping using remote sensing and GIS in citarum watershed, Indonesia. Proceedings of the 32rd Asian conference on remote sensing, Taipei, Taiwan

Sugiura A, Fujioka S, Nabesaka S, Tsuda M, Iwami Y (2016) Development of a flood forecasting system on the upper Indus catchment using IFAS. J Flood Risk Manage 9(3):265–277

Sugiura A, Fujioka S, Nabesaka S, Tsuda M, Iwami Y (2014a) Development of a flood forecasting system on upper Indus catchment using IFAS. Proceedings of 6th International Conference on Flood Management. Sao Paulo, Brazil, pp 1–12

Sugiura A, Fujioka S, Nabesaka S, Sayama T, Iwami Y, Fukami K, Tanaka S, Takeuchi K (2014b) Challenges on modelling a large river basin with scarce data: a case study of the Indus upper catchment. J Hydrol Environ Res 2(2014): 59–64

Astuti HP (2014) Rainfall-runoff modelling of Bengawan Solo catchment area with distributed model using integrated flood analysis system. Master Theses of Civil Engineering, RTS 551.48 Ast p, 2014. Institut tenkologi Sepuluh Nopember, Indonesia

Aziz F, Satofuka Y (2015) Flood runoff analysis and water availability for integrated river basin management planning in Panj-e-Amu, Afghanistan. Master thesis of Engineering

Chow MF, Jamil MM (2017) Review of development and applications of integrated flood analysis system (IFAS) for flood forecasting in insufficiently-gauged catchment. J Eng Appl Sci 12(special issue 11), 9210–9215

Chinh DD, Thuan NTT, Van PT, Thanh TN, Manh VV (2014) Research the applicability of IFAS model in flood analysis (pilot at Bang Giang river basin in Cao Bang Province). Proceedings of the 28th EnviroInfo 2014 Conference, Oldenburg, Germany, pp 1–8

FFC (2017) Annual flood report for the year 2016. Federal Floods Commission, Government of Pakistan

Fukami K, Herath S (2009) Flood risk management demonstration project under the Asian water cycle initiative for the global earth observation system of systems (FRM/AWCI/GEOSS). Final report for APN project. ARCP2009–01CMY-Fukami

Hafiz I, Sidek LM, Basri H, Fukami K, hanapi MN, Livia L, Jaafar AS (2014) Integrated flood analysis system (IFAS) for Kelantan river basin. Proceedings of IEEE 2nd international symposium on telecommunication technologies (ISTT) (2014)

Hafiz I, Nor NDM, Sidek LM, Basri H, Hanapi MN, Livia L (2013) Application of integrated flood analysis system (IFAS) for Dungun river basin. Proceedings of 4th international conference on energy and environment (ICEE 2013). IOP Conf. Series: Earth and Environmental Science 16 (2013)
ICHARM (2011) IFAS system instruction guidebook, pp 43
IPCC (2014) Climate change 2014: synthesis report. Contribution of working groups I, II and III to the fifth assessment report of the intergovernmental panel on climate change. In: Pachauri RK, Meyer LA (eds) Core Writing Team, vol 151, IPCC, Geneva, Switzerland, pp 1–112
Fukami K, Fujiwara N, Ishikawa M, Kitano M, Kitamura T, Shimizu T, Hironaka S, Nakamura S, Goto T, Nagai M, Tomita S(2006) Development of an integrated flood runoff analysis system for poorly-gauged basins. Proceeding of the 7th International Conference on Hydroinformatics, pp 2845–2852
Kimura N, Tai A, Chiang S, Wei H-P, Su Y-F, Cheng C-T, Kitoh A (2014) Hydrological flood simulation using a design hyetograph created from extreme weather data of a high-resolution atmospheric general circulation model. Water 6:345–366
Miyamoto M, Ono M, Nabesaka S, Okazumi T Iwami Y (2014) Applicability of a flood forecasting method utilizing global satellite information to an insufficiently-gauged river basin: a case of a river basin in the Philippines. Proceeding of 11th International Conference on Hydroinformatics (HIC), (2014), pp 1–9
Chow MF, Jamil MM, Che Ros F, Yuzir MAM, Hossain MS (2019) Evaluation of parameter regionalization methods for flood simulations in Kelantan River basin. In J Innov Technol Explor Eng 8(7S):313–318
NDMA (2010) Annual report of national disaster management authority. Government of Pakistan
Chuenchooklin S, Pangnakorn U (2015) Flood management tool for small catchment in the Nan river basin Thailand. J Appl Sci Res 11:1–6
Shahzad AS, Gabriel HF, Haider S, Mubeen A, Siddiqui MJ (2018) Development of a flood forecasting system using IFAS: a case study of scarcely gauged Jhelum and Chenab river basins. Arab J Geosci 11:383
Sugiura T, Fukami K, Inomata H (2008) Development of integrated flood analysis system (IFAS) and its applications. Proc World Environ Water Resourc Congress 1–10
Hendra Y, Fauzi M, Sutikno S (2015) Hybrid data hujan ARR dan Satelit guna peningkatan efektifitas model IFAS. Ann Civil Eng Sem 61–72

Chapter 6
Heat Vulnerability Index Development and Application in Medan City, Indonesia

Martiwi Diah Setiawati, Marcin Pawel Jarzebski, and Kensuke Fukushi

Abstract In Medan City, the Indonesian largest urban area in Sumatra Island, the extreme temperature events in the past 30 years have become more frequent. This phenomenon causes potential health thread to rapidly growing urban population. Moreover, the situation is likely to become more severe in the future under urban sprawl and climate change. Accordingly, it is imperative to understand the vulnerability of the city to extreme temperature with robust indices and spatial visualization in order to provide a baseline for planning responses to such risk. Thus, this research aims to assess heat vulnerability in Medan City, by developing and applying Heat Vulnerability Index (HVI) based on the commonly used health indicators and Principal Component Analysis (PCA). Spatial visualization of the HVI analysis demonstrated a higher vulnerability within the downtown areas of Medan City, compared with suburbs where the density of built-up areas contributes as the highest factor loading for the index. The positive effect of urban green area also can be shown by the assessment of PCA where the sign of factor loading was negative. Furthermore, our assessment also indicates that the HVI has positive correlation with number of hypertension patient and respiratory disease patient as 0.71 and 0.46, respectively. Though the impacts of heat stress are alarming, the integration of heat-related risk map into the master plan for spatial planning of Medan City have not been taken into consideration. This information will be very useful for local authorities when deciding on targeted campaign of urban climate adaptation, for example, increasing open green space by private sector including green roof top initiative. In addition, one of the HVI components (i.e., Universal Thermal Climate Index) also could assist the heat warning system in the future. However, some limitations existed in this study such as the analysis was at sub-district level instead of finer spatial resolution and the ideal study should include more heat-related illnesses but these data are not currently available in the study area.

Keywords Heat vulnerability index · Medan · Urban climate · Environmental health

---

M. D. Setiawati (✉) · M. P. Jarzebski · K. Fukushi
Institutes for Future Initiatives, The University of Tokyo, Tokyo, Japan
e-mail: martiwi1802@gmail.com; martiwi1802@ifi.u-tokyo.ac.jp

R. Djalante et al. (eds.), *Integrated Research on Disaster Risks*, Disaster Risk Reduction,
https://doi.org/10.1007/978-3-030-55563-4_6

6.1 Introduction

6.1.1 Increasing Heat Stress and Its Health Effects

Human health is highly sensitive to thermal environment (WHO 2008). The risk of heat exposure is a growing concern in the urban areas which have been already experiencing heat island effect, yet imminent to increase in the future (Tomlinson et al. 2011a, b). Furthermore, the increasing exposure to heat can potentially bring economic losses through reduced productivity (Kjellstrom and Meng 2016) and morbidity (Robine et al. 2008). Exposure to such heat, which adds stress to human health, is an extremely critical issue that needs to be tackled, as it causes a range of health impacts such as heatstroke, heat exhaustion, heat cramp, increased risk of kidney diseases, reduced concentration and work efficiency (Monazzam et al. 2014) (Jay and Kenny 2010). Vulnerability to heat stress, however, vary on physical and social conditions as other studies demonstrated (Reid et al. 2009; Coutts et al. 2007; Wolf and McGregor 2013; Kim et al. 2011).

Most developing countries are at risk against climate change impact, despite not contributing the majority of greenhouse gasses emission (Lundgren et al. 2013). As a prominent example is Indonesia, whereas its carbon dioxide emission rate is below the world average (World Bank 2016) and yet has been experiencing high impacts of extreme temperature events. Indonesia is a tropic country where inter-annual temperature variability is low. However, the number of very hot days is foreseen to increase the most (Hoegh-Guldberg et al. 2018). In Borneo and Sumatra islands, Indonesia, there was spatially consistent trend of increasing warm nights and days of climatic data between the late 1960 and 2003 (Alexander et al. 2006). Such abnormal events will be more frequent in the future. Under the most severe RCP8.5 scenario, Indonesia was projected to have three heat waves in the period of 2020–2052, and these heatwaves were projected as frequent as every two years during the period of years between 2068 and 2100 (Russo et al. 2014). Moreover, any increase in global temperature is projected to have a negative impact on human health including heat-related mortality and morbidity and cities are special concern which has the potential to be hotter due to urban heat islands effect (Hoegh-Guldberg et al. 2018). Furthermore, the heat exposure causes potential economic losses to Indonesia, with estimated loss amount of 250 billion US, equivalent to GDP loss of 6 percent, by the year 2030 (Kjellstrom and Meng 2016).

6.2 Approaches to Measure Heat Stress and Vulnerability

Heat stress is a condition when our body is unable to maintain a healthy temperature. The quantitative assessment of heat stress has been introduced in a form of heat stress indices. Existing variety of heat stress indices are applied as tools for assessing heat stress but they are limited to basic environmental parameters on human response in the computation (Monazzam et al. 2014). The most recent one is proposed by Blazejczyk et al. (2012), i.e., the Universal Thermal Climate Index (UTCI), a non-occupational heat stress index, which combines all the key climate factors into a single number. It is designed to evaluate the outdoor thermal environment in the context of public health, in reference to human thermo-physiology studies (Golbabaei et al. 2013). However, there are still many factors that have not yet counted in the heat stress indices, such as population characteristic and physical properties, which also affect heat stress level (Kjellstron et al. 2016). For this purpose, we developed Heat Vulnerability Index (HVI) which covered heat stress index, physical properties, and population characteristics. The HVI should be assessed by the range of determinants of heat-related health effects and through studying epidemiologic literature and consulting public health experts (Bao et al. 2015).

In urban areas, HVI calculation needs to consider more comprehensively the local climate variability particularly in relation to higher heat absorption of certain land use types (e.g., vegetation and building coverage), surface properties (e.g., emissivity, land surface temperature), and decreasing ventilation geometry of the urban area (Oke 1987; Stabler et al. 2005). These elements are responsible for micro-climate alternations in urban areas known as the Urban Heat Island (UHI). Previous studies clearly showed that alternation of green spaces into build areas already drastically heat up the environment and influences human health in urban areas (Kilbourne et al. 1982; Tan et al. 2007). Additional anthropogenic heat release, particularly from increasing energy consumption (i.e., electricity, gas, and oil) (Smith et al. 2009), contributes to temperature increase. Furthermore, population characteristic such as age, poverty level, health condition, and population density also affect heat-related illness (Kjellstron et al. 2016). Alongside the higher number of covariates needed to be included in HVI estimation that brings a challenge of reducing larger set of variables into a smaller one. Such methodological concern in heat vulnerability studies can be solved through a Principal Component Analysis (PCA) method (Bao et al. 2015). The PCA is a technique that allows an increase of the accuracy and objectively in analyzing large multivariate datasets and to reduce their dimensionality (Jolliffe 2014).

6.3 Study Objectives

A number of studies and approaches have been undertaken globally to evaluate heat-related risks. The vulnerability indices have been developed in the US (Reid et al. 2009), Europe (Wolf and McGregor 2013), Canada (Jay and Kenny 2010), Australia (Coutts et al. 2007), and China (Bao et al. 2015). The aforementioned studies were mainly conducted in the developed and mid-latitudes countries. However, the city characteristic and climate pattern (i.e., inter-annual temperature variability is low) in Indonesia are different from the aforementioned studies. Thus the specific HVI is necessary to be developed. Nevertheless, until the present time, there are no specific vulnerability indices developed specifically for Indonesia, despite, as described earlier, highly prone heat-related risks. Moreover, in the face of urban sprawls and substantially higher human exposure to heat (Basu and Samet 2002), the UHI should be incorporated into generating HVI (Rooney et al. 1998). Subsequently, the emission from anthropogenic sources as the contributor to the heat generation was previously not included into spatial risk assessment (Reid et al. 2009; Wolf and McGregor 2013).

Thus, this study aims to develop a tailor-made heat vulnerability index for Indonesia, and demonstrate its applicability through assessment in a selected city. The HVI refinement for the context of Indonesia is done through inclusions of the UTCI, which in previous studies often get overlooked, i.e., the UHI with its anthropogenic heat emission source and land-use changes in urban environment, and sensitivity adjustment (e.g., health, age, and income). The inter-linkages across elements will be evaluated through PCA to understand the degree of importance of each element. A spatial visualization of the HVI will be conducted to identify highly vulnerable areas. It will allow to design and undertake adaptation strategies, which are crucial to prevent or reduce undesirable impacts such as death, illness, loss of work productivity.

6.4 Methods

6.4.1 Study Area

To showcase the heat vulnerability in Indonesia and the HVI method applicability, Medan City as one of the pilot area of National Action Plan on Climate Change Adaptation (RAN-API), a rapidly growing urban area with significant exposure to health-related risks, has been selected. Medan City is the third-largest city in Indonesia, and the largest outside of Java, located on Sumatra Island, and the capital of North Sumatra Province (Fig. 6.1). The reasons for selecting the city by Indonesian Government and researcher team as pilot project was to consider as a model for replication to other growing cities in the country, to take various climate change adaptation measures. In the past three decades, Medan City urban population had increased by 718 thousand people and in the next three decades, the population is expected to increase by additional 1.4 million people (Statistical Agency 2015a, b). The increasing population is also responsible for settlement area increase, whereas in the past 14 years had doubled its area and consequently green open space was reduced for nearly two-third (Lubis et al. 2014). In addition, annual average temperature in Medan City showed increasing trend in the past 30 years and with increasing extreme temperature events frequency (Meteorological agency of Indonesia 2016). Medan City experienced increasing number of days with extreme temperature, defined as above 34 °C, in years 1980–2014 (Handayani 2010) as can be seen in Fig. 6.2. These changing environmental conditions are presumed to cause higher vulnerability to the heat risk in the future to Medan City, whereas the current state of governmental preparedness to such heat has been demonstrated as inadequate yet (Russo et al. 2014). Thus,

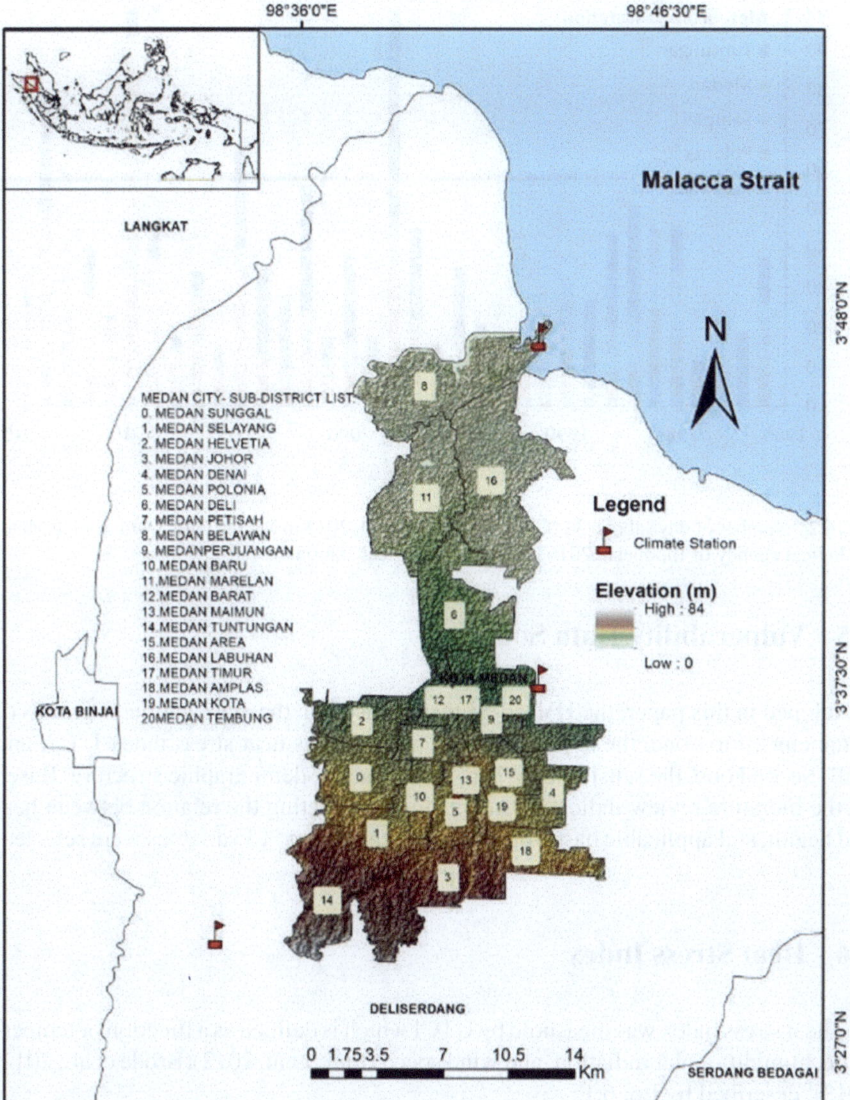

Fig. 6.1 Location of study area—Medan City, Indonesia

determining the areas more prone to heat risk and prevention measures are necessary for the future risk management and adaptation planning strategies.

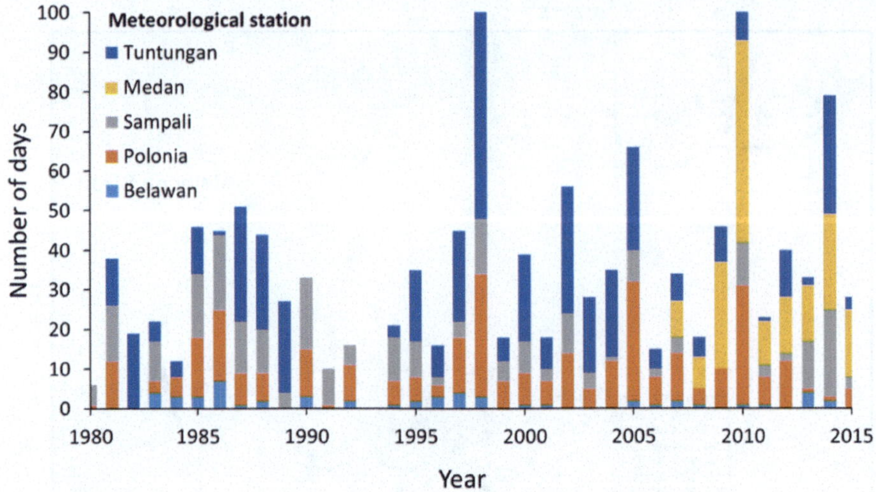

Fig. 6.2 Number of days above 34 °C per year from 1980–2015 in Medan City. *Data source* Meteorological agency of Indonesia 2016, five meteorological stations

6.5 Vulnerability Data Sources

Developed in this paper, the HVI is estimated based on the aggregation of groups of parameters. First one, the exposure to the heat such as heat stress index UTCI and UHI. Second one, the sensitivity to heat such as socio-demographic structure. Based on the literature review indicators satisfactorily reflecting the relation between heat and health, and applicable based on existing data sources, 13 variables were selected.

6.6 Heat Stress Index

The heat stress index was measured by UTCI which is defined as a function of temperature, humidity, solar radiation, and wind speed (Fiala et al. 2012)(Bröde et al.; 2012, 2013), described by Eq. 6.1:

$$\mathrm{UTCI} = f(t; \mathrm{RH}; \mathrm{Tmrt}; v) = t + offset(t; \mathrm{RH}; \mathrm{Tmrt}; v) \tag{6.1}$$

Note: f means function

the original equation is very long, you can access here: https://drive.google.com/file/d/1zsySSC8J8t1wAe5eVsz1WASXHJo9aRTo/view?usp=sharing or here: https://www.ladybug.tools/ladybug-comfort/docs/_modules/ladybug_comfort/utci.html. All calculation was conducted by Bioklima package 2 where t is maximum temperature (°C), RH is relative humidity (%), Tmrt is mean radiant temperature (°C), and v is wind speed (m/s) (Fiala et al. 2012). UTCI is using the following assumptions: walk 4 km/h, weight 74 kg, and metabolic rate 135 W/m^2.

Calculating the UTCI was conducted in few steps. Firstly, solar radiation was calculated (i.e., conversion from t, sunshine duration and RH to solar radiation) by equation given by Garg and Garg (1983), and the solar radiation was used to calculate Tmrt with Bioklima package 2.6 (https://www.igipz.pan.pl/Bioklima-zgik.html). Using the same package, the overall UTCI was calculated. For tropical countries, UTCI was divided into five classes; no thermal stress (9–26 °C), moderate heat stress (26–32 °C), strong heat stress (32–38 °C), very strong heat stress (38–46 °C), and extreme heat stress (above 46 °C) (Fiala et al. 2012).

6.7 Urban Heat Island

The Urban Heat Island was estimated by six representative variables, namely vegetation coverage, building coverage, Land Surface Temperature (LST), electricity and gas consumptions from several sources (i.e., industry, residential, office, department store and school), and oil consumption from land transportation (i.e., car, motorcycle, bus, and truck). Building contributes to the UHI because of their large thermal capacity, which is exposed to the sun through the roof and the wall, and through the heat dissipated because of space conditioning, electricity loads, and metabolic heat generation from the occupants (Phelan et al. 2015).

6.8 Vegetation and Building Coverage

High building density is associated with an increased heat-rellated illness and deaths (Hondula et al. 2012). It is caused by unshaded area increasing exposure to direct sunshine, and reduced outdoor space limiting natural air ventilation. Vegetation and building coverage area were sourced from satellite images captured by Landsat Operational Land Imager (OLI) 8 data and acquired from United States Geological Survey (http://www.usgs.gov). The images were processed in ArcMap 10.3 (ESRI), and utilized Normalized Difference Vegetation Index (NDVI) approach as explained in Eq. 6.2.

$$\text{NDVI} = \frac{\rho_{\text{NIR}} - \rho_{\text{Red}}}{\rho_{\text{NIR}} - \rho_{\text{Red}}} \tag{6.2}$$

where ρ_{NIR} is near-infrared Band of Landsat 8 (0.85–0.88 μm) and ρ_{Red} is red band of Landsat 8 (0.64–0.67 μm). After calculating NDVI, threshold for each land cover was defined in reference of land cover map from the Indonesian Geospatial Agency (BIG), i.e., accessible here: https://tanahair.indonesia.go.id.

6.9 Land Surface Temperature

Land Surface Temperature (LST) is commonly used to measure the magnitude of UHI at the surface level (Tomlinson et al. 2011a, b). Satellite thermal data are able to represent the spatial gradient of radiometric surface temperature (Coll et al. 2010). The satellite platform of LANDSAT 8 with 30 m resample of spatial resolution was used to estimate the LST (Gluch et al. 2006). Two input variables were brightness temperature (Tb) and emmisivity (ε). The brightness temperature calculation followed method prescribed by LANDSAT user guide (United Stated Geological Survey (USGS) 2016). The emissivity was assessed by vegetation proportion of NDVI (Sobrino et al. 2008; Artis and Carnahan 1982). The output unit of this calculation is in Celsius degree (°C) unit.

The LST calculation procedure is presented in Fig. 6.3.

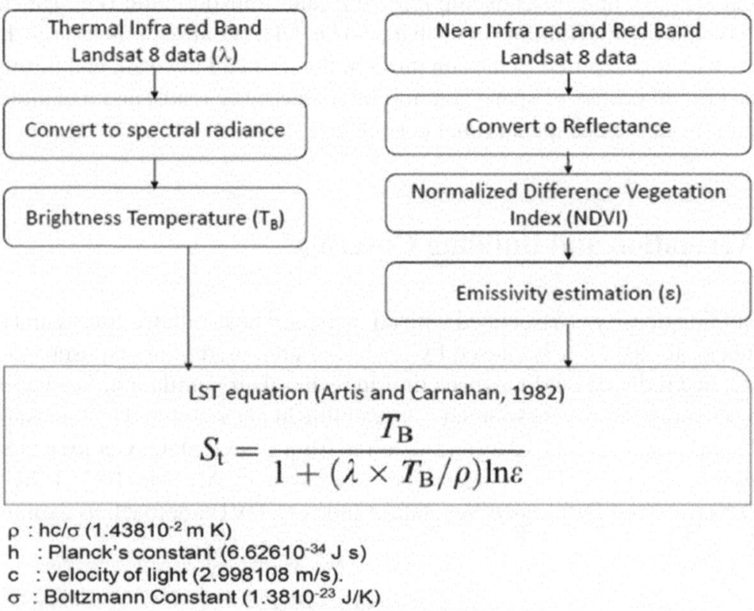

Fig. 6.3 Land surface temperature calculation (*Source* authors)

6.10 Electricity and Gas Emission

Electricity and gas consumptions in 2015 were collected from statistical agency of Medan City (https://medankota.bps.go.id). In this paper, electricity and gas consumptions were assumed as having the same rate of emission, hence its consumption was converted to emission using method applied by the Japanese Environmental Agency (https://www.env.go.jp/air/report/h16-05). The conversion factors are presented in Appendix 1 and 2. 2.3.4 Gas Emission from land transportation oil consumption.

Due to lack of data of vehicles number per sub-district, the data on gas emission were derived for each sub-district by using congestion hotspot number and total number of vehicle in the city (Medan Transportation Bureau 2016). Annual oil consumption and subsequently its emission were estimated by the assumption of 10,000 km/year travel distance per each vehicle. Furthermore, we assumed consumptions were same as emission.

6.11 Sensitivity to Heat

Sensitivity to heat was conducted based on number of demographic and socioeconomic variables. There were multiple factors considered as increasing risk of health, namely population density related to tendency of increased population density in the central city districts with higher heat exposing (Coutts et al. 2007), age related to higher mortality among elderly people and children (Reid et al. 2009; Coutts et al. 2007; Wolf and McGregor 2013), income level as poverty frequently increase heat exposure in various aspects of life (Kim et al. 2011). This includes those with pre-existing illness or impaired physical or mental health (Kaiser et al. 2001; Reid et al. 2009). In the study area, the lower income people are often living in houses with aluminum roof and without cooling system, and their work is located outdoors (i.e., peddler, construction worker, public transportation driver). Residents of lower income with respiratory, cardiovascular or nervous system problem are at increased risk of mortality as they are unable to take care for themselves or they have limited mobility (Tomlinson et al. 2011a, b; Wolf and McGregor 2013). In this study, the number of patients with hypertension and respiratory patients were included into analysis as one of the illness triggering the heart attack (Pan et al. 1995) but also due to the data availability.

6.12 Heat Vulnerability Index

The Heat Vulnerability Index was created based on 13 indicators as shown in Table 6.1.

Table 6.1 Indicators for HVI

Parameter group	Variable	Data source
Exposure to heat		
Heat stress index	Temperature (1980–2014)	Data online BMKG 2016 (Indonesia meteorological agency) (5 stations within Medan City)
	Solar radiation (1980–2014)	
	Relative humidity (1980–2014)	
	Wind speed (1980–2014)	European Center for Medium-Range Weather Forecasts (ECMWF)
Urban Heat Island (UHI)	Land surface temperature	Landsat Oli 8
	Transportation emission	Transportation agency
	Electricity consumption	Statistical agency of Medan City
	Gas emission	Statistical agency of Medan City
	Building density	Landsat Oli 8 and geospatial agency
	Vegetation density	Landsat Oli 8 and geospatial agency
Sensitivity to heat		
Socio-demographic	Population density	Statistical agency of Medan City
	Age <5	Statistical agency of Medan City
	Age >65	Statistical agency of Medan City
	Illness	Statistical agency of Medan City
	Outdoor worker	Statistical agency of Medan City

Vulnerability to heat was conceptualized as a function of exposure to heat and sensitivity of people. Exposure to heat was described by the UTCI and UHI as a combination from LST, land cover, and anthropogenic sources. The index value was normalized to relative position between values 0 and 1, where 0 is the lowest vulnerable and 1 is the most vulnerable to heat area. Sensitivity of people was described by socio-demographic factors such as population density, illness, elderly people, kids, poor family, and people who works outdoor (e.g., Traffic police, military, market trader, and construction worker). The approach taken to developing the heat vulnerability index presented here is an inductive one, in which PCA is used to identify groups of covariant heat risk factors as represented by a number of principal components. The HVI was built using PCA to reduce the indicators to component level, adding them together, and weighting them according to the variance.

6.13 Spatial Data Processing

To produce a spatial risk assessment index and map, the data set was specially prepared in the ArcMap 10.3, through unifying the projection coordination system into Universe Transfer Mercator (UTM) 47 North of World Geodetic System (WGS) 1984, and converting vector/polygon data into raster data, maintaining 30 m spatial resolution of the vector/polygon data. The Heat Vulnerability Index has been computed and projected spatially indicating areas of different levels of vulnerability.

6.14 Results

6.14.1 Heat Stress Level in Medan City

In 1982, 1997, and 2014 the spatial distribution of UTCI has drastically increased, with maximum 4°C rise between the year 1982 and 2014. In 2014, there were two levels of UTCI stress level in Medan City, namely strong heat stress (32–37 °C) and very strong heat stress (38–46 °C) as presented in Fig. 6.4. The increase is highly related to urban area sprawl. The frequency of days with UTCI at very strong heat stress level has been rising especially in the urban area, whereas the rural areas the frequency has been decreasing (Fig. 6.5).

6.15 Land Surface Temperature and Urban Heat Island in Medan City

The Land Surface Temperature in Medan City varied from 21 to 35 °C at the pixel level (Fig. 6.6). In general, higher LST was located in the central part of Medan City and generally lower in the northern and southern part. The distribution of the high LST was related to building cover (Fig. 6.6), such as northern part of Medan with dense building coverage indicated high LST, and areas of lower LST were concentrated in higher vegetation areas (i.e., vegetation, river, ponds) (Fig. 6.6).

6.16 Heat Vulnerability Map

Principal Component Analysis was described in Table 6.2. These explained 77.5% of the variance. The percentage variance value of each PCA was used as the weights in the calculation of HVI value. Consideration of the component loading matrix indicates that component 1 related all variables except transportation emission. Furthermore, component 2 only related to transportation emission. The negative sign of

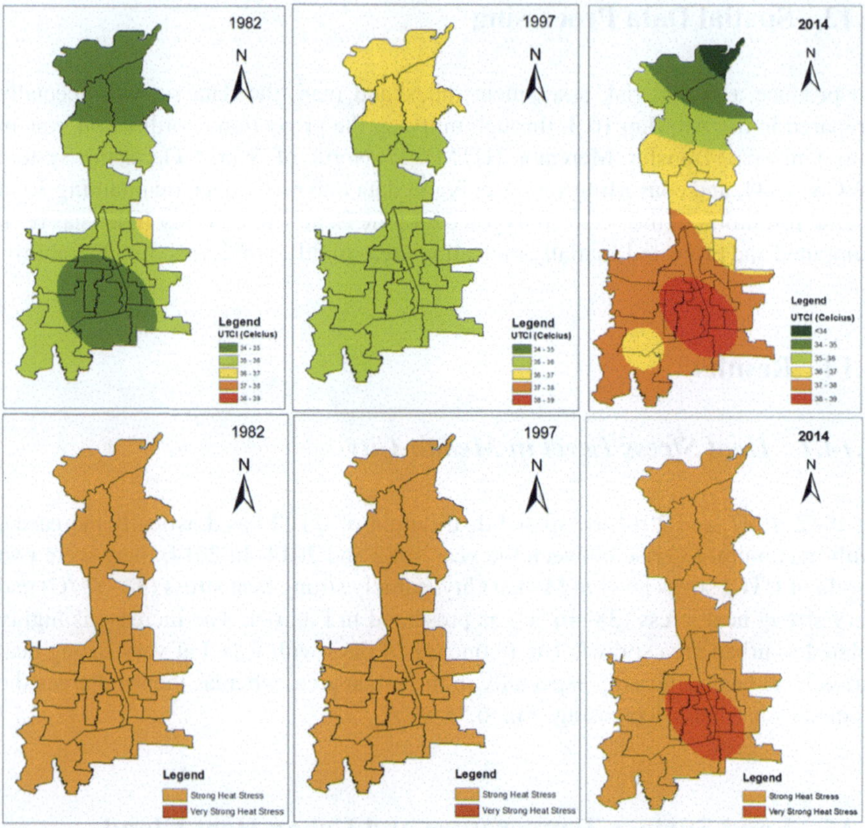

Fig. 6.4 Spatial distribution of heat stress level (UTCI) in Medan City (*Source* authors)

factor loading in the component 1 indicates that vegetation density has a strong cooling effect where it can reduce the HVI index.

Based on the interview with local health agency, the most cases of heat-related injury was heat exhaustion, but it rarely needs hospital treatment. Thus, there is no data reported yet regarding heat-related illness in the study area. In 2016, one heatstroke case was reported in Bali-Indonesia and caused mortality (Satya et al. 2018). Satya et al. (2018) stated that heatstroke is an uncommon case in Indonesia, so the possibility of a diagnostic error by a doctor will be very high. Since there is no reliable data of heat-related injuries, we tried to confirm our result with the most vulnerable people of heat-related mortality that is people with cardiovascular diseases and respiratory illness (Reid et al. 2009; Curriero et al. 2002). However, cardiovascular diseases cases were also not reported by local health agency, thus, hypertension cases were used as the most important risk factor for cardiovascular disease.

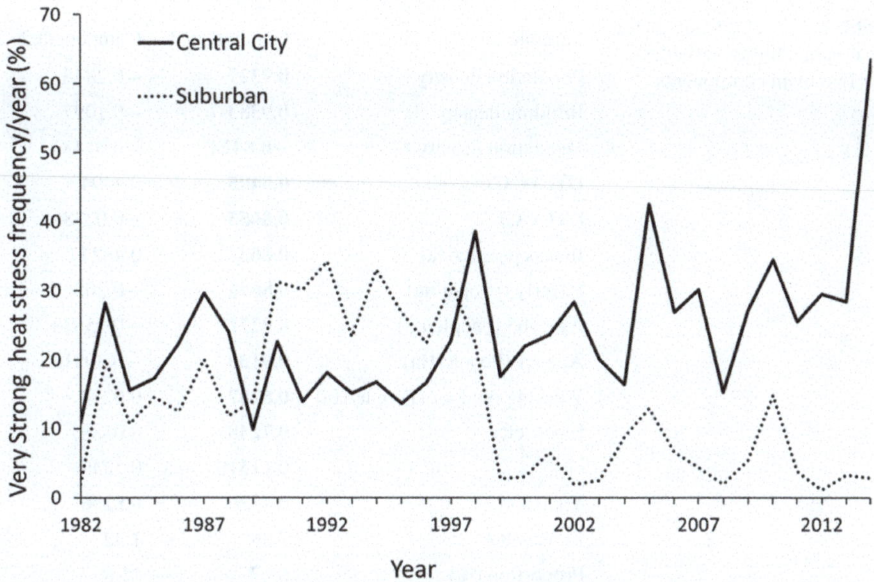

Fig. 6.5 Percentage of very strong heat stress frequency (UTCI ≥ 38 °C) in a year (*Source* authors)

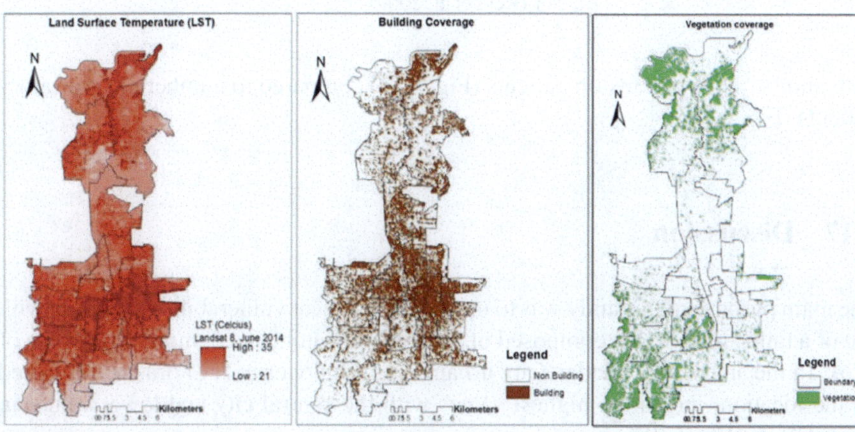

Fig. 6.6 Land surface temperature, building coverage, and vegetation coverage in Medan city (*Source* authors)

Upon confirming the determinants of heat vulnerability and aggregating it into one index (Fig. 6.7), the heat vulnerability map was overlaid with hypertension and respiratory prevalence in the city. Overall, higher vulnerability was seen in the center part of Medan City, which indicated with the gradation from orange to red color, and cases of hypertension and respiratory were found to be prevalent at these particular areas (Fig. 6.7a, b). The highest positive correlation with HVI was found

Table 6.2 Factor loadings for heat vulnerability variables for two main components

Variable	Component 1	Component 2
Population density	**0.9327**	−0.2848
Building density	**0.9383**	−0.1097
Vegetation density	**−0.8428**	−0.0143
UTCI (°C)	**0.6405**	0.4494
LST (°C)	**0.8653**	−0.0238
Illness people/Ha)	**0.663**	0.48277
Poverty (people/ha)	**0.8596**	−0.2086
Age >65 (people/ha)	**0.9271**	−0.25628
Age <5 (People/Ha)	**0.9130**	−0.30516
Working outdoor (People/Ha)	**0.8287**	0.0274
Electricity	**0.7248**	0.04887
Gas	0.11518	0.5249
Transport	0.4326	**0.8296**
Eigenvalue	7.88	1.42
Proportion (%)	65.7	11.8
Cumulative (%)	65.7	77.5

Note Absolute values >0.4 are the most significant loadings on that factor (Reid et al. 2009)

with number of hypertension patients (Fig. 6.8a) compared to number of respiratory patients (Fig. 6.8b).

6.17 Discussion

The main purpose of the study was to investigate the heat vulnerability in Medan with use of a tailor-made HVI, composed of reliable indicators and computed with use of PCA. In the analysis of Medan City urban area in Indonesia, HVI drastically varied on the location, indicating highest values with the central city and lower values in the outskirts (Fig. 6.7).

In our analysis, higher vulnerability was seen within the downtown areas compared with suburban areas. The higher vulnerability of city centers have been also found in UK as a direct consequence of UHI phenomenon (Tomlinson et al. 2011a, b; Wolf and McGregor 2013).

Our analysis is an approach similar to methodologies used for spatial risk assessment (Reid et al. 2009; Wolf and McGregor 2013). Previous approaches in reviewed studies did not include the emission from anthropogenic sources in the heat estimation. However, this study incorporated three sources of heat emissions, namely electricity use, gas, and fuel combustion. The spatial analysis showed that the spatial

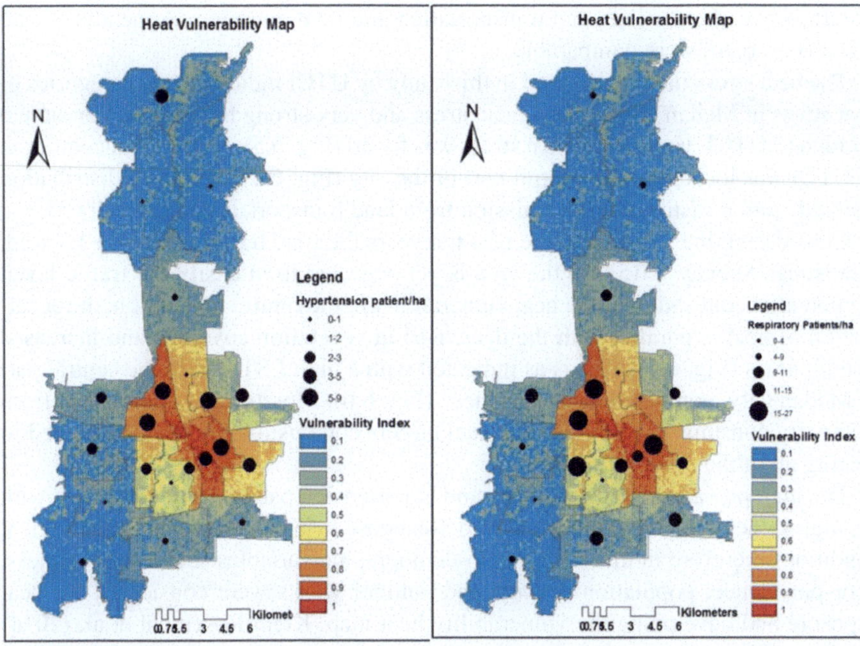

Fig. 6.7 Heat vulnerability map and number of hypertension and respiratory patients in Medan City (*Source* authors)

Fig. 6.8 Correlation between HVI and **a** Hypertension patients number and **b** Respiratory patients number (*Source* authors)

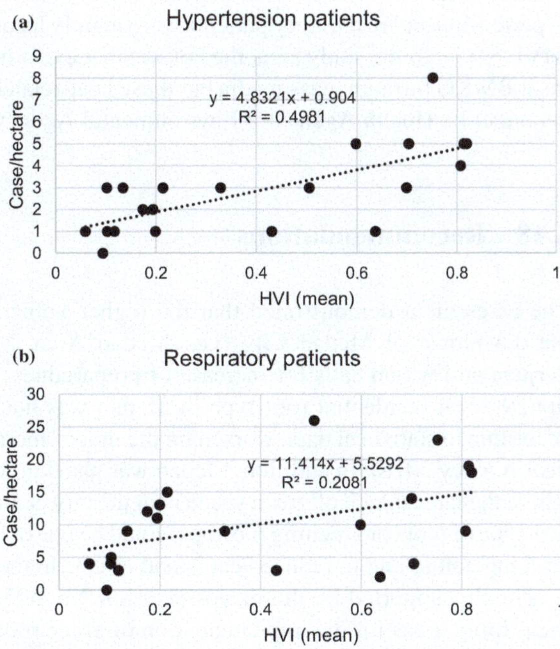

pattern between emission from transportation and total emission (Appendix 3) and UTCI (Fig. 6.3c) were comparable.

The heat stress index measured in this study by UTCI indicated two categories of heat stress in Medan City: strong heat stress and very strong heat stress. For annual average of UTCI, very strong heat stress was found (Fig. 6.5). Highest concentration of UTCI was located in the central part of the city (Fig. 6.4) and UTCI distribution denoted similar distribution to emission from land transportation (Appendix 3).

Considering the fact that in the past ten years the road traffic increased 1.7 fold, (Statistical Agency 2015a, b) the area is yet expected to intensify the traffic level, the potential contributor to the heat vulnerability in the future. In addition, the acceleration of heat is parallel with the decreased in vegetation coverage and increased of built areas (Fig. 6.5). This is as indicated with a high LST within the central part of Medan City, compared to its periphery (Fig. 6.6). Moreover, heat emission from electricity consumption and gas and fuel engine combustion greatly contributed to heating up of the micro-climate.

The high prevalence of respiratory and hypertension patients was consistent with the high value of HVI (Figs. 6.7 and 6.8). Response to the heat vary on physiological conditions therefore in this study socio-demographic variables such as age, illness, poor prevalence, population density, and outdoor works were considered for heat exposure and developing the vulnerability heat map. Klein Rosenthal et al. (2014) found significant increase of mortality rate in high poverty areas of New York City, especially among the elderly during extreme temperature days, due to poor housing conditions without air-conditioning and altered land cover increasing high surface temperatures. The similar trend was also found in Medan city where high number of hypertension and respiratory patients were mainly located in the high vulnerable area (HVI $\geq$ 0.7). In the study area, the hot weather event information was announced by local BMKG through mass media but these heat-related risks have not been reported yet either by Health Agency or Environmental Agency.

6.18 Recommendations

The assessment demonstrated that the higher vulnerability areas were located in the downtown of Medan City (i.e., Medan Area, Medan Kota, Polonia, Medan Perjuangan) which calls for increase of preparedness for the heat shocks. In addition, 90% of residential roof type in Medan was made of aluminum and asbestos according to statistical data, worsening the indoor thermal conditions (Medan Statistical Agency 2018). In addition, Medan was also known as Kota Ruko ("city shop") where the largest lack of green space is in the city center. Thus, gradually promoting new type of roofs or greening rooftops would be one of the solutions to cool the indoor air temperature and to reduce heat island effects in the limited space. This initiative is actually supported by the mayor decision No. 522.4/1553.X/IX 2013 regarding the priority areas for the implementation of green rooftop. Furthermore, it was also mentioned in the newest version of regional spatial planning called as RTRW 2011–2031 (Medan Spatial Planning Agency 2011) and in the regional detailed spatial planning called as RDTR 2015–2035 (Medan Spatial Planning Agency 2015a, b) that the green rooftop is one of open green space component in the city. However, the municipal target, technical guideline, and seedling assistance were not clearly mentioned in those aforementioned documents. In addition, this initiative will be applicable only for building with concrete roof type where only 5% for residential roof and commercial building.

Re-establishing green spaces in the city, as it was shown in this study to be drastically reduced thought the time, would be another measure reducing the land surface temperature in the central part of the city. As a reference, the increase of the green coverage in New York City and Phoenix City showed significant improvement in the city micro-climate (Phelan et al. 2015). In the mid-term regional planning document of Medan City (RPJMD), increasing open green space was one of the strategies to achieve a clean and healthy urban environment. Based on national regulation Law No. 26 of 2007 on spatial management stated that the proportion of the open green space in the urban region must be at least 30% of the total urban area where 20% is public space and 10% is private space. However, the actual area of open green space in Medan City was less than 10% (RDTR 2015–2035). This number is far from the minimum requirement of national regulation. This condition also partly due to the regional cluster in the RTRW document which stated that the downtown area was focused on trade and services region while the southern region was focused on green open land. Even though open green space is part of the adaptation to reduce heat stress, but there is not yet a policy or legally binding instruments specifically designated for the implementation of urban climate adaptation measures, including Jakarta as the capital city and benchmark development of Indonesia. As shown in Figs. 6.2 and 6.5, the number of extreme temperatures and the frequency of very strong heat stress were gradually increased, particularly during strong El-Nino phenomenon. This condition will be worse in the future when there is no policy regarding urban climate adaptation. Urban climate including urban heat island is not yet included in the national adaptation action plan (RAN-API) document. It is simply because the scale of RAN-API is very broad and assesses climate change in general. For regional planning document such as RTRW, RDTR, RPJMD, and REISTRA (mid-term strategic master plan), there is no concrete urban climate adaptation intervention yet.

The heat warning systems using UTCI considering especially for vulnerable groups and the key actors (i.e., health agency, meteorological agency, spatial planning agency, disaster management authority, and scientist) employing the HVI to design measures against extreme heat are recommendable for the city as part of disaster preparedness plan. The UTCI also can be used as a quantitative indicator to monitor the effectiveness of urban greening to reduce urban heat stress levels. In addition, setting the warning threshold should be conducted in future research in order to estimate when to issue a warning. There are four key elements of early warning system: risk knowledge, technical monitoring and warning service, communication and dissemination of warning, and response capability (UNISDR 2006). The risk knowledge component is very crucial in order to employ early warning system as part of development master plan for the city. We conducted several time workshop in the study area, however, UHI effect and heat-related risk have not been a major consideration in their development planning. Satya et al. (2018) warn that early symptom of heatstroke was communication chaos in the community, and it was associated with mystical matters, resulting in delays in handling patients. Thus, knowledge and awareness of the community about symptoms of heat-related risk, especially heatstroke will be very important to avoid heat-related mortality. In addition, the most important prevention is to educate the vulnerable groups. Since there is not yet a policy or legally binding regarding urban climate adaptation measures and lack of knowledge of heat-related risk of the municipal government and community, the heat early warning will be difficult to implement in the Medan city.

Employing at the city level performed in this study indicators give more information about local vulnerability. The methodology demonstrated in this paper can be used with variety of local data regardless of the geographical location thus it can be a universal tool for the heat vulnerability assessment, and with incorporation of future climate scenarios it can be useful for future heat vulnerability prediction. The method could be also focused on other local contexts through using data from household survey such as child and elderly actual health condition against their living standards, etc.

6.19 Conclusion

Increasing heat exposure in the urban environment has become a hazardous to human health. Using refined HVI, an index calculated by PCA method based on exposure to the heat and sensitivity to the heat, 13 variables in total, for the spatial assessment of heat vulnerability level was demonstrated as one of the possible to create response to the heating environment problem. In rapidly growing Medan City in the Indonesian province of North Sumatra, it has been shown that over the past three decades the heat tŕess was increasing. In addition, recent rapid growth of energy consumption and traffic, together with the drastic land-use change from green to build area affected in high vulnerability level, measured in this study through HVI.

The number of respiratory and especially hypertension health problem cases that are related to high temperature and under these conditions may lead to heart attacks and eventually morality was found spatially prevalent at the areas of high HVI. In this study, the employed novel approach to map the vulnerability to extreme temperature events allows to develop warning systems and to respond to climate change promptly. In longer perspective, the spatial distribution of high vulnerability areas also allows to introduce various spatial planning measures such as relocation of heat-emitting industrial areas, re-allocation of green areas back to the city to decrease surface temperature, and rising awareness of heat-related risk. Improving airflow through the city by considering climate data on the wind will improve cooling of the especially congested areas.

Acknowledgements We would like to thank Ministry of the Environment Japan for funding the research under the theme of Climate Change Adaptation in Indonesia in collaboration with Ministry of National Development Planning of the Republic of Indonesia (BAPPENAS), and to Indonesian authorities (BMKG and BPS) for providing climatic and statistical data. Also, this research was partially supported by the Japan Society for Promotion of Science (JSPS) through Kakenhi Program (i.e., Award no. 19H0114, PI: Kensuke Fukushi) and JSPS's core-to-core program, Center of Excellence in Health Risk Assessment for Adaptation to Climate Change.

Appendix 1: Energy Consumption in Non-Residential Building (MJ/m^2/year)

Type	Electricity	Urban gas
Office (including factory)	788	213
Department store	1458	438
Commodity retail	1421	185
Other retail	1421	185
Restaurant/cafe	1418	2013
Hotel	1400	1431
School	377	191
Hospital	996	846
Others	836	664

Appendix 2: Energy Consumption Per Residential Building (MJ/year)

	Electricity (MJ/year)	Urban gas (MJ/year)
Single person	10,497.44882	8524.069917
Multiple people	13,672.69638	10,784.191

Appendix 3: Anthropogenic Resources of Heat

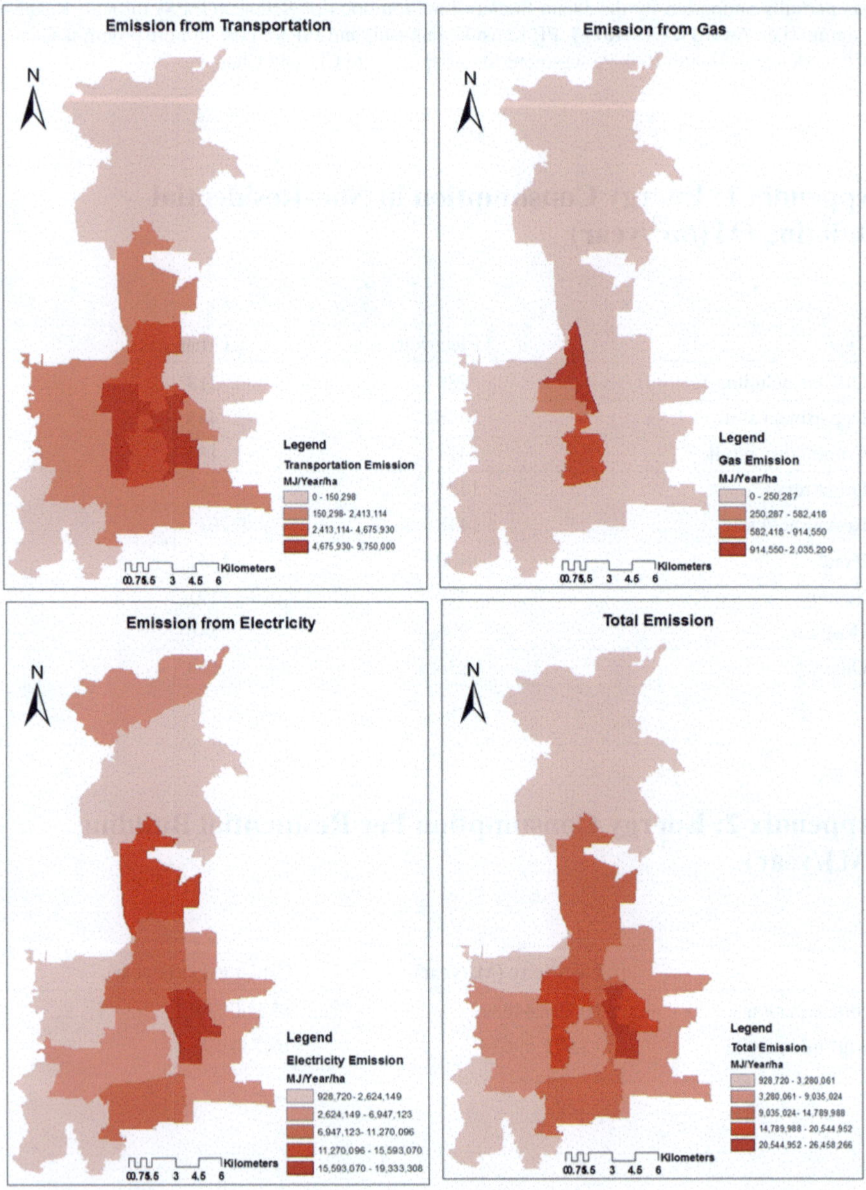

Appendix 4: Heat Vulnerability Index Map

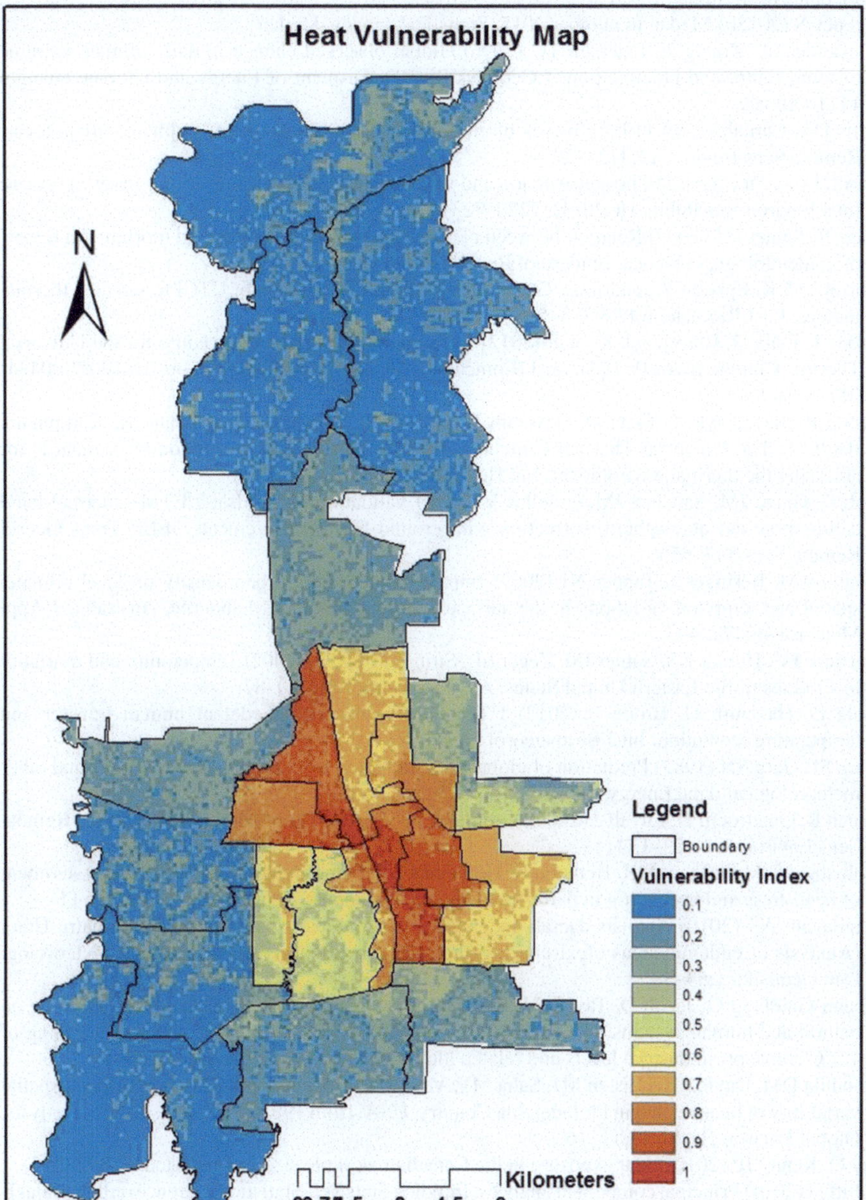

References

Agency MSP (2015a) Rencana detail tata ruang dan peraturan zonasi kota medan tahun 2015–2035. Medan, North Sumatra
Agency S (2015b) Medan in number 2015. Statistical agency, Medan
Alexander LV, Zhang X, Peterson TC (2006) Global observed change in daily climate extreme of temperature and precipitation. J Geophys Res (Department of Energy and Climate change) 1111:D05109
Artis DA, Carnahan WA (1982) Survey of emissivity variability in thermography of urban areas. Remote Sens Environ 12:313–329
Bao J, Li X, Yu C (2015) The construction and validation of the heat vulnerability index, a review. Int J Environ Res Public Health 12:7220–7234
Basu R, Samet JM (2002) Relation between elevated ambient temperature and mortality: a review of epidemiologic evidence. Epidemiol Rev 24:190–202
Blazejczyk K, Epstein Y, Jendritzky G, Staiger H (2012) Comparison of UTCI to selected thermal indices. Int J Biometeorol 56:515–535
Bröde P, Fiala D, Błażejczyk K, et al (2012) Deriving the operational procedure for the Universal Thermal Climate Index (UTCI). Int J Biometeorol 56: 481–494. https://doi.org/10.1007/s00484-011-0454-1
Bröde P, Błazejczyk K, Fiala D, Havenith G, Holmér I, Jendritzky G, Kuklane K, Kampmann B (2013) The Universal Thermal Climate Index UTCI compared to ergonomics standards for assessing the thermal environment. Ind Health 51:16–24
Coll C, Galve JM, Sanchez JM, Caselles V (2010) Validation of Landsat-7/ETM+ thermal-band callibration and atmospheric correction with ground base measurements. IEEE Trans Geosci Remote Sens 547–555
Coutts AM, Beringer J, Tapper NJ (2007) Impact of increasing urban density on local climate: spatial and temporal variations in the surface energy balance in Melbourne, Australia. J Appl Meteorol 46:477–493
Curriero FC, Heiner KS, Samet JM, Zeger SL, Strug L, Patz JA (2002) Temperature and mortality in 11 cities of the Eastern United States. Am J Epidemiol 155:80–87
Fiala D, Havenith G, Brode P (2012) UTCI-Fiala multi-node model of human transfer and temperature regulation. Int J Biometeorol 56:429–441
Garg SP, Garg SN (1983) Prediction of global solar radiation from bright sunshine hours and other meteorological data. Energy Convers Mgmt 23(2):113–118
Gluch R, Quattrochi D, Luvall J (2006) A multi-scale approach to urban thermal analysis. Remote Sens Environ 104:123–132
Golbabaei F, Monazzam MR, Hematjo R, Hosseini M, Farhang Dehghan S (2013) The assessment of heat stress and heat strain in pardis petrochemical complex, Tehran, Iran. IJOH 5:5–11
Handayani AS (2010) Analisis daerah endemik bencana akibat cuaca ekstrim di Sumatra Utara (Analysis of endemic areas ofextreme weather disaster in North Sumatra). Jurnal Meteorologi Dan Geofisika 1:54–58
Hoegh-Guldberg O, Jacob D, Taylor M, Bindi M (2018) Impacts of 1.5°C of global warming on natural and human systems. In: An IPCC Special Report on the impacts of global warming of 1.5°C above pre-industrial levels and related global greenhouse gas emission pathway
Hondula DM, Davis RE, Leisten MJ, Saha MV, Veazey LM, Wegner CR (2012) Fine-scale spatial variability of heat-related in Philadelphia Country, USA, from 1983–2008: a case-series analysis. Global Environ Health 11:11–16
Jay O, Kenny JP (2010) Heat exposure in the Canadian workplace. Am J Ind Med 53:842–853
Jolliffe I (2014) Principal component analysis. In Wiley StatsRef: Statistics reference online, Balakrishnan N, Colton T, Everitt B, Piegorsch W, Ruggeri F, Teugels JL (Eds). https://doi.org/10.1002/9781118445112.stat06472
Kaiser R, Rubin C, Henderson A (2001) A heat related death and mental illness during the 1999 Cincinnati heat wave. The Am J Forensic Med Pathol 303–307

Kilbourne E, Keewhan C, Jones T, Thacker S (1982) Risk factor of heat stroke, a case control study. J Am Med Assoc 247:3332–3336
Kim YM, Kim S, Cheong HK, Kim EH (2011) Comparison of temperature indexes for the impact assessment of heat stress on heat-related mortality. Environ Health Toxicol 1–9
Kjellstrom T, MMeng. (2016) Impact of climate change condition on occupational health and related economic losses: a new feature of global and urban health in the context of climate change. Asia Pacific J Public Health 28(2):28–37
Kjellstron T, Briggs D, Freyberg C, Lemke B, Otto M, Hyatt O (2016) Heat, human performance, and occupational health: a key issue for the assessment of global climate change impacts. Annu Rev Public Health 37:97–112
Klein Rosenthal J, Kinney PL, Metzger KB (2014) Intra-urban vulnerability to heat-related mortality in New York City, 1997-2006. Health & place 30:45–60. https://doi.org/10.1016/j.healthplace.2014.07.014
Lindley SJ, Handley JF, Theuray N, Peet E, McEvoy D (2006) Adaptation strategies for climate change in the urban environment: assessing climate change related risk in UK urban areas. J Risk Res 9:543–568
Lubis DP, Widhiastuti R, Syahrin A (2014) Management of model green open space towards the construction of the green city (case study in medan-indonesia). Int J Educ Res 2:1–4
Lundgren K, Kuklane K, Gao C, Holmer I (2013) Effects of heat stress on working populations when facing climate change. Ind Health 51:3–15
Medan Spatial Planning Agency (2011) Rencana tata ruang wilayah kota medan 2011–2031. Medan
Medan Statistical Agency (2018) Indikator Kesejahteraan Rakyat Kota Medan (Welfare Indicator of Medan Society). Medan Statistical Agency, Annual Government Report, Medan
Medan Transportation Bureau (2016) Area Traffic Control System (ATCS). Medan
Meteorological agency of Indonesia (2016) Data online of Indonesia Meteorological agency. http://dataonline.bmkg.go.id/. Accessed 18 Jul 2016
Monazzam MR, Golbabaei F, Hematjo R, Hosseini M, Nassairi P, Farhang DS (2014) Evaluation of DI, WBGT and Swreq/PHS heat stress indices for estimating the heat load on the employees of a petrochemical industry. IJOH 6:6–10
Oke TR (1987) Boundary layer climate, 2nd edn. Routledge, London and New York
Pan W-H, Li L-A, Tsai M-J (1995) Temperature extremes and mortality from coronary heart disease and cerebral infarction in elderly Chinese. The Lancet 345:353–355
Phelan PE, Kaloush K, Miner M, Golden J, Phelan B, Silva Ill H, Taylor RA (2015) Urban heat island: mechanism, implications, and possible remedies. Annu Rev Environ Res 285–307
Reid CE, O'neill MS, Gronlund CJ, Brines SJ, Brown DG, Diez-Roux AV, Schwartz J (2009) Mapping community determinant of heat vulnarability. Environ Health Perspectives 1730–1737
Robine J-M, Cheung SL, Van Le RoyS H, Oyen CG, Michel J-P (2008) Death toll exceeds 70000 in Europe during the summer of 2003. CR Biol 22:670–683
Rooney C, Mcmichael AJ, Kovats RS, Coleman MP (1998) Excess mortality in England and Wales, and in Greater London, during the 1995 Heatwave. J Epidemiol Community Health 52:482–486
Russo S, Dosio A, Graverssen RG, Sillman J, Carrao H, Dunbar MB, Singleton A, Montagna P, Barbola P, Vogt JV (2014) Magnitude of extreme heat waves in present climate and their projection in warming world. J Geophys Res: Atmos 119:12500–12512
Satya IMH, Wiryana IM, Sutawan IBKY (2018) Exertional heatstroke: sebuah laporan kasus. Medicina 49:303–307
Smith C, Lindley S, Levermore G (2009) Estimating spatial and temporal patterns of urban antropogenic heat fluxes for UK cities: the case of Manchester. Theor Appl Meteorol 98:19–35
Sobrino JA, Jimenez-Munoz JC, Soria G, Romagueira M, Guanter L et al (2008) Land surface emissivity retrieval from different VNIR and TIR sensors. IEEE Trans Geosci Remote Sens 46:316–327
Stabler L, Martin C, Brazel A (2005) Microclimates in desert city were related to land use and vegetation index. Urban for Urban Green 3:137–147

Tan J, Zheng Y, Song G et al (2007) Heat wave impacts on mortality in Shanghai, 1998 and 2003. Int J Biometeorol 51:193–200

Tomlinson CJ, Chapman L, Thornes JE, Baker CJ (2011) Including the urban heat island in spatial heat health risk assessment strategies: a case study for Birmingham, UK. Int J Health Geographics 10(42):1–14

Tomlinson CJ, Chapman L, Thornes JE, Baker CJ (2011) Including the urban heat island in spatial heathealth risk assessment strategies: a case studyfor Birmingham, UK. Int J Health Geographics 2–14

United Stated Geological Survey (USGS) (2016) LANDSAT 8 (L8) data user handbook. https://landsat.usgs.gov/documents/Landsat8DataUsersHandbook.pdf (March 29). Accessed 10 Jan 2016

UNISDR (2006) Developing early warning system: checklist. Bonn. https://www.unisdr.org/files/608_10340.pdf

WHO (2008) Heat-health action plan: guidance. World Health Organisation Regional Office for Europe, Copenhagen, Denmark

Wolf T, G MCGregor. (2013) The development of heat wave vulnerability index for London, United Kingdom. Weather and Clim Extremes 1:59–68

Chapter 7
Comparative GIS-Based Assessment of Landslide Susceptibility of Chepe River Corridor, Gandaki River Basin, Nepal

Kripa Shrestha, Udhab Raj Khadka, and Mandira Singh Shrestha

Abstract Loss of life and property due to human induced and natural landslides are common in Himalayan region of Nepal. Intense precipitation over a short period of time, unplanned road construction, degradation of forests, and change in land use are expected to increase the number and intensity of landslides in the already fragile mountainous tracts of the country. The landslide susceptibility assessment using GIS and remote sensing tools identifying hazard/susceptibility, vulnerability and risk are very useful for disaster risk reduction and management. In this chapter, Statistical Index Model and Logistic Regression Model were compared for performance through Geographical Information System (GIS), to derive landslide Susceptibility map of the Chepe River corridor. Eleven factors (slope, aspect, geology, distance to road, land use, rainfall, elevation, relief, drainage density, plan curvature, and profile curvature) were considered as possible key factors for the landslide susceptibility assessment. To validate the models, Receiver Operating Characteristic (ROC) was used. The result shows the Logistic Regression Model has 82% prediction accuracy, whereas Statistical Index Model has 63% prediction accuracy. Based upon the accuracy assessment, the logistic regression model seems to have better applicability in the Chepe River corridor in comparison to statistical index method using same eleven triggering factors.

Keywords Landslide · Susceptibility assessment · GIS · Chepe River · Nepal

---

The original version of the book was inadvertently published with an incorrect layout in Chapter 7. The layout has been corrected and the book has been updated to reflect this. The correction to this chapter is available at https://doi.org/10.1007/978-3-030-55563-4_20

K. Shrestha (✉) · M. Singh Shrestha
International Centre for Integrated Mountain Development (ICIMOD), Khumaltar, Lalitpur, Nepal
e-mail: k.kripashrestha@gmail.com; kripa.shrestha@icimod.org

M. Singh Shrestha
e-mail: mandira.shrestha@icimod.org

U. R. Khadka · M. Singh Shrestha
Central Department of Environmental Science (CDES), Tribhuvan University, Kirtipur, Nepal

R. Djalante et al. (eds.), *Integrated Research on Disaster Risks*, Disaster Risk Reduction,
https://doi.org/10.1007/978-3-030-55563-4_7

7.1 Introduction

Nepal's varied topography makes it susceptible to climate-related disasters and it experiences a range of natural hazards, some of which occur annually, e.g., floods and landslides (UNDP 2009). Disasters triggered by natural hazards cause heavy loss of life, property, and unparalleled threat to sustainable development. With more than 7,000 deaths, Nepal ranked 23rd in terms of total natural hazard-related deaths globally between 1988 and 2007 (Baruwal 2014); 11th in the world in terms of vulnerability to earthquakes; 30th in terms of water-induced hazards such as flood and landslides (UNDP 2009), and 20th for the most multi-hazard prone countries in the world (Dangal 2011). MoHA and DpNet-Nepal (2015) stated that the various studies and reports over the last 33 years have shown that each year, floods, landslides, fires, avalanches, and epidemics kill hundreds of people and destroy property worth billions of rupees and the extreme weather events associated with heavy rainfalls are the principal cause of cascading natural disasters in Nepal. Landslides in Nepal are the country's costliest and deadliest type of natural disaster, but their management is still seen as low priority (IRIN 2013). Landslide susceptibility, vulnerability, and risk maps are vital for disaster management and for planning development activities in the mountainous country like Nepal. Despite high importance of landslide risk evaluation for decision-making, comparatively minimal efforts have been made to establish and systematically test methods for landslide risk assessment and understand their advantages and limitations (Guzzetti 2012).

Ministry of Forest and Environment (MoFE 2010) stated that the overall vulnerability of Lamjung district in terms of disaster is very high with the index ranging from 0.100 to 0.787. The district is considered to be highly vulnerable to landslide, rainfall, temperature, and glacial lake outburst flood (GLOF). The studies on landslide in the river corridor should not be underestimated as it can cause a huge destruction. The major purpose of this chapter is to assess the landslide susceptibility along the Chepe River corridor within the boundary of Lamjung and Gorkha districts. Since this area is unexplored in term of landslide assessments, this chapter will be a baseline for the landslide assessments.

7.2 General Description of Chepe River Corridor

The study was conducted in the Chepe River, which lies in the border between the two districts Gorkha and Lamjung (Fig. 7.1). The river originates from the Dudhpokhari at an approximate elevation of 5300 m above sea level. The name of Chepe River is linked to Nepal's political history and the formation of the nation state, which offers a rich context of institutional responses to landslides. It has a total catchment area of 308 km^2 at the confluence of Marshyangdi River. The total area of the river corridor is about 47.301 km^2 considering the buffer of 500 m on both sides of the river. The terrain is very rugged, precipitation distribution pattern seems to be very much

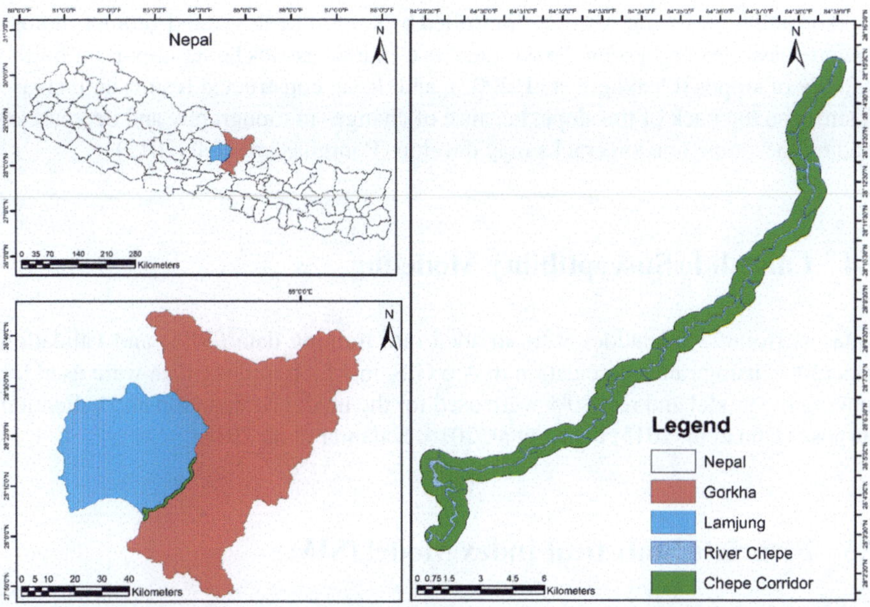

Fig. 7.1 Study area: Chepe River Corridor

influenced by the spotty convective activities, so the pattern from one station to the other nearby station is also different and complex (Pokhrel 2003). The Chepe River corridor touches five municipalities, i.e., two from Lamjung district (Rainas and Dudhpokhari) and three from Gorkha district (Palungtar, Siranchowk, and Ajirkot). Specifically, the landslide occurrence location is 28° 3′ 30.40" N and 84° 28′ 40.49" E to 28° 15′ 40.78" N and 84° 39′ 2.26" E.

7.3 Methodology

Fieldwork was carried to take the coordinates of the landslides, for the field verification of landslide mapped through google earth and to get the preliminary information about the occurrence of landslide in between January and June 2018. Arc GIS 10.2, R studio, Google Earth Engine (GEE), and JMP were used for the study. For the analysis, 11 parameters, i.e., slope, aspect, rainfall, distance to road, geology, elevation, plane curvature, profile curvature, relief, drainage density, and land use were taken into consideration on the basis of literature review. Various factors are responsible for the occurrence of the landslide, i.e., Slope is the measure of an angle between a location in the ground surface and the horizon (Ohlmacher 2006); Aspect giving the direction of the slope (Tian et al. 2010); Curvature controlling the convergence or divergence of landslide material and water in the direction of landslide motion

(Carson and Kirk 1972); Altitude controlled by several geologic and geomorphological processes (Pourghasemi 2008); Land use characteristics being important for the stability of slopes (Ocakoglu et al. 2002); and Road constructed results an increase in stress on the back of the slope, because of changes in topography and decrease of load on toe, some tension cracks may develop (Pourghasemi et al. 2012).

7.4 Landslide Susceptibility Modeling

Datasets from 73 landslides were divided into training data (70%) and validation data (30%) using random selection in Arc GIS, in which training data were used for running the model and rest 30% were used for the model comparison and validation purpose (Dou et al. 2015; Fayez et al. 2018; Kalantar et al. 2018).

7.5 Bivariate Statistical Index Model (SIM)

The SIM was used for the landslide analysis (Van Westen 1997):

$$W_{ij} = ln\left(f_{ij}/f\right) = ln\left(\left(A_{ij}{}^{*}/A_{ij}\right) \times \left(A/A^{*}\right)\right) = ln\left(\left(A_{ij}{}^{*}/A^{*}\right) \times \left(A/A_{ij}\right)\right)$$

where

W_{ij} = weight given to class i of parameter j.
F_{ij} = landslide density within class i of parameter j.
F = Landslide density within entire map.
$A_{ij}{}^{*}$ = area of landslide in class i of parameter j.
A_{ij} = area of a class i of parameter j.
A^{*} = total area of landslide in entire map.
A = total area of entire map.

7.6 Landslide Susceptibility Index (LSI)

$$\text{LSI} = \sum_{j=1}^{n} W_{ij}.$$

where

W_{ij} = weight of class i of parameter j.
n = number of parameter.

7.7 Multi-variate Logistic Regression Model (LRM)

Considering p independent variables, $x_1, x_2, \ldots, x_p$, affecting landslide occurrences, we define the vector $X = (x_1, x_2, \ldots, x_p)$. The independent variables are with values of 1 (presence) or 0 (absence).

The conditional probability that a landslide occurs is represented by $P(y = 1/X)$. The logit of the multiple LRM (Hosmer and Lemeshow 2000) is

$$\text{Logit}(y) = b_0 + b_1 x_1 + b_2 x_2 + \ldots + b_p x_p$$

where b_0 is the constant of the equation, and $b_1, b_2, \ldots, b_p$ are the coefficients of variables $x_1, x_2, \ldots, x_p$.

The probability $P(y = 1/X)$ can be expressed in the LRM:

$$P\left(y = \frac{1}{X}\right) = \frac{1}{1 + e^{-(b_0 + b_{1x_1} + b_{2x_2} + \ldots\ldots + b_{px_p})}}$$

where "e" is the constant 2.718.

Higher the value of coefficient, higher will be the weightage.

7.8 Landslide Susceptibility Index Classification

For the differentiation of different susceptibility class, susceptibility index classification was done using R studio. Both the landslide and non-landslide points were taken for the landslide susceptibility index classification. The susceptibility index having 25% of landslide was classified as low susceptibility, 25–50% as medium susceptibility, 50–75% as high susceptibility, and 75–100% as very high susceptibility (Lee and Pradhan 2007).

7.9 Landslide Susceptibility Model Validation and Comparison

It is very important to check the efficiency or the validation of the landslide susceptibility model. In this study, Receiver Operating Characteristic (ROC) index has been used for the validation of the model (Pontius and Schneider 2001). A good fit model has Area Under Curve (AUC) values that range from 0.5 to 1, while values below 0.5 represent a random fit (Youssef et al. 2016). The ROC of SIM and LRM using 30% (n = 18) of landslides were obtained to check the accuracy and reliability of the model.

7.10 Results and Discussion

7.10.1 Landslide Mapping

A total of 73 landslides were detected from the field visit and google earth. A series of field visits also helped with checking the sizes and shapes of landslides and material involved like construction materials, rocks, boulders, muds, etc. The landslides with the area ranging from 3.87 to 30,600.79 m^2 are heterogeneously distributed over the area. Some landslides are located near the confluence with Marshyangdi River, some along the middle section of the river, and some close to the river source (Fig. 7.2).

Many factors are responsible for the landslide occurrence. Generally, as slope increases, the probability of landslide occurrence also increases (Meten et al. 2015). However, Chen et al. (2001) state that there is no appropriate relationship between the steepness of a slope and the probability of occurrence of landslides. Landslides are prone to occur on slopes having a particular range of steepness. The slope class from 10 to 50° showed an increasing trend of landslide, whereas from 50 to 70°, it is decreasing. Aspect (slope orientation) affects the exposure to sunlight, wind, and precipitation thereby indirectly affecting other factors such as soil moisture,

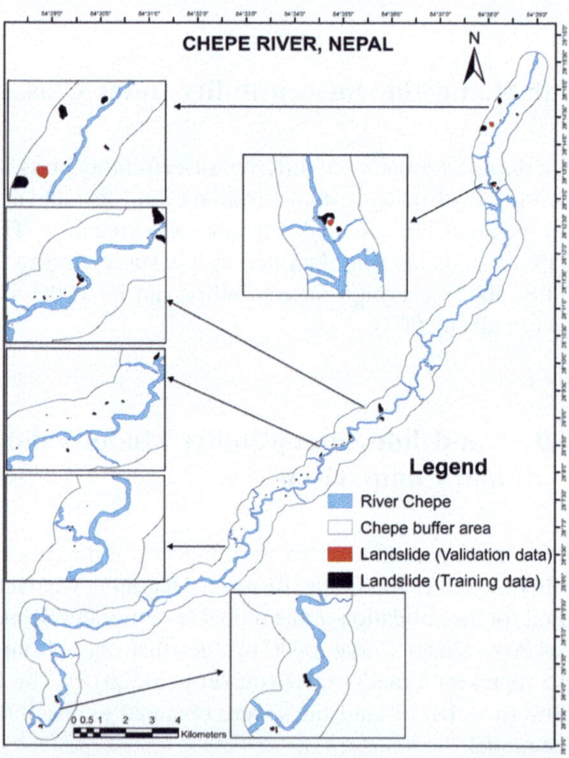

Fig. 7.2 Landslide inventory map showing the distribution of landslides in the Chepe River corridor

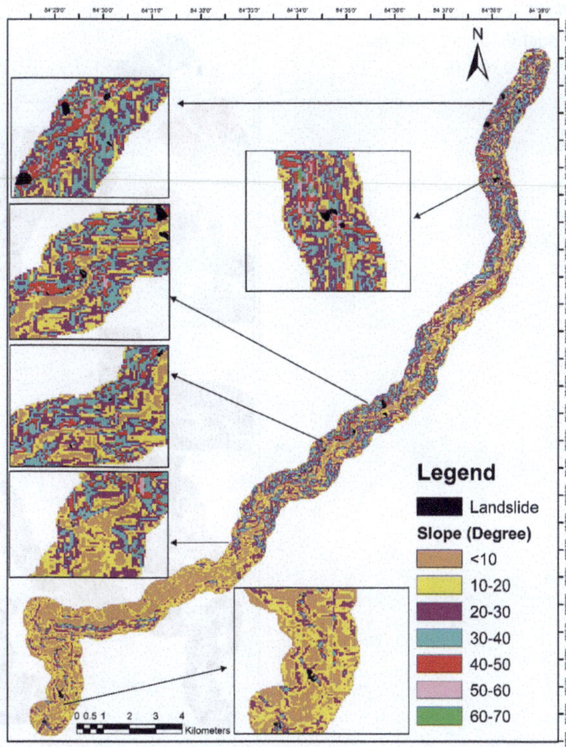

Fig. 7.3 Distribution of landslide along with different slopes

vegetation cover, and soil thickness that contribute to landslides (Clerici et al. 2006). Aspect toward the southeast and south is found mostly to contribute to landslides in the study area. This might be because the south-facing slopes are generally steep and anti-dipping slopes and many south-facing facets are on the windward side of the summer monsoon rain (Ghimire 2011) (Figs. 7.3, 7.4, 7.5, 7.6, 7.7, 7.8, 7.9, 7.10, 7.11, 7.12, 7.13 and 7.14).

The land use also has a significant role in the stability of soil slope. Landslide was found to occur more in the forested area than in the barren lands which are similar to the findings in Chure area (TU-CDES 2016). Based on the field survey, the maximum area covered by landslides is in the forest. This may be due to factors like haphazard road construction, broader girth of older trees, spring source creating gullies, etc. Moreover, opening in canopy due to small scale chronic disturbances such as lopping of branches for fodder and fuelwood reduces interception of rain as a result in events of high precipitation soil erosion takes place along leading to mass wasting of soil eventually leading to landslides. As per legal jurisdiction and definition the area may fall into forest zone, nevertheless the disturbance and human activities continues in terms of lopping, litter removal. Apart from this unplanned developmental activity in already fragile belt of the Himalayas also takes huge toll on the geology which further aggravates the phenomenon. SIM has also stated that slope (40–50°), distance

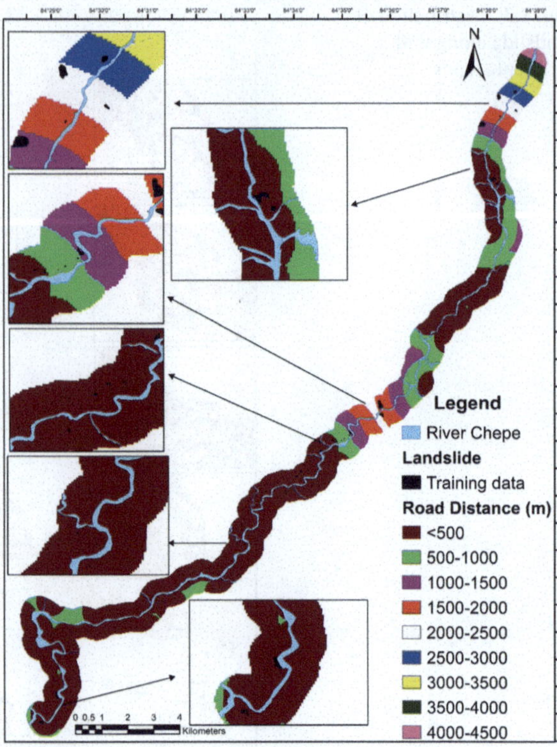

Fig. 7.4 Distribution landslide along with the distance to road

to road (less than 500 m), geology (Himal group, Ranimatta and Ulleri formation), and rainfall (2800–2900 mm) are mostly responsible for the occurrence of landslide which also might be the reason for the observation of maximum landslide in the forest areas as there are many associated factors responsible for occurrence of landslide. Most of the area of the corridor falls in the forest which might be another reason for maximum landslides observed in the forested areas. Trees and forests can make a positive contribution in various situations; however, it also increases landslide risk by imposing load on unstable slopes and via wind-related effects; they are unlikely to prevent or minimize deep landslides or slides on very steep slopes (FAO 2018).

According to Bhattarai and Pradhan (2013), the construction activities like roads are preferentially built along the same relief and are therefore landslide hazards in an area are observed more or less on the same relief. The study area has the maximum area of landslides in the relief of less than 30 m. Additionally, Most of the area of Chepe River falls within the elevation range of 400–800 m where the area of occurrence of landslide is maximum in the elevation range of 800–1200 m. According to Ghimire (2011), there is no exact relation between occurrence of landslide and elevation and therefore elevation alone cannot explain the occurrence of landslides. Elevation along with other linked parameters like aspect, slope is interlinked for causing landslides in any place.

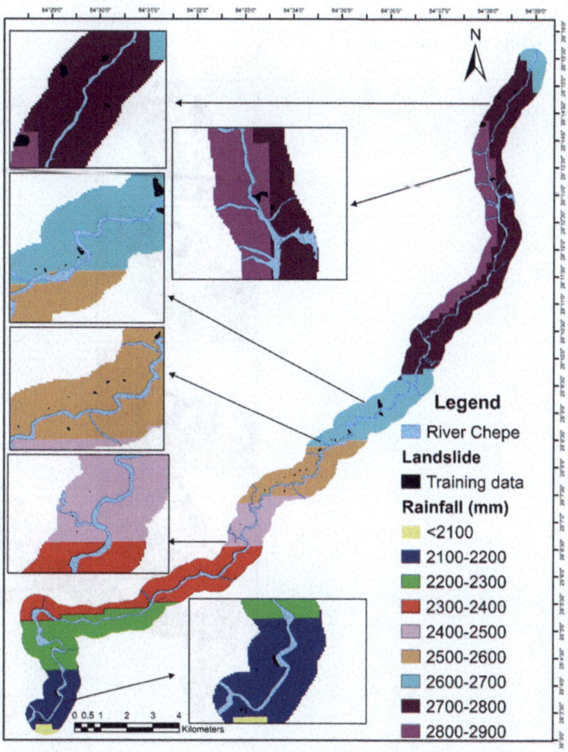

Fig. 7.5 Distribution landslide along with different rainfall intensities

The roads built on the slopes cause the loss of toe support. The change of the topography and loss of support lead to increase in strain behind the slope and development of cracks. It leads to the instabilities occurring in the slope because of the negative effects such as water infiltration (Devkota et al. 2013) and with that the road segment may act as a barrier, a net source, a net sink, or a corridor for water flow, and depending on its location in the area, it usually serves as a source of landslides (Pradhan and Lee 2010). The road network also increases surface flows which contribute to landslides. Maximum landslides were found to be near to the road, i.e., <500 m. With the increasing distance from the road, the landslide is also less which shows that the landslide might be triggered due to the haphazard road construction. The similar result was reported by Devkota et al. (2013) in the Mugling–Narayanghat road section which stated that road construction is one of the reasons behind the landslide occurrence. Implementation of improved standards for road construction needs to be undertaken immediately to reduce landslide risk.

The curvature values represent the morphology of the topography (Lee et al. 2004; Erener and Duzgun 2010). Devkota et al. (2013) concluded that the concave curvature is most susceptible to the landslide, whereas Lee and Pradhan (2007) observed that convex curvature is more susceptible to landslide which indicates that there may be other factors like rainfall, slope, geology, land use, etc. along with curvature type

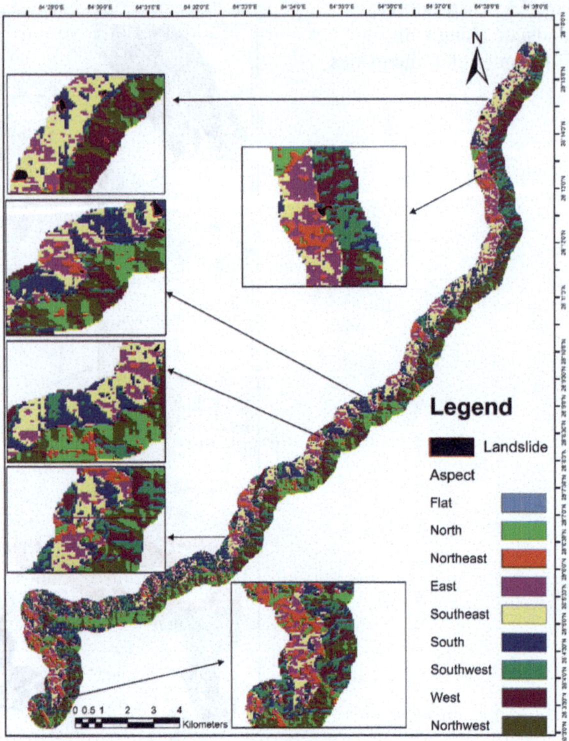

Fig. 7.6 Distribution of landslide with different aspects

responsible for the landslide. In this river corridor, landslide was observed maximum in the planar curvature.

Geology plays an important role in causing landslide failure. Most of the landslide in the study area is found to contain Himal group and Ranimatta formation. Himal group consists of rocks like marble, gneiss, magmatite, whereas Ranimatta formation consists of phyllite, quartzite, metasandstone, and conglomerate (Ojha 2009). Thin beds of phyllite between quartzite beds is one of the causes of the landslide. Phyllite has less friction angle and higher weathering rate which gets weathered during long environmental exposure, due to which the landslides are active. Himal group consists of gneiss which contains more amount of feldspar and weathering rate of feldspar is higher which undergoes chemical weathering with water. Due to which land sliding is common in gneiss (Gasim et al. 2015).

Rainfall plays an important role in landslide triggering effects. Rainfall triggered landslides are reported every year in different countries. Annual monsoon rainfall in Nepal ranges from as low as 160 mm in the northwestern region to as high as 5500 mm in some of the isolated areas of central Nepal (Chalise and Khanal 2001). The rainfall of intensity 2800–2900 mm is observed to have more landslide and the area of landslide was observed to be increasing with the increase in the intensity of rainfall in the Chepe River corridor. The distribution of the daily precipitation

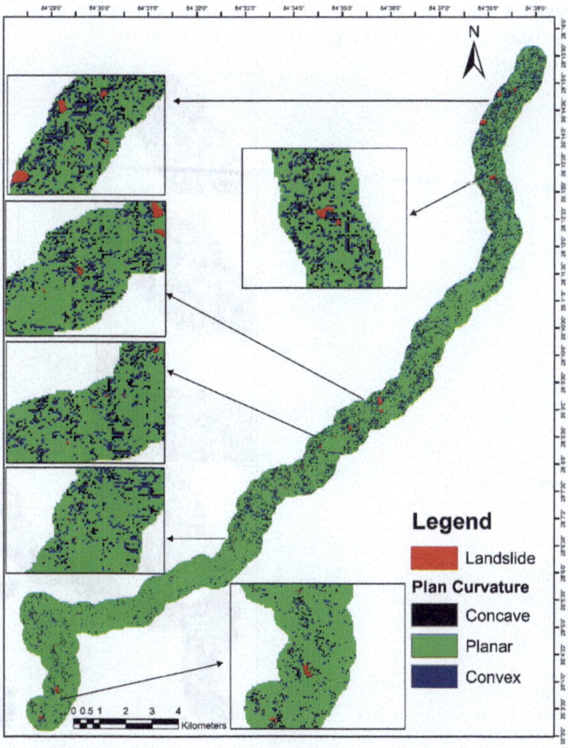

Fig. 7.7 Distribution of landslide along with plan curvature

in the monsoon in the context of Nepal is also uneven where sometimes 10% of total annual precipitation can occur in a single day (Alford 1992) and 50% of total annual precipitation is often recorded within 10 days of monsoon period (Dahal and Hasegawa 2008). Such kind of rainfall pattern plays a vital role in the occurrence of landslide in Nepal. The study by Dahal and Hasegawa (2008) on the rainfall threshold for landslide in Nepal Himalaya, stated that on an average, a rainfall of 10 h or less requires a rainfall intensity in excess of 12 mm h^{-1} to trigger failure, but a rainfall duration of 100 h or longer with an average intensity of 2 mm h^{-1} can also trigger landslides in the Himalaya. However, considering the daily rainfall scenario, this study concludes that when daily rainfall amount exceeds 144 mm, there is always risk of landslides in Himalayan slopes (Dahal and Hasegawa 2008). The result of the model showed slope, distance to road, and geology are the major contributing factor of the landslide in the river corridor (Table 7.1).

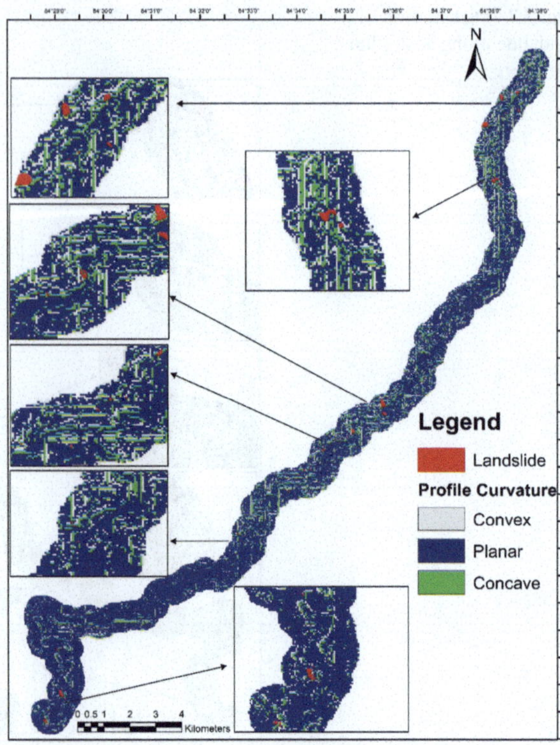

Fig. 7.8 Distribution of landslide along with profile curvature

7.11 Landslide Susceptibility Index (LSI) from Statistical Index Model

The result from the landslide susceptibility index classification showed the maximum value of 1 and the minimum value of 0.1 for susceptibility index value. By this method, both the landslide distribution and susceptibility index are used for the classification purpose of susceptibility index class (Figs. 7.15 and 7.16, Table 7.2).

7.12 Landslide Susceptibility Index (LSI) from Multi-variate Logistic Regression Model

Logistic regression model, one of the proposed favorable models to deal with the problem of combination of heterogeneous data, has been widely used for mapping landslide susceptibility. It is based on the point-based analysis of landslide. LRM shows that there is greater significance of land use in the occurrence of the landslides, whereas geology and profile curvature have very less significance. There is

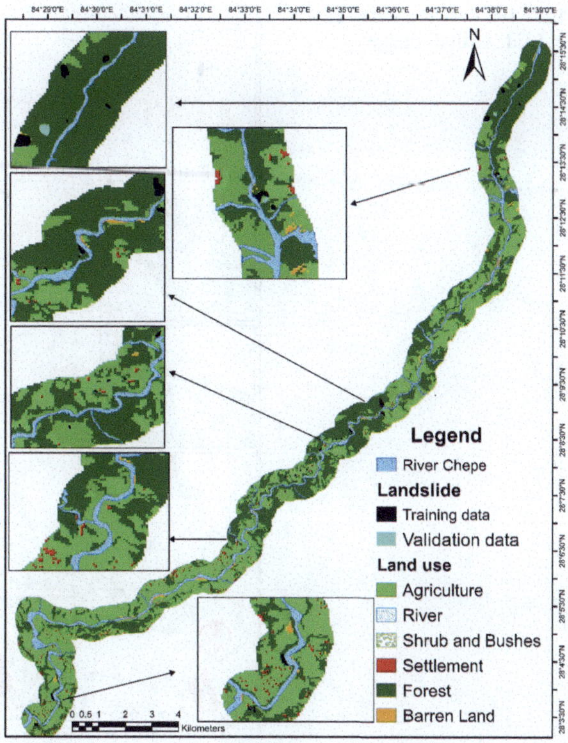

Fig. 7.9 Distribution of landslide with different land use patterns

a general consensus that geological information is one of the most decisive parameters regarding landslide manifestation and thus included in Landslide Susceptibility Models (LSMs) (Rossi et al. 2010; Pourghasemi et al. 2012). However, the study by Paudel (2016) also revealed that the necessity of using geological information in addition to topographical parameters is not always high (Table 7.3).

Using both landslide and non-landslide points, the following equation was obtained from JMP software (Fig. 7.17 and Table 7.4).

$Z =$ Float $(-20.4030 - 0.0190 \times$ Elevation $+ 2.2773 \times$ Drainage $- 0.0028 \times$ Road $+ 0.0631 \times$ Profile Curvature $+ 1.1295 \times$ Plan Curvature $- 0.0335 \times$ Relief $+ 0.0092 \times$ Rainfall $+ 0.0069 \times$ Aspect $+ 0.0967 \times$ Slope $+$ Con ("Landuse" $= = 1, -3.2927$, Con ("Landuse" $= = 2, 1.079$, Con ("Landuse" $= = 3, 2.3956$, Con ("Landuse" $= = 5, -0.1821, 0)))) +$ Con ("Geology" $= = 2, -13.5925$, Con ("Geology" $= = 4, -10.4906$, Con ("Geology" $= = 5, 11.6915$, Con ("Geology" $= = 8, 12.3917, 0)))))$.

Map prepared was converted into the log expression:

$$\mathrm{P}(y = 1/X) = \frac{1}{(1 + \mathrm{Power}(2.718 - \mathrm{rastermap})}$$

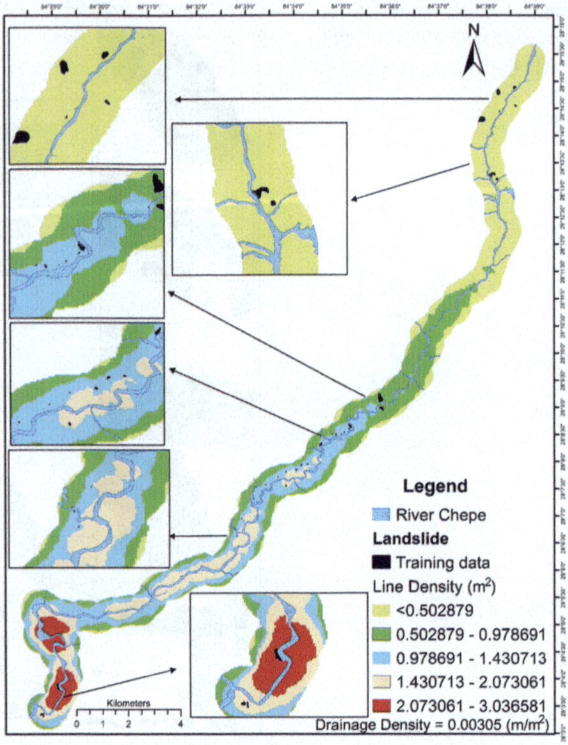

Fig. 7.10 Distribution of landslide with drainage density

The whole model test shows that the-loglikelihood has dropped to 30.59 (from 68.62) and the difference in model by including the variable is significant with Prob > ChiSq (<0.0001) and chi-square statistics (76.05009). This means models are compared and hypothesis is tested which showed model improvement has been achieved.

The result of landslide susceptibility index classification showed the maximum value of 0.7309 and the minimum value of 0.5. The result obtained showed that about 25% of the landslide has the LSI less than 0.62, about 50% of landslide is in the range of 0.62–0.64, about 75% of landslide had at the range of 0.64–0.645 and 100% of landslide had the range of 0.645–0.7309 (Figs. 7.18 and 7.19, Table 7.5).

7.13 Comparing LSI from Both: Receiver Operating Characteristic (ROC)

Landslide Susceptibility Models after the preparation should be validated before using to ascertain their reliability. According to Vakhshoori and Zare (2018), the validity of LSMs had simply been presented by simple statistics such as landslide

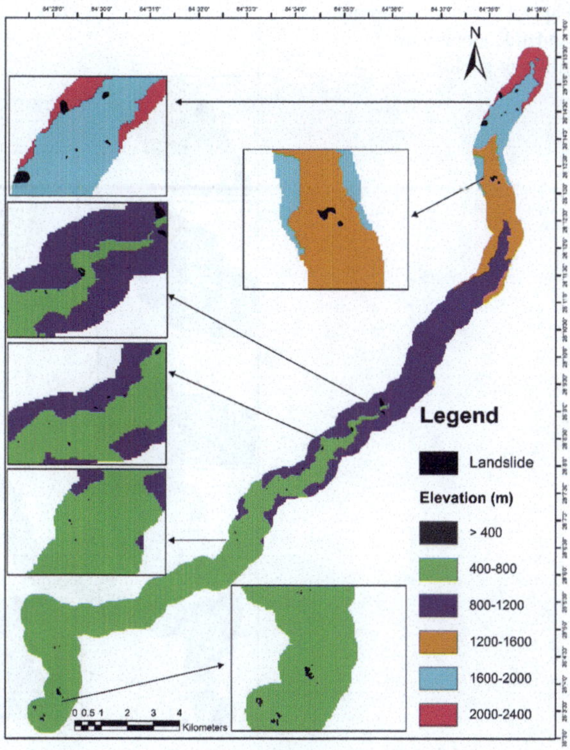

Fig. 7.11 Distribution of landslide with different elevations

percentage per susceptible zones and model efficiency. Owing to the inefficiency of these simple statistics (Provost and Fawcett 1997; Provost et al. 1998), threshold-independent methods, like ROC, have been recommended for validation (Fielding and Bell 1997; Begueria 2006; Akgun (2012); Corominas et al. 2014). The landslide susceptibility assessment in this study was carried out using two different models, i.e., SIM (bi-variate model) and LRM (multi-variate model). Furthermore, the results were validated using the ROC analysis to evaluate the correlation between the landslide susceptibility maps and landslide inventory points as well as to compare the effectiveness of model in landslide susceptibility mapping of Chepe River corridor (Figs. 7.20 and 7.21).

The results obtained shows that a value of Area Under Curve (AUC) for SIM was 0.6296 and the prediction accuracy was about 63%. Similarly, AUC for LRM was 0.8209 and the prediction accuracy was 82%. The results obtained from ROC indicate that the LRM looks to be more accurate in terms of the performance of landslide susceptibility mapping and has better prediction accuracy than the SIM in the study area. Similary few studies like Pourghasem et al. (2013) prepared the

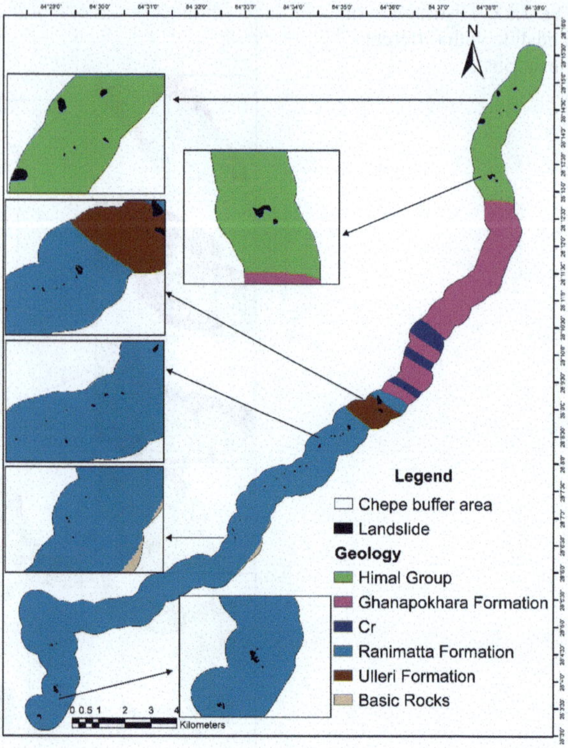

Fig. 7.12 Distribution of landslide along with different geology

landslide susceptibility mapping of Iran by binary logistic regression, analytical hierarchy process, and SIMs along with this they also compared the performances of model, which also shows that LRMs is better than SIMs; Similar result has been observed by Lim et al. (2011) in Penang island, Malaysia and Dou et al. (2015) in Osado Island, Japan where LRM was found to be more reliable than SIMs in the study of landslide susceptibility. Akgun (2012) also compared LRM with other models and the result showed LRM has a better accuracy than other models. Similary a study by Acharya and Pathak (2017) in Marsyangdi River Basin (Chepe River is a tributary of Marshyangdi River), West Nepal also used LRM for the assessment which showed 74 and 77% accuracy.

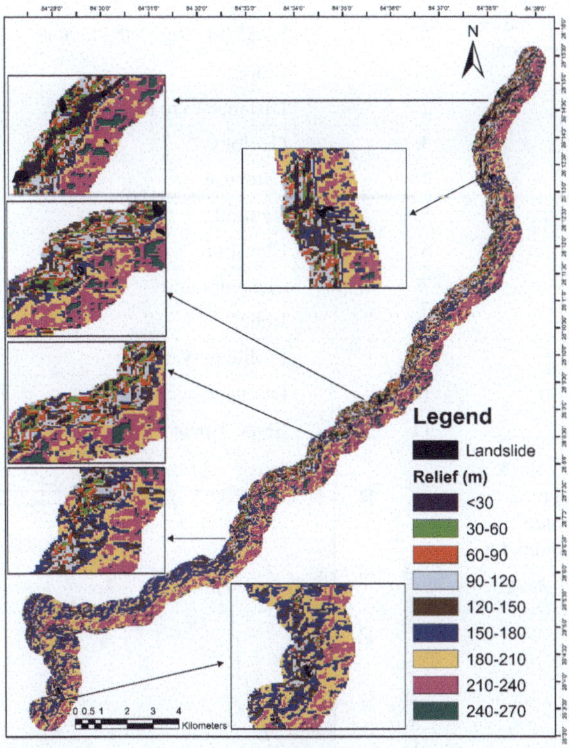

Fig. 7.13 Distribution of landslide along with different relief

Fig. 7.14 Created gully resulting in the landslide

Table 7.1 Weightage values for the triggering factors of landslide

S. no	Landslide triggering factors	Weightage
1	Slope	5.5006
2	Distance to road	5.1733
3	Geology	2.2722
4	Land use	1.798
5	Rainfall	1.7502
6	Elevation	1.7061
7	Plan curvature	1.5474
8	Relief	1.1621
9	Profile curvature	0.9767
10	Drainage density	−1.8401
11	Slope aspect	−3.6418

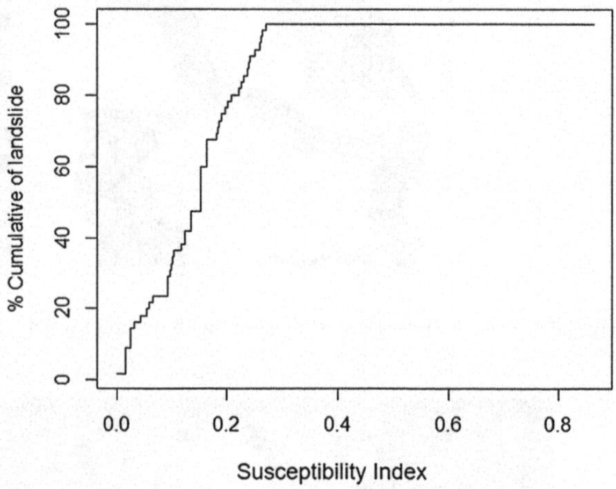

Fig. 7.15 Percentage cumulative of landslide with susceptibility index for SIM

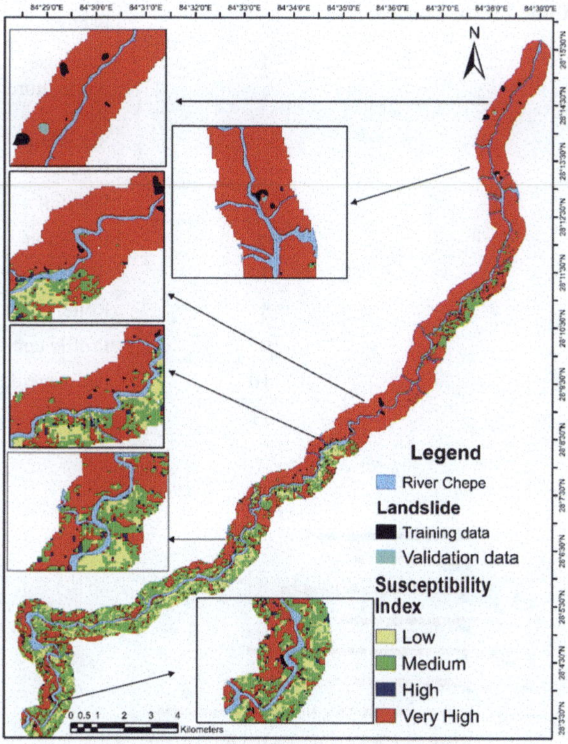

Fig. 7.16 Landslide susceptibility map prepared by SIM

Table 7.2 Landslide susceptibility index classification

S. no	Landslide susceptibility index	Susceptibility class
1	<0.1	Low
2	0.1–0.19	Medium
3	0.19–0.2	High
4	0.2–1	Very high

Table 7.3 Coefficient value of factors in LRM

S. no	Factors	B
1	Slope	0.0967
2	Distance to road	- 0.0028
3	Geology Ghanapokhara formation Cr Ranimatta formation Ulleri formation	−13.5925 −10.4906 11.6915 12.3917

(continued)

Table 7.3 (continued)

S. no	Factors	B
4	Land use Agriculture River Shrubland Forest	−3.2927 1.079 2.3956 −0.1821
5	Rainfall	0.0092
6	Elevation	−0.0190
7	Plan curvature	1.1295
8	Relief	−0.0335
9	Profile curvature	0.0631
10	Drainage density	2.2773
11	Aspect	0.0069

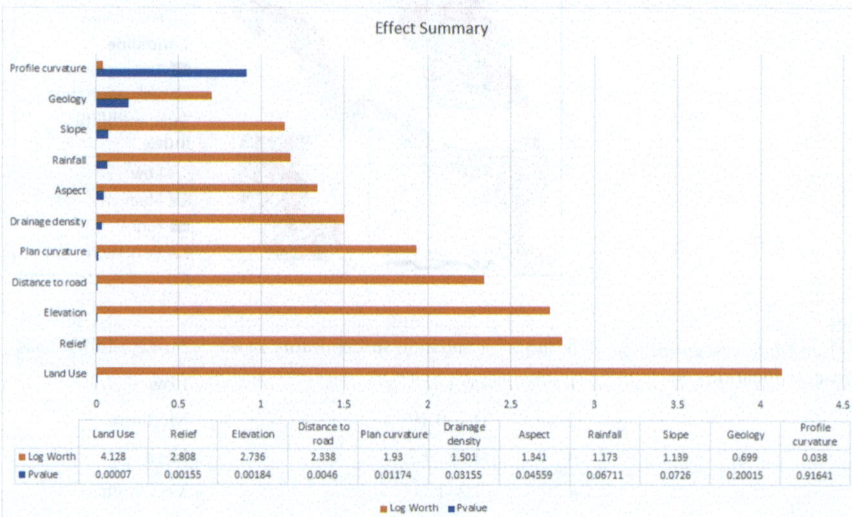

Fig. 7.17 Probability value for each triggering factor of landslide

Table 7.4 Whole model test

Model	LogLikelihood	DF	ChiSquare	Prob > ChiSq
Difference	38.025046	15	76.05009	< 0.0001*
Full	30.591474			
Reduced	68.616520			
RSquare (U)		0.5542		
AICc		99.8171		
BIC		134.705		
Observations (or Sum Wgts)		99		

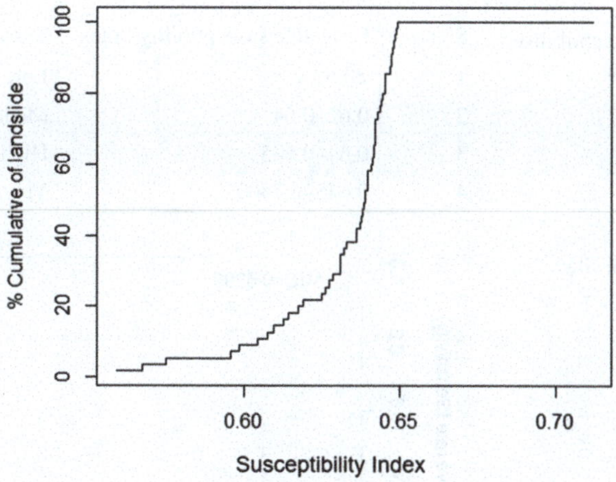

Fig. 7.18 Percentage cumulative of landslide with the susceptibility index using LRM

Fig. 7.19 Landslide susceptibility map prepared by LRM

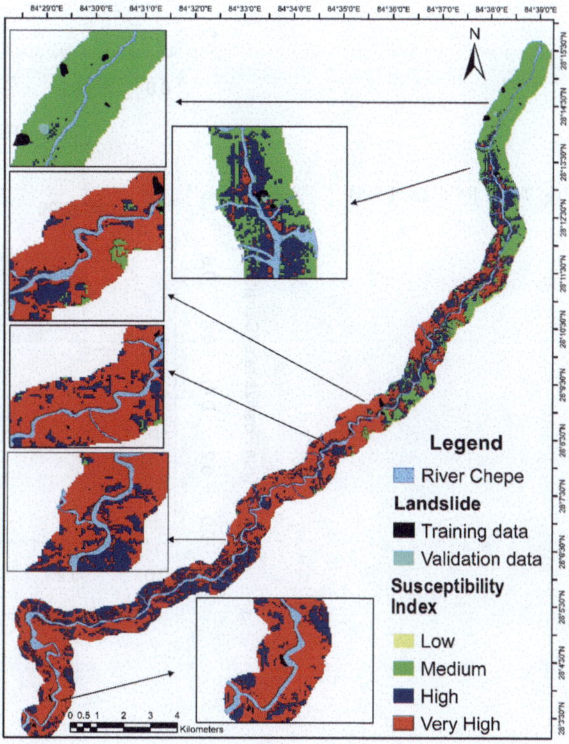

Table 7.5 Landslide susceptibility index classification

S. no	Landslide susceptibility index	Susceptibility class
1	<0.62	Low
2	0.62–0.64	Medium
3	0.64–0.645	High
4	0.645–0.7309	Very high

Fig. 7.20 ROC for SIM

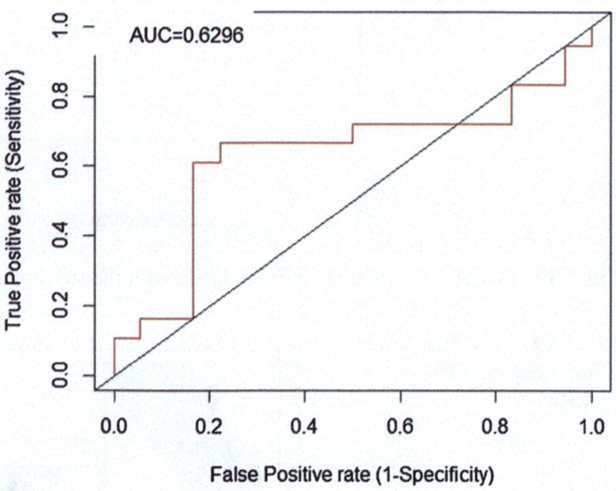

Fig. 7.21 ROC for LRM

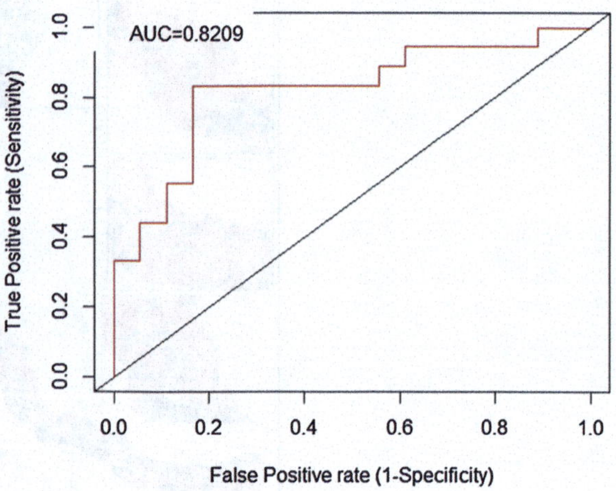

7.14 Conclusion

Most of the development activities taking place in the Himalayas are done with utter disregard to the mountain environment and its characteristics. Unplanned and unscientific development is making the Himalayas more prone to disasters like landslides. There is great need to bridge the gap between scientist, policy-makers, and those implementing the projects in order to come up with an inclusive plan and implementation strategy with regard to fragile areas like the Himalayan belt. Based upon the accuracy assessment, the LRM seems to have better applicability in the Chepe River corridor in comparison to statistical index method using same eleven triggering factors. Due to the time constrains, these models are not tested in the other parts of Nepal, however, on the basis of the literature review, LRMs has better performance, better applicability, and accuracy than the SIMs (Akgun 2012; Dou et al. 2015; Lim et al. 2011; Mondal and Mandal 2018; Mousavi et al. 2011; Pourghasem et al. 2013). For the landslide control, prioritization should be given to the proper land use management to control the landslide of the area. Development is necessary; however, haphazard road construction is increasing the likelihood of occurrence of landslide and its negative impact can be predicted to increase in the coming days if not managed carefully.

This study has identified the landslide-prone areas of the Chepe River corridor which can be helpful to raise awareness about the risk in the area and take a step for the mitigation. Proper Land Use planning (according to Land Use Policy of Nepal 2015) and its implementation should be done in order to use the land sustainably. And proper baseline study on the hazard and risk zonation should be done prior to the construction of roads, houses, and the other infrastructures. The science plan of IRDR focusing on the hazards and its effect can be implemented in the climate change affected country like Nepal. The government of Nepal has prepared Disaster Risk Reduction National Strategic Action Plan (NSAP) (2018–2030) in which DRR activities have been prioritized corresponding to the four priorities of the Sendai Framework For Disaster Risk Reduction (SFDRR) (2015-2030) and 18 priority actions. This Strategic Action Plan has also set the targets for reducing disaster loss considering the targets set by the Sustainable Development Goals 2030 in terms of substantially reducing the impact of disasters nationally. It aims to reduce the disaster mortality rate, reduce the number of disaster affected people, reduce the direct economic loss in GDP from disasters, reduce the damage to critical infrastructure and disruption of basic services enhancing resilience, develop Disaster Risk Management DRM strategies and action plans at the provincial and local levels (MoHA 2019).

Acknowledgements On behalf of all the authors, I am thankful to the Central Department of Environmental Science, Tribhuvan University, Nepal for giving this opportunity and HI-AWARE, ICIMOD for the technical and financial support to carry out this research. I am pleased to express my deepest gratitude to Dr. Deo Raj Gurung, Aga Khan Development Network (AKDN), Tajikistan, Dr. Uttam Paudel, Housing Recovery and Reconstruction Platform (HRRP) for their invaluable guidance throughout my study. I would also like to thank Mr. Ajay Shrestha, Mr. Anish Gurung, Mr. Kesav Paudel, Mr. Nabin Bhandari, Mrs. Radhika Maharjan and Mr. Salar Saeed Dogar for their support in this study.

Disclaimer The views and interpretations in this publication are those of the authors and they are not necessarily attributable to their organizations.

References

Acharya S, Pathak D (2017) Landslide hazard assessment between Besi Sahar and Tal area in. Int J Adv Remote Sens GIS, 5(1):29–38

Akgun A (2012) A comparison of landslide susceptibility maps produced by logistic regression, multi-criteria decision, and likelihood ratio methods: a case study at ˙Izmir, Turkey. Landslides 9:93–106

Alford D (1992) Hydrological aspects of the Himalayan region. Occasional Paper No. 18. ICIMOD. Kathmandu, p 68

Baruwal A (2014) Disaster profile of Nepal. Monographic issue. Emergency and disaster reports 1(3):3–49

Begueria S (2006) Validation and evaluation of predictive models in hazard assessment and risk management. Nat Hazards 37(3):315–329

Bhattarai BC, Pradhan AMS (2013) A case study on landslide hazard mapping in Changunarayan VDC, Nepal. In: Landslide science and practice. Springer, Berlin, Heidelberg, pp 551–559

Carson MA, Kirk MJ (1972) Hillslope form and process. Cambridge University Press, London, p 475

Chalise SR, Khanal NR (2001) Rainfall and related natural disasters in Nepal. In: Tianchi L, Chalise SR, Upreti BN (eds) Landslide hazards mitigation to the Hindukush–Himalayas. ICIMOD, Kathmandu, pp 63–70

Chen HH, Chiou CR, Kang TS (2001) The application of GIS on studying the characteristics of landslide areas in the Shi-Meng Reservoir watershed. J Exp Natl Taiwan Univ 15(3):3–222

Clerici A, Perego S, Tellini C, Vescovi P (2006) A GIS-based automated procedure for landslide susceptibility mapping by the conditional analysis method: the Baganza valley case study (Italian Northern Apennines). Environ Geol 50(7):941–961

Corominas J, Van Westen C, Frattini P, Cascini L, Malet JP, Fotopoulou S, Pitilakis K (2014) Recommendations for the quantitative analysis of landslide risk. Bull Eng Geol Env 73(2):209–263

Dahal RK, Hasegawa S (2008) Representative rainfall thresholds for landslides in the Nepal Himalaya. Geomorphology 100(3):429–443

Dangal R (2011) Country profile Nepal. Disaster risk management: policies and practices in Nepal. Kobe: Asian Disaster Reduction Center. Available in: https://www.adrc.asia/countryreport/NPL/2011/FY2011B_NPL_CR.pdf. Assessed 5 June 2018

Devkota KC, Regmi AD, Pourghasemi HR, Yoshida K, Pradhan B, Ryu IC, Althuwaynee OF (2013) Landslide susceptibility mapping using certainty factor, index of entropy and logistic regression models in GIS and their comparison at Mugling-Narayanghat road section in Nepal Himalaya. Nat Hazards 65(1):135–165

Dou J, Bui DT, Yunus AP, Jia K, Song X, Revhaug I, Zhu Z (2015) Optimization of causative factors for landslide susceptibility evaluation using remote sensing and GIS data in parts of Niigata, Japan. PLoS ONE 10(7):e0133262

Erener A, Duzgun HSB (2010) Improvement of statistical landslide susceptibility mapping by using spatial and global regression methods in the case of More and Romsdal (Norway). Landslides 7(1):55–68

Fayez L, Pazhman D, Pham BT, Dholakia MB, Solanki HA, Khalid M, Prakash I (2018) Application of frequency ratio model for the development of landslide susceptibility mapping at part of Uttarakhand state, India. Int J Appl Eng Res 13(9):6846–6854

Fielding AH, Bell JF (1997) A review of methods for the assessment of prediction errors in conservation presence/absence models. Environ Conserv. 24(01):38–49

Gasim MB, Toriman ME, Juahir H (2015) Phenomenon of slope failure occurrences along Gerik-Jeli highway. Science Alert, Malaysia

Ghimire M (2011) Landslide occurrence and its relation with terrain factors in the Siwalik Hills, Nepal: case study of susceptibility assessment in three basins. Nat Hazards 299–320

Guzzetti F (2012) Landslide inventory maps: new tools for an old problem. Earth Sci Rev 42–66

Hosmer DW, Lemeshow S (2000) Special topics. Applied logistic regression, 2nd edn., pp 260–351

https://www.fao.org/asiapacific/forestry-publication. Accessed 25 Sep 2018

IRIN (2013) No early warning for Nepali's deadly landslides. http://www.irinnews.org/report/98381/no-early-warning-nepal%E2%80%99s-deadly-landslides

Kalantar B, Pradhan B, Naghibi SA, Motevalli A, Mansor S (2018) Assessment of the effects of training data selection on the landslide susceptibility mapping: a comparison between support vector machine (SVM), logistic regression (LR) and artificial neural networks (ANN). Geomatics, Nat Hazards Risk 9(1):49–69

Lee S, Pradhan B (2007) Landslide hazard mapping at Selangor, Malaysia using frequency ratio and logistic regression models. Landslides 4(1):33–41

Lee S, Ryu JH, Won JS, Park HJ (2004) Determination and application of the weights for landslide susceptibility mapping: using an artificial neural network. Eng Geol 71:289–302

Lim KW, Tay LT, Lateh H (2011) Landslide hazard mapping of Penang island using probabilistic methods and logistic regression. In: Proceedings of the IEEE international conference on imaging systems and techniques. Penang, Malaysia, pp 273–278, 17–18 May 2011

Meten M, Prakashbhandary N, Yatabe R (2015) Effect of landslide factor combinations on the prediction accuracy of landslide susceptibility maps in the Blue Nile Gorge of Central Ethiopia. Geoenviron Disasters 2:9. https://doi.org/10.1186/s40677-0150016-7

MoFE (2010) National Adaptation Programme of Action (NAPA). Government of Nepal, Ministry of Forest and Environment (previously MoPE). Kathmandu, Nepal

MoHA (2019) Nepal Disaster Report. Ministry of Home Affairs. Government of Nepal

MoHA, DPNet-Nepal (2015) Nepal Disaster Report. Ministry of Home Affairs, Government of Nepal and Disaster Preparedness Network-Nepal

Mondal S, Mandal S (2018) RS & GIS-based landslide susceptibility mapping of the Balason River basin, Darjeeling Himalaya, using logistic regression (LR) model. Georisk: Assess Manage Risk Eng Syst Geohazards 12(1):29–44

Mousavi SZ, Kavian A, Soleimani K, Mousavi SR, Shirzadi A (2011) GIS-based spatial prediction of landslide susceptibility using logistic regression model. Geomatics, Nat Hazards Risk 2(1):33–50

Ocakoglu F, Gokceoglu C, Ercanoglu M (2002) Dynamics of a complex mass movement triggered by heavy rainfall: a case study from NW Turkey. Geomorphology 42(3):329–341

Ohlmacher C (2006) The relationship between geology and landslide hazards of Atchison, Kansas, and vicinity. Kansas Geol Surv

Ojha TK (2009) Magnetostratigraphy, topography and geology of Nepal Himalaya: a GIS and Paleomagnetic Approach. Ph.D. dissertation. University of Arizona

Paudel U (2016) Landslide susceptibility assessment using machine learning with emphasis on scaling and topographic representation issues. Ph.D. dissertation. Department of Natural Environmental Studies, Graduate School of Frontier Sciences, The University of Tokyo (unpublished)

Pokhrel N (2003) Study of areal precipitation distribution pattern in the Chepe catchment, Nepal. Butwal Power Company Ltd, Kumaripati Lalitpur, GPO 11728. Kathmandu, Nepal, pp 1–7

Pontius RJ, Schneider LC (2001) Land-cover change model validation by an ROC method for the Ipswich watershed, Massachusetts, USA. Agr Ecosyst Environ 85:239–248

Pourghasemi HR (2008) Landslide hazard assessment using fuzzy logic (Case Study: A part of Haraz watershed), M.Sc. Thesis, Tarbiat Modarres University International Campus. Iran, p 92

Pourghasemi HR, Moradi HR, Aghda SF (2013) Landslide susceptibility mapping by binary logistic regression, analytical hierarchy process, and statistical index models and assessment of their performances. Nat Hazards 69(1):749–779

Pourghasemi HR, Pradhan B, Gokceoglu C, Moezzi KD (2012) Landslide susceptibility mapping using a spatial multi criteria evaluation model at Haraz Watershed, Iran. In: Terrigenous mass movements. Springer, Berlin, Heidelberg, pp 23–49

Pradhan B, Lee S (2010) Landslide susceptibility assessment and factor effect analysis: backpropagation artificial neural networks and their comparison with frequency ratio and bivariate logistic regression modelling. Environ Model Softw 25(6):747–759

Provost FJ, Fawcett T (1997) Analysis and visualization of classifier performance: comparison under imprecise class and cost distributions. In: Heckerman D, Mannila H, Pregibon D, Uthurusamy R (eds) Proceedings of the 3rd international conference on knowledge discovery and data mining (KDD-97). Menlo Park (CA), AAAI Press, pp 43–48, 14 Aug 1997

Provost FJ, Fawcett T, Kohavi R (1998) The case against accuracy estimation for comparing induction algorithms. In: Shavlik J (ed) Proceedings of the ICML 1998. San Francisco (CA), Morgan Kaufmann, pp 445–453, 24 July 1998

Rossi M, Guzzetti F, Reichenbach P, Mondini AC, Peruccacci S (2010) Optimal landslide susceptibility zonation based on multiple forecasts. Geomorphology 114:129–142

Tian H, Nan H, Yang Z (2010) Select landslide susceptibility main affecting factors by multi objective optimization algorithm. In: Proceedings of the 6th international conference on natural computation (ICNC '10), pp 1830–1833

TU-CDES (2016) Landslide inventory characterization and engineering design for mitigation works of Chure Area in ten districts. Central Department of Environmental Science, Tribhuvan University and Government of Nepal, President Chure-Tarai Madhesh Conservation Development Board, Kathmandu

UNDP (2009) Found on August 10, 2009. http://www.undp.org/mdtf/UN-REDD/overview.shtml

Vakhshoori V, Zare M (2018) Is the ROC curve a reliable tool to compare the validity of landslide susceptibility maps? Geomatics, Nat Hazards Risk 9(1):249–266

Van Westen CJ (1997) Statistical landslide hazard analysis. Application Guide, ILWIS 2.1 for Windows. Enschede

Youssef AM, Pourghasemim HR, Pourtaghi ZS, Al-Katheeri MM (2016) Landslide susceptibility mapping using random forest, boosted regression tree, classification and regression tree, and general linear models and comparison of their performance at Wadi Tayyah Basin, Asir Region, Saudi Arabia. Landslides 13(5):839–856

Chapter 8
Level of Disaster Resilience and Migration Patterns in Japanese and Foreign Residents

Hiroaki Matsuura

Abstract Building community resilience is one of the central priorities of the SFDRR 2015–2030. However, no agreement exists over the exact measurement of the progress of building disaster resilience at community level. This chapter aims to contribute to the literature of resilience measurement in two ways. First, the chapter develops a Japanese version of the Resilience Capacity Index (RCI), which was ranked at the highest sub-national level resilience index in terms of maturity according to the UNDP's systematic review (2014). Second, it examines how high and low community resilience measured by RCI affects inter-municipality migrations among Japanese nationals and foreign residents. There are four findings in this chapter; first, in areas unaffected by earthquakes, Japanese people are more likely to move-out from municipalities with low resilience, while they are less likely to move-out from municipalities with high resilience. Second, in areas severely affected by earthquakes, municipalities with low resilience suffer additional out-migration, while no additional out-migration can be observed in municipalities with middle or high resilience. Third, in-migration shows that having high resilience does not always benefit municipalities. Municipalities with high resilience (community connectivity) tend to form a closed set of networks within the community unconnected to larger society. Fourth, foreigners do not respond to the level of resilience in both earthquake-affected and non-earthquake affected municipalities. They seem to behave as if there is a separate community for non-Japanese people. The results of this paper suggest the need for a strategy to reduce the number of municipalities with low resilience and encourage more inclusive and participatory communities for foreigners.

Keywords Resilience capacity index · Community level · Inter-municipality migrations · Inclusiveness · Japanese · Foreigners

H. Matsuura (✉)
Shoin University, Kanagawa, Japan
e-mail: hiroaki.matsuura@shoin-u.ac.jp

R. Djalante et al. (eds.), *Integrated Research on Disaster Risks*, Disaster Risk Reduction,
https://doi.org/10.1007/978-3-030-55563-4_8

8.1 Introduction

The Sendai Framework for Disaster Risk Reduction 2015–2030 sets out seven targets and four priorities for global action against disasters (UNDRR 2015). The word 'resilience' is included in one of the seven targets: 'Substantially reduce disaster damage to critical infrastructure and disruption of basic services, among them health and educational facilities, including through developing their resilience by 2030'. The Sendai Framework is the successor of the Hyogo Framework for Action 2005–2015, which sets the first global plan to help nations work towards disaster resilience by strengthening community capabilities (UNDRR 2007). Learning from the Hyogo Framework, the Sendai Framework emphasizes local level actions and the importance of collaboration at global, national and local levels (UNDRR 2015).

The Sendai Framework for Disaster Risk Reduction 2015–2030 was also the first agreement of the post-2015 development agenda (UNDRR 2016). Implementing the Sendai Framework is necessary for achievement of the Sustainable Development Goals, which set the framework for global development policy over 15 years. In turn, progress on the SDGs, especially Goal 11 to 'make cities and human settlements inclusive, safe, resilient and sustainable', can substantially develop community resilience in the face of disasters. Building resilience for future disasters is critical not only to help save lives and properties from potential disasters, but also to enhance sustainable economic growth and poverty reduction not disrupted by disasters.

The concept of resilience has been widely accepted in literature covering disaster management. There is also an increasing trend of developing a wide variety of quantitative metrics measuring for the level of resilience (Quinlan et al. 2016). Despite the popularity of concept, however, no agreement exists over the exact measurement of resilience against natural disasters. In 2014, the United Nations Development Programme (2014) systematically surveyed the index of disaster resilience and classified the maturity of disaster resilience measurements into six phases. These range from early models yet without a developed indicator framework (Phase 1) to models that already have institutionalized disaster resilience measurements and are collecting data frequently (Phase 5) (Winderl 2014). No single resilience index reached phase 6, where the relevance of measurement has empirically verified. With the other four indicators (Hyogo Framework for Action Monitor, World Risk Index, Global Focus Model and Socio-Economic Resilience Index), the Resilience Capacity Index (RCI) was categorized as being at phase 5. Unlike the other four indicators that are calculated at national level, the Resilience Capacity Index is calculated at sub-national level. The RCI was originally developed by Foster (2007) and is a compound of 12 variables aggregated into three categories: regional economic capacity, socio-demographic capacity and community connectivity capacity (Foster 2011). The RCI was originally applied to rank the level of resilience in 361 metropolitan statistical areas in the United States, but it has also been used for other countries sub-national units such as 52 counties in the northern Gulf of Mexico region, 290 municipalities in Sweden, 40 localities in Sri Lanka and 79 Slovak districts (Reggiani et al. 2016; Reams et al. 2012; Östh et al. 2015; Abenayake et al. 2016; Psycharis et al. 2014).

This chapter aims to contribute to the literature of resilience measurement in two ways. First, the chapter develops a Japanese version of the Resilience Capacity Index (RCI), which was the most promising sub-national level resilience index according to the UNDP's systematic review in 2014. The second is to examine how high and low community resilience measured by RCI affects inter-municipality migrations among Japanese nationals and foreign residents in both earthquake-affected and non-earthquake affected areas.

The rest of the chapter is organized as follows. Section 2 describes the data used in the analysis and procedure that were used to calculate the Resilience Capacity Index at municipality level in Japan. Section 3 presents the empirical strategy to estimate the association between community resilience and inter-municipality migration among Japanese and foreign residents. Section 4 presents the results of the empirical analysis. Finally, Sect. 5 summarizes the discussion and concludes the chapter.

8.2 Method

8.2.1 Data Sources

This chapter uses four different types of variables. The first type of variable is 12 indicators used to calculate the RCI. The second type of variable is the variable of inter-municipality migration, which comes from the Report on Internal Migration compiled by the Statistics Bureau of Japan from 2012 to 2016. The third type of variable is the number of deaths due to earthquakes. This variable is used to classify the earthquake-affected and non-affected municipalities. The last type of variable is municipality-level control variables used in the regression analysis.

8.2.2 Resilience Capacity Index in Japan

The original RCI developed by Foster (2007) has been applied to the 361 metropolitan regions in the United States. In this paper, I calculated the RCI of 1,741 municipalities in Japan. Data collected are from the period 2010–2016: some indicators are only available from the longer version of the Census every ten years, while other indicators are available every year. Local election results, Housing and Land Survey and the Census are also available every five years. As previously noted, the RCI is a composite measure of 12 different indicators across three different capacity domains: regional economic capacity, socio-demographic capacity and community connectivity capacity.

Regional Economic Capacity reflects (1) income equality, (2) diversification of the economic employment, (3) regional affordability measured by housing costs and income levels and (4) dynamic business environment. The first indicator is income equality. RCI assumes that the more equal a municipality's distribution of income, the more cohesive the community response to disasters. An inverse of Gini coefficient was used to measure the degree of income equality in both original RCI and the RCI calculated in this paper. The Gini Coefficient was first calculated using the municipality-level income distribution data in Housing and Land Survey. This survey is conducted every five years; thus, I used the latest available dataset from 2013. Then I subtracted the Gini Coefficient from 1 to obtain the measure of income equality. The second economic indicator is economic diversification. RCI assumes that economic diversity increases community resilience by reducing the economic risk of the municipality. The original data come from the Economic Census for Business Frame in 2014. In order to calculate economic diversification, I first calculated Herfindahl–Hirschman Index (HHI), a measure of concentration to the particular sector, by using the number of employees across 22 different sectors in each municipality. HHI is equal to 1 if all employees work in one of these 22 sectors. On the other hand, if employees work equally spread across 22 different sectors, it is $22 * \left(\frac{1}{22}\right)^2 = 0.046$. I took an inverse of this value after the z-transformation since a low value means more economic diversity. The third economic indicator is regional affordability. Regional affordability is an indicator of economic security, measured by the proportion of income not spent on rent. The data on average monthly rent comes from the 2013 Housing and Land Survey. I divided this number by per capita monthly income calculated from data on taxable income, compiled by Ministry of Internal Affairs and Communications, and then subtracted it from 1. In the original RCI, regional affordability was measured by the percentage of households who spend less than 35% of their income on housing. However, I was unable to obtain these data at municipality level in Japan. Therefore, I calculated average housing expenditure as a percentage of household income in each municipality. The fourth economic indicator is business environment indicators. RCI assumes that the more dynamic the local business environment, the more adaptable and resourceful, and thus more resilient against disasters. The original RCI used the Indiana Business Center's 2010 Innovation Index, which measured economically dynamic areas by using multiple indicators. No such index exists in Japan. However, the Japan Productivity Center has been publishing labour productivity statistics since 1958. The literature argues that labour productivity is related to competitive environment, innovative human capital and flexible workplace (Schnabel 1997; Bauer 2003). For this reason, I simply used labour productivity as a proxy for the dynamic business environment—obtaining data on labour productivity in 2016.

Socio-demographic indicators reflect the capacity of the municipality on the basis of social and demographic characteristics, including (1) educational attainment, (2) disability, (3) poverty and (4) health care access. The first socio-demographic indicator is the percentage of adults aged 15 and older who have completed a bachelor's or higher degree, which comes from the large-scale Census and is surveyed every ten years. I used the latest survey available from 2010. The second indicator is without disability. RCI assumes that municipalities with a higher percentage of people with disabilities are more vulnerable to disasters. Unfortunately, the data on the percentage of people with disability are not available since the Japanese government has not implemented a comprehensive survey of people with disabilities at municipality level. Some municipalities disclose their information on the number of people with disability certificates, but not all. Therefore, I used data on the number of disability-related institutions per 1,000 population, including institutions for support of the disabled, rehabilitation homes for the physically disabled, institutions for the protection of persons with intellectual disabilities, social rehabilitation facilities for mentally disabled persons and institutions supporting social activities for the physically disabled. I used the latest available data from 2011. I took an inverse of this value after the z-transformation. The third indicator is out of poverty. Poverty imposes a constraint on households and communities to respond to and recover from a disaster. Out of poverty is measured by the inverse of per capita public assistance expenditure. The data on per capita public assistance expenditure comes from the System of Social and Demographic Statistics. Again, I take an inverse of this value after the z-transformation. The last socio-economic indicator is health care access. The original RCI used health insurance coverage as a proxy for the general capacities useful for effectively responding to and recovering from a disaster. However, this makes little sense in the context of Japan because Japan achieved universal health coverage through the social insurance system in 1961 and the entire population is virtually covered by a social health insurance plan. However, there is still a disparity in access to health care due to a shortage of physicians and their regional maldistribution. Therefore, I used the data on number of physicians per 1,000 population, which comes from the System of Social and Demographic Statistics.

Community Connectivity Indicators reflect (1) the number of civic organizations, (2) metropolitan stability measured by the percentage of long-term residents in the community, (3) homeownership and (4) voter participation rates. RCI assumes the density of civic organizations as a proxy for community engagement. Data on the number of non-profit organizations per 10,000 residents come from the NPO Hiroba Database. Last access to the database was on 24 December 2015. I divided the number of NPOs by the number of total population in each municipality. The second indicator is the measure of metropolitan stability, which assesses rootedness in community. I used the share of population that remained resident in the municipality over a five-year period in the Census 2015. The third indicator, homeownership, provides a measure of attachment and commitment to place. I used data from the Housing and Land Survey to calculate the proportion of owner-occupied housing units in total occupied housing units. Finally, I measured civic engagement by using voter participation. Higher voter participation indicates trust in the democratic process in local governance. The dataset comes from the Japanese Local Elections Dataset, compiled by Horiuchi and others (Horiuchi 2019). The share of voter-eligible population who voted in the last election (in 2015) was calculated in each municipality. There are missing observations because there are fewer candidates than seats available and no election is held in 633 municipalities.

Due to the different measurement units of these 12 indicators, z-score transformation was employed to convert all the indicators to a common scale with a mean of zero and a standard deviation of one. As noted above, I took an inverse of HHI, the number of disability-related institutions per 1,000 population, and per capita public assistance expenditures after the z-transformation. Then I took an average of these 12 indicators to calculate the Japanese-version RCI score. Each of the three dimensions of RCI, Economic capacity attributes, Socio-Demographic capacity attributes and Community Connectivity attributes were also calculated in order to understand which attributes contribute to resilience against disasters. I assign missing values to each attribute if more than two out of four indicators are missing in each attribute. This results in six missing observations in RCI (four missing observations in Economic capacity attributes and five missing observations in Community Connectivity attributes). These six municipalities are located in the evacuation zones of the Fukushima nuclear accident.

Once RCI was calculated, I constructed two dummy variables for municipalities with high and low resilience. The dummy variable for municipalities with high (low) resilience is equal to 1 if a municipality is in the top (bottom) quartile of RCI. Otherwise it is 0. I also constructed high and low economic capacity attributes, high and low socio-demographic capacity attributes and high and low community connectivity attributes in a similar manner. Appendix shows the top ten and bottom ten of the RCI (and its components) ranking.

8.2.3 Migration Data of Japanese and Foreign Population

The in-migration and out-migration data come from a Report on Internal Migration compiled by the Statistics Bureau of Japan. The information is derived from the Basic Resident Registry (BRR) system of Japan. The BRR is a registry of citizens maintained by municipality governments. It contains personal information including name, current address, date of birth and gender for each citizen. The Basic Resident Registration Act of 1967 requires citizens to report a change of address to the local authority. The date of move and the place of past residence are recorded in the BRR. As a result, I can track the total number of in- and out-migration across municipalities in Japan.

Based on a revision of the Basic Resident Registration Act, foreigners who have stayed or will stay in Japan legally for more than three months are subject to the Basic Resident Registration System (from 2012). I calculated the in-migration and out-migration rate of foreign residents in each municipality separate from Japanese nationals by using the same data.

I constructed the out-migration rate by dividing the total number of out-migrations of Japanese nationals and foreigners by the total number of Japanese nationals and foreigners in each municipality, respectively. I also constructed the in-migration rate in a similar manner.

The chapter uses average in- and out-migration rate in the period 2012–2016. Figure 8.1 shows the time trends of in- and out-migration rate from 2010 to 2016. The rapid increase in in- and out-migration rate in 2012 was as a result of the 2011 Tohoku

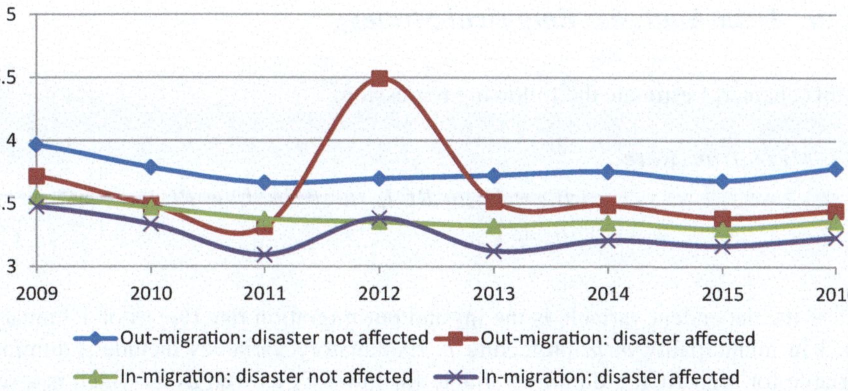

Fig. 8.1 Time trends of in- and out-migration rate in earthquake-affected and non-earthquake affected municipalities

earthquake and its subsequent tsunami and nuclear disaster. For the robustness check, I also excluded the year 2012 from the sample, but it did not change the results much.

8.2.4 Municipalities Severely Affected by Earthquakes

In 2011, a magnitude-9.0 earthquake, the largest earthquake in Japanese history, struck in the Tohoku region, triggering a massive tsunami and a nuclear meltdown. The earthquake and its resulting tsunami caused the deaths of 15,897 people, spread across 124 municipalities. I constructed a dummy variable for municipalities severely affected by earthquakes equal to 1 if somebody died in the municipality in 2011. Otherwise it is 0. No additional deaths occurred outside of the 124 municipalities between 2011 and 2015 before the 2016 Kumamoto Earthquake.

8.2.5 Other Control Variables

Logged per capita income, logged school density (per square kilometre) and logged population density (per square kilometre) are used as control variables in the regression model of this chapter. The data come from taxable income data, which are compiled from 2013 information by the Ministry of Internal Affairs and Communications, the Basic Resident Registry (BRR) and the School Basic Survey.

8.2.6 Data Analysis: Empirical Strategy

In this chapter, I estimate the following regression.

$$\begin{aligned} Migration\ Rate_m &= \beta_1 * High\ RCI_m + \beta_2 * Low\ RCI_m + Z_\gamma \\ &+ \theta_p + \varepsilon_m \end{aligned} \tag{1}$$

where the dependent variable is the in- and out-migration rate (%) as of 1 January 2013 in municipality *m in* prefecture *p*. Explanatory variables include a dummy variable for high RCI, a dummy variable for low RCI as well as Z, which is a set of other control variables that includes logged per capita income, logged number of schools per square kilometre and logged population density. θ_p is the prefecture fixed effects.

In addition to the specification above, I test whether high (low) resilience affects differently in municipalities affected and not affected by an earthquake. I estimate the following regression.

$$\begin{aligned} Migration\ Rate_m &= \delta_1 * (DIS_m * High\ RCI_m) + \delta_2 * (DIS_m \\ &Low\ RCI_m) + \delta_3 * DIS_m + \beta_1 * High\ RCI_m + \beta_2 \\ &* Low\ RCI_m + Z_\gamma + \theta_p + \varepsilon_m \end{aligned} \tag{2}$$

where DIS is a dummy variable for earthquake-affected municipalities (at least one earthquake-related death). I am particularly interested in the sign of δ_1 and δ_2, which are the coefficient of the interaction of dummy variables for earthquake-affected municipalities with high RCI and low RCI, respectively.

8.3 Results

8.3.1 Level of Resilience Capacity Index in Japan

Table 8.1 shows summary statistics of the inputs for each of the 12 indicators used to calculate the RCI. The single variable with the most missing observations was voter participation rate because there were fewer candidates than seats available, thus no election was held in these municipalities. Income equality, regional affordability and homeownership data also contain a relatively large number of missing observations since the Housing and Land Survey of 2013 excludes towns and villages with a population of 15,000 or less. The average out-migration rate per Japanese population

Table 8.1 Summary statistics

Variable	Year	N	Mean	Std. Dev	Min	Max
Income equality (Gini Coefficient)	2013	1126	0.37	0.02	0.3	0.47
Economic diversification (Herfindahl–Hirschman Index)	2014	1738	0.14	0.05	0.08	1
Regional affordability* (Average share of housing expenditure)	2013	1115	0.18	0.03	0.06	0.39
Business environment (Labour productivity)	2016	1738	3811	1516	119	38,232
Educational attainment (share of adults who had completed a bachelor's or higher degree)	2010	1741	0.74	0.11	0.37	0.97
Without disability* (# of disability-related institutions per 1,000 population)	2011	1707	0.07	0.11	0	1.62
Out of poverty* (per capita public assistance expenditure)	2013	1741	39.99	18.65	15.02	212.3
Health care (# of physicians per 1,000 population)	2013	1741	1.54	1.73	0	35.11
Civic infrastructure (# of non-profit organizations per 10,000 residents)	2015	1736	46.5	182.5	0	5555.6
Metropolitan stability (share of residents more than five years of residence)	2015	1736	0.89	0.04	0.01	0.94
Homeownership rate	2013	1126	72.86	13.07	35.6	98.2
Voter Participation rate	2015	1085	0.54	0.12	0.25	0.92
Resilience capacity index	–	1735	−0.03	0.32	−3.35	2.65
Economic capacity		1737	−0.03	0.53	−6.19	9.34
Socio-demographic Capacity		1741	0.00	0.66	−4.98	5.31
Community connectivity		1736	0.01	0.53	−9.70	2.71
Municipalities with any deaths, 2011–2015	2011–2015	1741	0.07	0.26	0	1
Out-migration rate: Japanese	2012–2016 (average)	1741	3.73	1.37	1.29	18.08

(continued)

Table 8.1 (continued)

Variable	Year	N	Mean	Std. Dev	Min	Max
Out-migration rate: Foreigner		1738	24.42	37.02	0.00	714.47
In-migration rate: Japanese		1741	3.33	1.54	0.76	17.20
In-migration rate: Foreigner		1738	33.22	37.57	0.00	712.03
Per capita taxable income	2013	1741	2747	481	1213	9017
School density	2013	1741	0.27	0.56	0	8.60
Population density	2013	1741	1026	2356	2	19,367

*There are 1741 municipalities in Japan as of 2016. I calculate Resilience Capacity Index (RCI) unless each of the three dimensions (Economic capacity, Socio-demographic Capacity, and Community connectivity) of RCI has missing values in more than two variables. As a result, RCI is calculated for the 1735 municipalities

is 3.7%, while an average of 24.4% of the foreign population moved out within a single year. The average in-migration rate per Japanese population is 3.3%, while an average of 32.2% of the foreign population moved out within a single year. Foreigners move-in and move-out much more frequently than Japanese nationals.

8.3.2 *Pattern of Migration and Level of Resilience*

Third, in-migration shows that having high resilience does not always benefit municipalities. Municipalities with high resilience (community connectivity) tend to form a closed set of networks within the community unconnected to larger society. Fourth, foreigners do not respond to the level of resilience in both earthquake-affected and non-earthquake affected municipalities. They seem to behave as if there is a separate community for non-Japanese people. The results of this paper suggest the need for a strategy to reduce the number of municipalities with low resilience and encourage more inclusive and participatory communities for foreigners.

Table 8.2 presents the average out-migration and in-migration rate by level of resilience. For both Japanese nationals and foreigners, there is a clear pattern that

Table 8.2 Migration by level of resilience

By resilience category	Out-migration (%)		In-migration (%)	
	Japanese	Foreigner	Japanese	Foreigner
Low resilience	4.3	30.7	3.7	38.4
Mid resilience	3.6	23.3	3.3	32.7
High resilience	3.3	20.7	3.1	29.5

more people move-out from and move-in to municipalities with low resilience, while less people move-out from and move-in to municipalities with high resilience.

Table 8.3 presents the results of multivariate regressions of in- and out-migration rate (specification 1). In the first column, the dependent variable is out-migration rate, and the explanatory variables include prefecture fixed effects as well as a dummy variable for high resilience, a dummy variable for low resilience, logged per capita income, logged school density and logged population density. A dummy variable for high resilience is associated with a 0.22% reduction in out-migration rate, while a dummy variable for low resilience is associated with a 0.45% increase in out-migration rate. Column two shows the same regression in the sample of foreigners. Unlike Japanese nationals, foreigners do not move-out more from municipalities with low resilience. Similarly, they do not move-out less from municipalities with high resilience.

The next two columns show the results of the analysis of in-migration among the Japanese and foreign population. The in-migration rate among Japanese nationals is 0.16% lower in municipalities with high resilience, compared to municipalities with middle resilience. This decrease in the in-migration rate is smaller than the decrease in out-migration (0.22%). Thus, higher resilience is associated with social increase. On the other hand, the in-migration rate is 0.38% higher in municipalities with low resilience, compared to municipalities with middle resilience. However, this increase in in-migration is smaller than the increase in out-migration (0.45%). Thus, low resilience is associated with social decrease. However, these associations are not observed in the sample of foreigners.

From columns five to eight, I disaggregated resilience in RCI's three sub-indices: regional economic capacity, socio-demographic capacity and community connectivity capacity. Column five replicates the analysis of the out-migration rate with Japanese nationals using the disaggregated RCI. High community connectivity is associated with a 0.33% decrease in out-migration, while low community connectivity is associated with a 0.99% increase in out-migration. In addition, low economic capacity is associated with a 0.21% increase in out-migration. Column seven replicates an analysis of the in-migration rate with Japanese nationals using disaggregated RCI. High community connectivity is associated with a 0.29% decrease in in-migration, while low community connectivity is associated with a 0.85% increase in in-migration. In addition, low economic capacity is associated with a 0.26% increase in in-migration, while low socio-demographic capacity is associated with a 0.21% decrease in in-migration. The results show that the main reason for low migration in municipalities with high resilience comes from high community connectivity, while the main reason for high migration in municipalities with low resilience is both low economic capacity and low community connectivity.

In columns six and eight, I replicate the same regressions by using the sample of foreigners. The coefficient of socio-demographic capacity, economic capacity and community connectivity are no longer significant in the sample foreign population.

Until this point, I examined the direct association of the Resilience Capacity Index with the migration behaviours of Japanese nationals and foreigners. Table 8.4 shows the result of estimating specification two and examines whether the presence

Table 8.3 Multivariate analysis of migration behaviour: the role of resilience

Variables	1	2	3	4	5	6	7	8
	Out-migration rate		In-migration rate		Out-migration rate		In-migration rate	
	Japanese	Foreigners	Japanese	Foreigners	Japanese	Foreigners	Japanese	Foreigners
High RCI	−0.224***	−1.817	−0.156**	−2.254				
	(0.062)	(1.727)	(0.071)	(1.890)				
Low RCI	0.449***	2.4	0.378***	0.524				
	(0.067)	(4.575)	(0.070)	(4.116)				
High socio-demographic Capacity					0.0405	−1.067	0.152	−2.26
					(0.073)	(2.625)	(0.106)	(2.661)
Low socio-demographic Capacity					−0.068	−1.579	−0.209**	−4.458
					(0.089)	(2.773)	(0.099)	(2.900)
High economic capacity					0.0516	−0.786	0.0248	0.158
					(0.062)	(2.660)	(0.060)	(2.119)
Low economic capacity					0.210***	−0.497	0.256***	1.275
					(0.049)	(2.419)	(0.072)	(2.649)
High community connectivity					−0.334***	−2.255	−0.293***	−3.076
					(0.052)	(3.061)	(0.057)	(3.606)
Low community connectivity					0.991***	2.22	0.845***	2.974*
					(0.063)	(1.470)	(0.072)	(1.726)
Logged per capita income	4.572***	9.589	5.753***	9.177	3.467***	6.017	4.692***	6.157
	(0.640)	(9.030)	(0.782)	(9.891)	(0.788)	(11.830)	(0.946)	(13.880)
Logged school density	0.909***	−0.374	0.870***	0.413	0.937***	0.3	0.912***	1.043

(continued)

Table 8.3 (continued)

Variables	1	2	3	4	5	6	7	8
	Out-migration rate		In-migration rate		Out-migration rate		In-migration rate	
	Japanese	Foreigners	Japanese	Foreigners	Japanese	Foreigners	Japanese	Foreigners
	(0.249)	(1.636)	(0.287)	(1.993)	(0.249)	(1.627)	(0.285)	(1.846)
Logged population density	−0.728***	−2.164*	−0.570**	−2.627*	−0.899***	−3.297**	−0.758***	−3.949**
	(0.212)	(1.078)	(0.242)	(1.319)	(0.221)	(1.546)	(0.248)	(1.824)
Prefecture fixed effect	Yes	Yes	Yes	Yes	Yes	Yes	Yes	Yes
Observations	1,733	1,730	1,733	1,730	1,733	1,730	1,733	1,730
R-squared	0.572	0.063	0.59	0.068	0.636	0.063	0.633	0.071

Note Standard errors are clustered at prefecture level. ***$p < 0.01$, **$p < 0.05$, *$p < 0.1$

Table 8.4 Multivariate analysis of migration behaviour: the role of earthquakes and resilience

Variables	Out-migration rate		In-migration rate	
	Japanese	Foreigners	Japanese	Foreigners
High resilience	−0.219***	−2.405	−0.170**	−3.21
	(0.067)	(2.038)	(0.072)	(2.359)
Low resilience	0.397***	3.007	0.361***	0.692
	(0.062)	(5.286)	(0.073)	(5.458)
Any deaths	−0.0629	−1.185	−0.0668	−0.294
	(0.106)	(2.200)	(0.111)	(2.183)
High resilience*any deaths	−0.015	8.218	0.168	7.685
	(0.131)	(5.910)	(0.196)	(6.524)
Low resilience*any deaths	0.670**	−9.947	0.237	−2.786
	(0.320)	(6.169)	(0.325)	(5.965)
Logged per capita income	4.561***	8.745	5.766***	8.582
	(0.645)	(8.337)	(0.797)	(9.620)
Logged school density	0.908***	−0.862	0.918***	−0.206
	(0.258)	(1.675)	(0.298)	(1.975)
Logged population density	−0.742***	−1.685	−0.609**	−2.254
	(0.222)	(1.146)	(0.248)	(1.349)
Prefecture fixed effect	Yes	Yes	Yes	Yes
Observations	1733	1730	1733	1730
R-squared	0.567	0.061	0.58	0.064

Note Standard errors are clustered at prefecture level. ***$p < 0.01$, **$p < 0.05$, *$p < 0.1$

of resilience prevents out-migration and promotes in-migration in the area affected by earthquakes in addition to the direct effect of resilience on migration behaviour.

In the first column of Table 8.4, a dummy variable for municipalities with any earthquake-related deaths and their interaction term with high and low RCI are added to the specification in the first specification of Table 8.3. Similar to the result in Table 8.3, low RCI is associated with a 0.40% increase in out-migration rate in municipalities with no earthquake-related deaths, but low RCI is associated with a further 0.67% reduction in the out-migration rate in municipalities with earthquake-related deaths. This indicates that the municipalities with low resilience lose additional people in municipalities affected by earthquakes because of out-migration. On the other hand, municipalities with high resilience did not experience less out-migration or more in-migration in areas affected by earthquakes.

8.4 Discussion and Conclusion

This chapter develops a Japanese version of the Resilience Capacity Index (RCI), which was the most promising sub-national level resilience index according to the UNDP's systematic review of resilience index maturity at municipality level. The chapter then investigated the role of disaster resilience, measured by the Resilience Capacity Index, in the migration behaviours of Japanese and non-Japanese residents. The results of the empirical analysis found that (1) Even in the areas unaffected by earthquakes, Japanese people are more likely to move-out from municipalities with low resilience, while they are less likely to move-out from municipalities with high resilience. (2) In the areas severely affected by earthquakes, municipalities with low resilience suffered additional out-migration, while no additional out-migration was observed in municipalities with middle or high resilience. (3) The analysis of in-migration shows that having high resilience does not always benefit municipalities. People are less (more) likely to move-in to the municipalities with high (low) resilience due to the high (low) within-community connectivity that excludes newcomers. This is the negative side of community connectivity. Municipalities with high resilience (community connectivity) tend to form a closed set of networks within the community unconnected to larger society, making these municipalities isolated from the outside world. (4) Foreigners do not respond to the level of resilience in both earthquake-affected and non-affected municipalities. They behave as if there is a separate community for non-Japanese people and do not seem to care about the level of resilience among the Japanese community.

There are several limitations in this study. First, the Resilience Capacity Index is the highest ranked sub-national level resilience index in terms of maturity according to the UNDP's systematic review in 2014. However, its relevance has not yet been empirically verified. This chapter finds that municipalities with low resilience lose additional residents, but municipalities with high resilience do not necessarily prevent out-migration in areas affected by earthquakes. One interpretation of this is that the RCI successfully identifies municipalities with no resilience against earthquakes, but fails to identify municipalities highly resilient to earthquakes. There is a possibility of underestimating the effect of having high resilience. Unfortunately, I cannot guarantee the relevance of the RCI. Future work should examine the relevance of the RCI as a measurement of the level of resilience.

The second limitation is the robustness of the Resilience Capacity Index. In many of the RCI's 12 indicators, there are more than two candidate variables to potentially measure each of the RCI's indicators, and I need to choose one from these. The choice of variable in each of the RCI's 12 indicators certainly affects the value of RCI. For income equality, when Gini coefficient is calculated, I assume that the average income of the households with an annual income of more than 15 million yen is 20 million yen. I checked the robustness of this assumption by calculating alternative Gini coefficients by assuming that the average income of the household in this category is 17.5 million or 22.5 million yen. For without disability, I also chose the number of disability-related institutions per 1,000 population instead of the proportion of people with a disability certificate, which can have a lot of missing values due to the institutional background. For health insurance coverage, I used the number of physicians per 1,000 population instead of the proportion of people covered by health insurance because I do not want to assign 100% to all municipalities. I checked the robustness of my results against the different choices of variables as much as possible, but I am unable to check all of the combinations.

There is also an issue of missing observations especially in voter participation rates, income equality, regional affordability and homeownership variables. I calculate RCI unless each of the three dimensions of RCI has missing values in more than two variables. As a result, the RCI in some municipalities is calculated from fewer variables than the others. Among all, RCI in four municipalities in Fukushima (Idate-mura, Kawauch-mura, Hirono-machi and Naraha-machi) was calculated from only seven indicators, while RCI in 278 municipalities was calculated from eight indicators.

Third is the measurement of post-disaster recovery. In this study, I primarily focus on in- and out-migration because population recovery is an essential part of the post-disaster recovery. However, in- and out-migration are not the only measures for the post-disaster recovery. Various economic, infrastructure and transportation metrics can be also used to measure the degree of post-disaster recovery (Aldrich 2012). I leave this for future studies.

Fourth, I limited the sample up to 2016, which was before 2016 Kumamoto Earthquake affected the in- and out-migration statistics. I also did not use information on the number of deaths caused by natural disasters other than earthquakes due to the limitation of data available. Again, I leave this to future studies.

Despite these limitations, the results of this chapter provide two important policy implications regarding the role of resilience in the post-disaster recovery process. First, reducing the number of municipalities with low resilience is an effective strategy to mitigate a post-disaster population decline. By doing so, I can avoid the negative side of high within-community connectivity associated with high resilience. Second, foreigners behave as if they have different communities from the local communities in which I measure the level of resilience using the Resilience Capacity Index. This is the issue of foreigner integration and community inclusion. Foreigners with a lack of access to information and resources are disproportionately affected by disasters. Fortunately, Goal 11 of the SDGs states the promotion of inclusive societies. The community inclusion of foreigners can substantially develop the resilience of all members in the community in the face of disasters.

Appendix: Ranking of Municipalities Based on the Resilience Capacity Index

Top 10	Resilience capacity index	# of missing indicators	Economic capacity	# of missing indicators	Socio-demographic capacity	# of missing indicators	Community connectivity capacity	# of missing indicators
1	Chiyoda-ku (Tokyo)	0/12	Oshino-mura (Yamanashi)	2/4	Chiyoda-ku (Tokyo)	0/4	Iitate-mura (Fukushima)	2/4
2	Oshino-mura (Yamanashi)	4/12	Tajiri-cho (Osaka)	2/4	Bunkyo-ku (Tokyo)	0/4	Chiyoda-ku (Tokyo)	0/4
3	Tajiri-cho (Osaka)	3/12	Erimo-cho (Hokkaido)	2/4	Eiheiji-cho (Fukui)	0/4	Kaneyama-machi (Yamanashi)	1/4
4	Bunkyo-ku (Tokyo)	0/12	Toyokoro-cho (Hokkaido)	2/4	Nagakute-shi (Aichi)	0/4	Kosuge-mura (Yamanashi)	1/4
5	Uchinada-machi (Ishikawa)	0/12	Toyota-shi (Aichi)	0/4	Uchinada-machi (Ishikawa)	0/4	Wakasa-cho (Fukui)	0/4
6	Shimotsuke-shi (Tochigi)	1/12	Kiyosato-cho (Hokkaido)	2/4	Shimotsuke-shi (Tochigi)	0/4	Kitayama-mura (Wakayama)	1/4
7	Kamiichi-machi (Toyama)	1/12	Chiyoda-ku (Tokyo)	0/4	Mibu-machi (Tochigi)	0/4	Neba-mura (Nagano)	1/4
8	Toyota-shi (Aichi)	1/12	Rokkasho-mura (Aomori)	2/4	Shinjuku-ku (Tokyo)	0/4	Ama-cho (Shimane)	1/4
9	Eiheiji-cho (Fukui)	0/12	Shintoku-cho (Hokkaido)	2/4	Chuo-shi (Yamanashi)	0/4	Mihara-mura (Kochi)	1/4
10	Minato-ku (Tokyo)	0/12	Shimukappu-mura (Hokkaido)	2/4	Minato-ku (Tokyo)	0/4	Shirataka-machi (Yamagata)	0/4
Worst 10								
1	Naraha-machi (Fukushima)	5/12	Iitate-mura (Fukushima)	2/4	Shimamaki-mura (Hokkaido)	0/4	Naraha-machi (Fukushima)	2/4
2	Shimamaki-mura (Hokkaido)	4/12	Satosho-cho (Okayama)	2/4	Shosanbetsu-mura (Hokkaido)	0/4	Hirono-machi (Fukushima)	2/4

(continued)

(continued)

Top 10	Resilience capacity index	# of missing indicators	Economic capacity	# of missing indicators	Socio-demographic capacity	# of missing indicators	Community connectivity capacity	# of missing indicators
3	Shosanbetsu-mura (Hokkaido)	3/12	Izumizak-mura (Fukushima)	2/4	Yusuhara-cho (Kochi)	0/4	Kawauchi-mura (Fukushima)	2/4
4	Susuto (Hokkaido)	4/12	Goka-machi (Ibaraki)	2/4	Nishiokoppe-mura (Hokkaido)	0/4	Onagawa-cho (Miyagi)	1/4
5	Tomari-mura (Hokkaido)	4/12	Hinoemata-mura (Fukushima)	2/4	Rikubets (Hokkaido)	0/4	Otoineppu-mura (Hokkaido)	2/4
6	Kawauchi-mura (Fukushima)	5/12	Mishima-mura (Kagoshima)	2/4	Furubira-cho (Hokkaido)	0/4	Wako-shi (Tochigi)	0/4
7	Minamifurano-cho (Hokkaido)	4/12	Toshima-mura (Kagoshima)	2/4	Suttsu-cho (Hokkaido)	0/4	Aogashima-mura (Tokyo)	0/4
8	Iitate-mura (Fukushima)	5/12	Yamatsuri-machi (Fukushima)	2/4	Nanmoku-mura (Nagano)	0/4	Minamisanriku-machi (Miyagi)	0/4
9	Kuromatsunai-cho (Hokkaido)	4/12	Hakone-machi (Kanagawa)	2/4	Minamifurano-cho (Hokkaido)	0/4	Otsuchi-cho (Iwate)	1/4
10	Furubira-cho (Hokkaido)	4/12	Ryuo-machi (Shiga)	2/4	Shimokawa-cho (Hokkaido)	0/4	Horonobe-cho (Hokkaido)	2/4

References

Abenayake C, Yoshiki M, Marasinghe A, Takashi Y, Masahiro I (2016) Applicability of extra-local methods for assessing community resilience to disasters: a case of Sri Lanka. J Env Assess Policy Manag 18:1650010

Aldrich DP (2012) Building resilience: social capital in post-disaster recovery. University of Chicago Press, Chicago, IL

Bauer TK (2003) Flexible workplace practices and labor productivity

Foster K (2011) Resilience capacity index: data, maps and findings from original quantitative research on the resilience capacity of 361 US metropolitan regions. https://brr.berkeley.du/rc

Horiuchi Y (2019) The Japanese local elections dataset. V1 ed.: Harvard Dataverse

Östh J, Reggiani A, Galiazzo G (2015) Spatial economic resilience and accessibility: A joint perspective. Comput Environ Urban Syst 49:148–159

Psycharis Y, Kallioras D, Pantazis P (2014) Economic crisis and regional resilience: detecting the 'geographical footprint' of economic crisis in Greece. Reg Sci Policy Pract 6:121–141

Quinlan AE, Berbés-Blázquez M, Haider LJ, Peterson GD (2016) Measuring and assessing resilience: broadening understanding through multiple disciplinary perspectives. J Appl Ecol 53:677–687

Reams MA, Lam NSN, Baker A (2012) Measuring capacity for resilience among coastal counties of the US northern Gulf of Mexico region. Am J Clim Change 1:194–204

Reggiani A, Hudec O, Siserova M (2016) Resilience capacity and vulnerability: the case of Slovakia

Schnabel M (1997) International competitiveness: labor productivity leadership and convergence among 14 OECD countries. Department of Commerce WP ESA/OPD, 97–94

UNDRR (2007) The Hyogo framework for action 2005–2015: building the resilience of nations and communities to disasters. UNDRR, Geneva

UNDRR (2015) Reading the sendai framework for disaster risk reduction 2015–2030. UNDRR, Geneva

UNDRR (2016) Implementing the Sendai Framework to achieve the Sustainable Development Goals. UNDRR, Geneva

Winderlss T (2014) Disaster resilience measurements: stocktaking of ongoing efforts in developing systems for measuring resilience

Part II
Governance and Management of Disaster Risks

Chapter 9
Understanding Social-Mediated Disaster and Risk Communication with Topic Model

Xianlin Jin

Abstract To aid mitigating disaster and risk, this chapter concentrates on identifying the complex disaster and risk communication patterns with a particular focus on open discussions about disasters on social media. It reviews the literature on disaster and risk communication along with the introduction of the Crisis and Emergency Risk Communication model (CERC). This chapter highlights the importance of utilizing big data tools to unpack the conundrum of social-mediated disaster and risk communication. Lastly, it includes a case study that aims to understand the communication patterns that emerged from the Hurricane Maria resolution stage. Topic modeling, semantic analysis, content analysis, and word-cloud were integrated to capture five topics related to Hurricane Maria: food support, mental and physical health, fatalities, government's responses, and water supply. Findings suggest that the public was concerned about inadequate support for the victims of Hurricane Maria and worried about how the affected community could recover from the disaster. Government and emergency management agencies should advance social media use in future management plans to initiate interactive communication. These institutions should utilize social media to recognize the public's concerns, communicate remediation, and offer relief responses in a timely manner.

Keywords CERC · Social media · Disaster and risk communication · Topic modeling

9.1 Introduction

As the Sendai Framework for Disaster Risk Reduction 2015–2030 posited, the ultimate goals of disaster risk reduction are to prevent new risk, reduce existing risk, and strengthen community resilience (United Nations Office for Disaster Risk Reduction 2015). This process requires stakeholders' (from states to communities) collaboration and a set of guidelines and principles for disaster risk reduction.

X. Jin (✉)
College of Communication and Information, University of Kentucky, 206 McVey Hall, 155 Graham Ave, Lexington, KY 40506, USA
e-mail: xianlin.jin@uky.edu

R. Djalante et al. (eds.), *Integrated Research on Disaster Risks*, Disaster Risk Reduction,
https://doi.org/10.1007/978-3-030-55563-4_19

Reducing threats and uncertainties caused by disasters demands emergency management professionals and government officials to provide immediate and accurate information (Seeger et al. 1998). When the public is engaged in this communication process, it becomes more important to interact with and disseminate information to the affected community. Accurate and timely information will not only reduce the public's uncertainty but also implement crisis coping strategies (Jin 2020). Social media offers real-time information (Morris et al. 2010; Sin and Kim 2013) and provides a platform where emergency management institutions and the public could have interactive communication.

With this in mind, this chapter focuses on disaster and risk communication in the social media context. In particular, this chapter highlights the importance of understanding disaster and risk communication by analyzing rich and real-time social media data with topic modeling. This chapter reviews the literature on disaster risk communication and introduces the Crisis and Emergency Risk Communication (CERC) model that is an emergency management guideline implemented in the United States. Furthermore, this chapter proposes to embrace big data tools in research and discusses the application of topic modeling in disaster and risk communication contexts. This article also presents a case study of communication surrounding Hurricane Maria on Twitter to illustrate the methodology, theoretical, and practical implications for disaster and risk communication.

9.2 Disaster and Risk Communication

Standard risk and emergency management can only help mitigate isolated risks (Jin and O'Hair 2020). Nevertheless, such standard management approach has limited effectiveness in reducing disaster risk that creates "serious disruption of the functioning of a community or a society involving widespread human, material, economic or environmental losses and impacts, which exceeds the ability of the affected community or society to cope using its own resources" (United Nations International Strategy for Disaster Reduction 2009, p. 9).

Disaster, risk, and crisis are interrelated. As the United Nations International Strategy for Disaster Reduction (UNISDR 2009) posited, disasters are often seen as outcomes of continuous risk conditions. Disaster risks contain various potential losses "in lives, health status, livelihoods, assets, and services" that are hard to determine (UNISDR 2009, p. 10). Risk and crisis are parts of a connected process; risk frequently escalates into a crisis that involves threat and requires urgent action (Reynolds 2002).

To reduce disaster and risk, the Sendai Framework highlighted four priorities: "Understanding disaster risk. Strengthening disaster risk governance to manage disaster risk. Investing in disaster risk reduction for resilience. Enhancing disaster preparedness for effective response and to 'Build Back Better' in recovery, rehabilitation, and reconstruction" (United Nations Office for Disaster Risk Reduction 2015, p. 14).

To achieve these goals, a number of studies have focused on disaster and risk communication (e.g., Jin, 2020; Lachlan et al. 2014; Lin et al. 2016a, b; Spence et al. 2015; Sutton et al. 2008). A majority of disaster and risk communication research has been primarily built on cost–benefit analysis (Jin and O'Hair 2020; Zinn and Taylor-Gooby 2006). In addition, social-mediated risk communication studies merely analyzed the tweets posted by official accounts that have been shown less active on social media (Jin, 2020; Lachlan et al. 2014; Lin et al. 2016a, b). Nevertheless, the knowledge gap between the public and emergency management experts leads to the public's uncertainty regarding management implementation. It is essential for disaster and risk management institutions to understand the public's concerns, uncertainty, and requests in a timely manner.

Particularly, the communication patterns of disaster vary across stages and hence require emergency management institutions to utilize different communication strategies (Reynolds and Seeger 2005). The United States Centers for Disease Control and Prevention (CDC) adopted the Crisis and Emergency Risk Communication (CERC) model to train emergency management and risk communication practitioners to build effective communication in emergencies, such as hurricanes. The next section reviews the CERC model's guidelines with a particular focus on the resolution stage's communication tasks.

9.2.1 Crisis and Emergency Risk Communication Model

The CERC model integrates disaster and risk communication with crisis communication and assists relevant stakeholders to manage emergencies (Reynolds and Seeger 2005). According to the CERC model (Reynolds 2002; Reynolds and Seeger 2005), there are five stages of crisis development: precrisis, initial event, maintenance, resolution, and evaluation. Each stage requires emergency management institutions to utilize different communication strategies. For example, the major tasks of disaster and risk communication in the precrisis stage include designing risk messages and warnings, as well as other preparations. The priority of the initial event stage is to reduce uncertainty and build reassurance. The maintenance stage requires more interactive communication between the public and emergency management institutions. For the resolution stage, disaster and risk communication highlights the importance of having open discussions about the causes of the crisis and relevant stakeholders' responsibility and responses to emergencies. Effective disaster and risk communication in the resolution stage will aid the public to fully understand the risk and help the relevant stakeholders to recover a positive reputation (Lachlan et al. 2016; Reynolds and Seeger 2005).

Previous studies have explored the social-mediated communication patterns of disasters during precrisis, initial event, and/or maintenance stages (Lachlan et al. 2016, 2014; Spence et al. 2015), while the open discussions of disasters in the resolution stage have been less studied. Inspired by the CERC model, this chapter concentrates on the social-mediated communication patterns of a disaster in the resolution stage. The following sections offer a review of how social media impacts disaster and risk communication.

9.2.2 *Disaster and Risk Communication on Social Media*

Social media has been shown to impact risk and crisis communication in multiple ways. For instance, social media has been used as information and news sources for communities (Cameron et al. 2012; Morris et al. 2010; Sin and Kim 2013). With social media, the public and governmental agencies can have "interactive, collaborative, conversational, and community-based" communication (Spence et al. 2015, p. 172). As Alexander (2014) emphasized, appropriate usage of social media in risk and emergency management can serve the following functions:

- Listening. Social media makes it possible to collect public opinions. Marken (2007) posited that social media enables ordinary citizens to voice their opinions and experiences. Recent research also suggests that crisis information influentials on social media can directly generate information and impact the information flow (Jin, 2020).
- Monitoring. Social media can be used to understand what the public is concerned about and how individuals respond to specific emergency events. With this approach, rumors can be deterred (Bird et al. 2012).
- Emergency planning. It is important to explore social-mediated disaster and risk communication in emergency management plans (Jin, 2020; Lachlan et al. 2014; Lin et al. 2016a, b). Recent research discovered that the discussion about air pollution on Sina Weibo (one Chinese micro-blog platform) impacted the Chinese government's responses to air pollution (e.g., releasing values of fine particles $PM_{2.5}$ and building $PM_{2.5}$ monitors across the country; see Jin and O'Hair 2020).
- Promoting social cohesion. According to an American Red Cross survey (Blanchard et al. 2010), 24% of the United States population and 31% of the internet users expressed that they would keep their family and friends updated about their safety through social media.

Despite the potential values of implementing social media usage during disasters, social media has not been fully utilized as a communication channel (Lin et al. 2016a, b). Previous research suggests that Twitter was largely used for emotional release rather than sharing disaster-related information during Hurricane Sandy. In addition, the majority of information shared on social media was from ordinary citizens, not governmental agencies (Lachlan et al. 2014).

Another challenge for current research is that chaotic systems of disasters are hard to be clarified. Previous researchers have distributed post-disaster surveys that are built on people's memory about their experiences during disasters (Sutton et al. 2008) and conduct content analysis with limited text samples from news reports or posts on social media (Lachlan et al. 2014; Spence et al. 2015). These studies often only examined partial communication patterns associated with disasters. The complete picture of crisis information emerging from social media remains unclear.

To disentangle the conundrum, data mining tools can be utilized to analyze real-time social media messages and allow researchers to study open discussions of disasters (Kavanaugh et al. 2012). In particular, this chapter proposes to utilize topic modeling to identify communication patterns surrounding disasters.

9.3 Topic Modeling and Social-Mediated Disaster and Risk Research

Social media provides a large volume of real-time data, which creates both opportunities and challenges for researchers to explore how social media impacts disaster and risk communication. On the one hand, researchers have offered some insights by analyzing retweet frequency of various stakeholders (Lovejoy et al. 2012), exploring the strategies to increase retweet numbers (e.g., URL and multimedia files, Lachlan et al. 2014), or conducting a content analysis of disasters related tweets with selected samples (Lin et al. 2016a, b; Spence et al. 2015). On the other hand, although these studies shed light on how social media impacts communication and information dissemination, it remains difficult to capture the chaotic disaster and risk communication patterns on social media. Because the limited sample size and traditional approach of exploring the linear relationship and group differences (Murphy 1996) can not fully handle the unstructured social media data. Facing this challenge, this chapter recommends bringing big data tools in disaster and risk communication research and particularly focuses on topic modeling.

9.3.1 Introduction of Topic Model

Topic modeling allows researchers to analyze a large volume of unstructured social media data and interpret meaning embedded in the data. Topic modeling has been used as a tool to detect text meaning for three decades (Rohani et al. 2016). Hofmann (2001) introduced the Latent Semantic Analysis (LSA) model in which the order of words is neglected while generating topics. Blei et al. (2003) developed the model of Latent Dirichlet Allocation (LDA) to identify topics by calculating the probability distribution over documents. A topic model in LDA specifies a simple probabilistic procedure to generate documents on the basis of latent (random) variables (Blei et al. 2003). It is also worth noting that topic models can capture polysemy as one word can be part of multiple topics. The goal of fitting a generative model is to find the best set of latent (random) variables that explain the observed data (Griffiths and Steyvers 2004; Steyvers and Griffiths 2007).

Although Twitter provides real-time data for researchers to study disaster risk communication, its word limits prevent researchers from accurately interpreting short posts (Tang et al. 2014). Previous research has applied and extended the LDA model in social media contexts. For instance, Hong and Davison (2010) trained standard topic models on micro-blog messages and compared the effectiveness of these models. They concluded that training the topic model on aggregated social media messages can lead to better performances of modeling social media text. With these implementation efforts, topic modeling has been seen as an appropriate tool for researchers to detect social media patterns.

With topic model analysis, researchers can identify major discussions associated with certain events. The next section reviews how topic model analysis has been used by researchers to unpack the patterns of social-mediated disaster and risk communication.

9.3.2 Applying Topic Model Analysis in Disaster and Risk Communication

Topic modeling has been used to identify health trends on social media (Asghari et al. 2018; Paul and Dredze 2014), explore disaster and risk communication patterns (Sadri et al. 2018), and assess disaster damage (Resch et al. 2018). For instance, to identify popular topics about Hurricane Sandy on social media, Sadri et al. (2018) conducted a topic model analysis with 763,000 English tweets posted by the top 4,029 users from October 14 to November 12, 2012. According to Sadri et al., storm prediction, storm watch, and preparedness were the major topics in the warning phase. In the response phase, the major concerns were about gas/fuel, food/water, and significant power outage. Nevertheless, Sadri et al. did not include tweets posted by users who did not tweet about Hurricane Sandy over 100 times. Such an approach may omit the opinions of non-frequent Twitter users including accounts of emergency management agencies, and these results may be exaggerated.

Another example of topic modeling's application in disaster and risk research is Resch et al.'s (2018) study that utilized LDA to analyze the tweets related to the Napa earthquake in August 2014. With this approach, Resch et al. detected earthquake footprints and generated a damage assessment mapping significant loss. Resch et al.'s findings largely focused on the general topics related to disaster losses, while open discussions about emergency management effectiveness, responsibility, and responses to disasters have not articulated. To aid disaster and emergency management, this chapter concentrates on disaster and risk communication that emerged from the resolution stage.

This chapter aims to bring topic modeling and latent semantic analysis to understand the complex disaster and risk communication patterns; as such, the chapter includes a case study of disaster and risk communication related to Hurricane Maria. In this case study, the author collected real-time data from Twitter with NodeXL

(Smith et al. 2010) and discovered five major topics that emerged from the resolution stage of Hurricane Maria: food support, mental and physical health, fatalities, government's responses, and water supply. The findings will help researchers, practitioners, and decision-makers understand the public's concerns, mitigate the public's uncertainty, and provide immediate disaster relief responses.

9.4 Case Study of Hurricane Maria Recovery Stage

In Fall 2017, Hurricane Maria hit multiple areas, such as Dominica, the United States Virgin Islands, and Puerto Rico. Hurricane Maria formed on September 16, 2017, became extratropical after September 30 and dissipated on October 2, 2017 (National Hurricane Center 2017). By the end of August 2018, the Puerto Rico government confirmed that estimated 2,975 people died because of Hurricane Maria (Santiago et al. 2018). This disaster has not only caused a loss but also created uncertainty for the public to be back to normal life.

With the purpose of understanding how the public discusses and communicates this disaster on social media, this case study used topic modeling, latent semantic analysis, content analysis, and word-cloud to discover the dominant topics of disaster and risk communication surrounding Hurricane Maria during the resolution stage. Specifically, this case study aims to answer the following question: What topics were associated with Hurricane Maria in the resolution stages?

9.4.1 Data Collection

The tweets related to Hurricane Maria were collected with the keyword searching function in NodeXL Professional (Smith et al. 2010), which connects to the public free Application Programming Interface of Twitter. Specifically, by searching the keywords "Hurricane Maria" and "HurricaneMaria," 12,146 tweets (original tweets and retweets) about Hurricane Maria were collected. These tweets were posted from November 6 to November 21, 2017. Specifically, this dataset contained 7,550 vertices. This data collection strategy enables the author to cover tweets with different hashtags and allows the case study to capture more open discussions surrounding Hurricane Maria.

9.4.2 Data Analyses

This study utilized topic model analysis, which is a generative model providing a probabilistic framework of text data. This case study has utilized JMP Pro13

(SAS Institute Inc. 2016) to conduct a series of topic modeling, semantic analysis, and word-cloud analyses. JMP Pro13 (SAS Institute Inc. 2016) uses a bag of words approach to conduct topic analysis in which the analysis is built on the word and phrase calculation. Specifically, topic analysis is performed by conducting a varimax rotated singular value decomposition of the document term matrix (DTM, see Klimberg and McCullough 2016).

Before generating the DTM and conducting a topic analysis, Regex tokenizing was applied to transform text to lowercase, identify the regular expressions, and remove the most punctuations. Furthermore, a stop-word list was created and applied to clean data. A Latent Semantic Analysis (LSA) was used to extract and represent terms and phrases from a corpus so that the DTM was reduced to a manageable size for further analysis. A Singular Value Decomposition (SVD) was generated to decomposes the DTM into three matrices:

$$DTM = D * S * T.$$

These matrices are defined as follows:

- D is an orthogonal document-document matrix of eigenvectors.
- T is an orthogonal term-term matrix of eigenvectors.
- S is a diagonal matrix of singular values (Klimberg and McCullough 2016, p. 338).

While generating the curated DTM, natural language processing techniques were used. Later, text clustering, principal components, and factor analysis were utilized to interpret the DTM (Klimberg and McCullough 2016). Lastly, topic analysis and word-cloud were generated. Topic analysis can calculate how much each tweet contributes to each topic. It can also discover the most popular topics of a specific event based on word frequency calculations. With this approach, this case study was able to identify the dominant topics regarding Hurricane Maria in its resolution stage.

9.5 Results

To capture the major topics associated with Hurricane Maria, this study found the top hashtags that were frequently used in the resolution stage (November 6–November 21, 2017). The top three hashtags were #puertorico (459), #theoriginals (381), and #hurricane (151). The top three URLs used in the tweets were identified as well. The first link was a denotation webpage created by the disaster relief and recovery program of UNIDOS (2017): https://hispanicfederation.org/unidos. The second link was about an article calling for volunteers for Puerto Rico, which is written by Navarro (2017): https://trib.al/6FI4dfx. The third link was an article (Dickerson 2017) focusing on the mental health crisis of Puerto Rico community after Hurricane Maria (https://www.nytimes.com/2017/11/13/us/puerto-rico-hurricane-maria-mental-health.html).

To further understand the communication patterns that emerged from the resolution stage of Hurricane Maria, this study combined topic modeling, semantic analysis content analysis, and word-cloud. To answer the research question, five topics were identified by conducting the topic model and semantic analyses: food support, mental health, fatalities, government responses, and water supply. The top terms that emerged in each topic were identified (Tables 9.1 and 9.2).

Among the five topics, the dominant topic was food support that explained the largest variance of the case (Table 9.3). The food support topic was about the quality

Table 9.1 Topic to terms of Hurricane Maria tweets in the resolution stage

Topic 1. Food support		Topic 2. Mental health		Topic 3. Fatalities	
Term	Loading	Term	Loading	Term	Loading
Barsand	0.94680	Stress	0.89880	CNN	0.80687
Pack	0.94212	Traumatic	0.89875	Funeral	0.77150
Supply	0.94172	Symptoms	0.89854	Homes	0.75646
Goods	0.94160	Showing	0.88119	Toll	0.69913
Candy	0.93222	Population	0.86482	Death	0.66586
Contracted	0.92940	Public	0.84666	Identify	0.55324
Cheez	0.92815	Health	0.78389	Asked	0.55298
Thread	0.85860	Officials	0.73956	Fatalities	0.55298
Company	0.84590	Nytimes	0.66908	Investigation	0.53307

Table 9.2 Topic to terms of Hurricane Maria tweets in the resolution stage (Continue)

Topic 4. Government's Responses		Topic 5. Water and Power Supply	
Term	Loading	Term	Loading
Realize	0.85927	Water	−0.6814
Needs	0.83000	Thousands	−0.6602
Governor	0.81293	Nowhere	−0.6494
Bill	0.79522	Bathe	−0.6493
Clinton	0.79377	Running	−0.6483
Supplies	0.72489	Juan	0.5745
Relief	0.59249	San	0.5717
Abcpolitics	0.58838	puerto rico	−0.5288
People	0.51363	Latinorebels	0.5108

Table 9.3 Variance explained by each topic

Topic	Variance	Percentage
Topic 1	9.1642	3.818
Topic 2	7.6685	3.195
Topic 3	7.0194	2.925
Topic 4	6.9874	2.911
Topic 5	6.8727	2.864

of free meals provided by the Federal Emergency Management Agency (FEMA) for victims. The food support topic contains "candy," "fema," "snack," and "meals." For example, on November 12, 2019, one twitter user criticized the government's deficient responses to the food needs of the impacted people in Puerto Rico: "It's been nearly two months since Hurricane Maria and thousands in Puerto Rico still don't have access to emergency food." Interestingly, on the same day, another user posted "Are Hurricane Maria Survivors Getting 'Skittle Meals' From The Government?".

The mental health topic concentrated on victims' trauma and mental health. The top terms in this topic contained "traumatic," "health," "stress," and "public." The New York Times' tweets on mental health have been shared by other twitter users. For example, the following tweets have been frequently retweeted: "After Hurricane, Signs of a Mental Health Crisis Haunt Puerto Rico" and "Much of Puerto Rico's population is showing symptoms of post-traumatic stress, public health officials say" (Dickerson 2017).

The fatalities topic was about the death toll result. People were also concerned about the fatalities caused by Hurricane Maria. The words such as "death," "toll," "funeral," "fatalities," and "investigation" contributed to this topic. For instance, the Cable News Network tweeted the Puerto Rico death toll news: "Puerto Rico has asked funeral homes to help identify Hurricane Maria fatalities after a CNN investigation into the death toll" (Sutter et al. 2017). This tweet has been retweeted 172 times and received 90 favorites.

The topic of the government's responses to the disaster focused on the inadequate responses provided by the government. This topic includes "governor," "needs," "supplies," "bill," and "Clinton." For example, the following tweet posted by the American Broadcasting Company News (ABC News) has been retweeted 491 times and received 1,248 favorites: "Puerto Rico governor says Bill Clinton's visit will help people realize the island still needs relief supplies more than 2 months after Hurricane Maria" (ABC News 2017).

The topic of water and power supply showed that Puerto Rico still did not have normal water and power supply. People complained about no access to power and no water for bath and drink. The water supply topic contains "running," "water," "bathe," and "puerto rico." By the time of data collection, the following tweet has been retweeted 2,740 times and received 2,031 favorites: "Nearly 6 weeks after Hurricane Maria, thousands in Puerto Rico still don't have running water, have nowhere to bathe or wash clothes. Now, it's making people sick." Similarly, a tweet posted by the New York Times has been retweeted 351 times: "After a month lost to Hurricane Maria, students return to class in San Juan, making do without electricity & water" (Alvarez 2017).

To provide a clear picture of mentioned within Hurricane Maria tweets, this study also visualized word frequency with word-cloud of JMP Pro13 (Fig. 9.1). The larger the word size, the more frequently it was mentioned. For details of the top terms and phrases associated with Hurricane Maria tweets in the resolution stage, please check Figs. 9.2 and 9.3.

Fig. 9.1 Word-cloud of Hurricane Maria tweets posted in the resolution stage

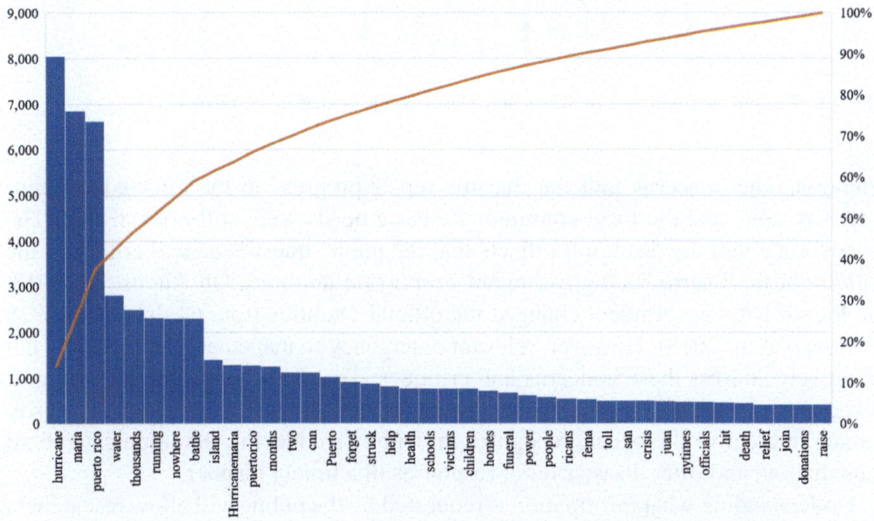

Fig. 9.2 Top word frequency of Hurricane Maria tweets posted in the resolution stage

9.6 Discussion

9.6.1 Implications

This case study identifies the major topics associated with Hurricane Maria in the resolution stage by analyzing real-time social media data. The most significant concerns about this disaster concentrate on the hazardous effects of this disaster and the critiques of inadequate disaster relief responses. Complains about a lack of water and power supply in Puerto Rico remain obvious even two months after Hurricane Maria dissipating on October 2, 2017. In addition, the mental health of affected community members after suffering this intense hurricane also attracts the public's

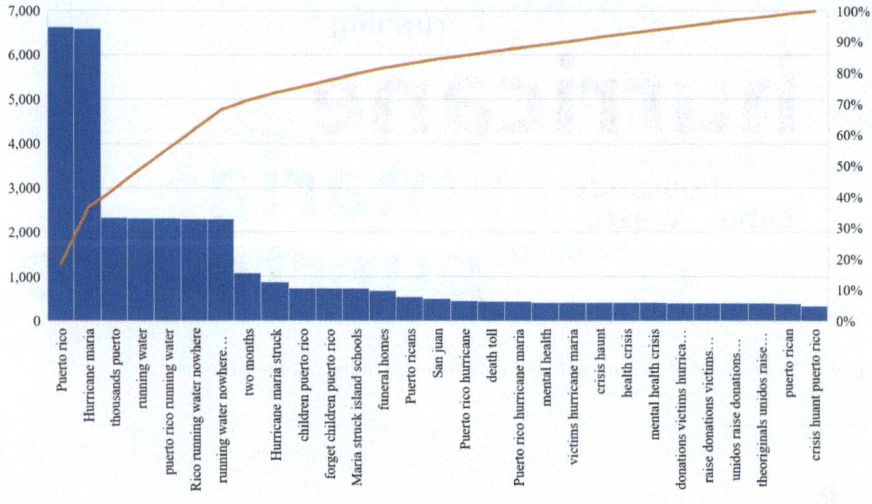

Fig. 9.3 Top phrase frequency of Hurricane Maria tweets posted in the resolution stage

attention. The concerns indicate that the repair progress in the affected community was slow, and the local community's basic needs were still not satisfied. The discussion about the death toll reflects that the public questioned and criticized the validity of the Puerto Rico government's report on fatalities. On August 28, 2018, the Puerto Rico government changed the official fatalities from 64 deaths to 2,975 (Santiago et al. 2018). However, relevant emergency management agencies did not effectively address these concerns and critiques. The case study's findings suggest that government and related agencies should employ advanced social media tools to initiate interactive dialogues, recognize the public's concerns, communicate about remediation, and offer disaster relief responses in a timely manner.

Understanding what information is requested by the public will allow researchers and emergency management practitioners to design effective messages to meet the need. Knowing the role of social media during disasters will allow emergency management practitioners to use the appropriate mediums to reach the public and distribute information about the threat, causes, and dangers, which help reduce the public's uncertainties. Understanding these patterns of disasters will also help practitioners calm an anxious public (Spence et al. 2015). Even though social media has been shown to impact crisis communication (e.g., Cameron et al. 2012; Jin, 2020; Lachlan et al. 2014; Morris et al. 2010), the FEMA's (2014) emergency response plan has not fully recognized social media's role in emergency management. As this case study's findings illustrate, it is important for emergency management institutions to understand the public's concerns, monitor the crisis communication and information flow, provide immediate responses to targeted audiences, and offer prompt disaster relief responses.

More importantly, this study posits the importance of bringing big data tools to address the complexity of disaster and risk communication. Traditional quantitative units exploring linear relationships and group differences may not effectively describe such chaotic systems (Murphy 1996). To unpack this conundrum, this chapter presents a case study to demonstrate that integrating topic modeling with content analysis can better capture the details of disaster and risk communication on social media and identify the public's concerns and requests. With this attempt, this chapter hopes to bring a discussion about what new methods can be adapted to handle the complexity of disaster risk. The knowledge will help researchers refine disaster risk framework and guide governments and emergency practitioners to provide effective communication and prompt disaster relief responses.

9.6.2 Future Directions

Disaster risk research faces many challenges because it requires multi-dimensional cooperation among various disciplines (e.g., public health, emergency management, and disaster risk communication) and different stakeholders (e.g., government, emergency management institutions, and communities). Although many emergency management institutions start utilizing social media to deliver messages to the public, institutional guidance and evidence remain unclear regarding how to take advantage of social media in preparation for a future crisis, monitor communication on social media, guide information flow, and respond to the public's opinions expressed on social media.

This chapter highlights the value of embracing the public's opinions in emergency management. The growing technology significantly impacts disaster and risk communication among emergency management agencies, mass media, and general citizens. The traditional one-way, top-down communication ignores the public's opinions (Jin and O'Hair 2020). In the future, more efforts should be devoted to examining disaster risk communication among the various stakeholders across platforms, such as governmental agencies' websites, newspapers, television, and social media.

Furthermore, this chapter encourages future researchers to utilize machine learning tools to address disaster risk problems. Machine learning, such as topic modeling, may be able to help researchers refine disaster risk frameworks. Such knowledge will provide implications for future theory building in disaster and risk, as well as practice in emergency management.

9.7 Conclusions

This book chapter highlights the importance of exploring open discussions about diasters on social media during the resolution period. It also proposes to apply big data tools in social-mediated disaster and risk communication. Particularly, this chapter posits that topic modeling can help identify chaotic systems that are hard to be examined because of the limited sample and reach of traditional methods exploring the linear relationship and group differences (Murphy 1996). From a practical standpoint, this chapter suggests the necessity of utilizing social media in future emergency management during disasters. Effective social media usages will assist emergency management institutions to monitor disaster and risk information dissemination on social media, understand the public's concerns, provide accurate information, build interactive communication with affected communities, and offer immediate relief responses.

References

Alexander DE (2014) Social media in disaster risk reduction and crisis management. Sci Eng Ethics 20:717–733. https://doi.org/10.1007/s11948-013-9502-z

Alvarez L (2017, October 24) Some Puerto Rico schools reopen, making do without power. The New York Times. https://www.nytimes.com/2017/10/24/us/puerto-rico-schools.html?hp&action=click&pgtype=Homepage&clickSource=story-heading&module=second-column-region&region=top-news&WT.nav=top-news

American Broadcasting Company News (2017, November 20) Puerto Rico governor says Bill Clinton's visit will help people realize the island still needs relief supplies more than 2 months after Hurricane Maria. American Broadcasting Company News. https://twitter.com/i/web/status/932867667349921792

Asghari M, Sierra-Sosa D, Elmaghraby A (2018) Trends on health in social media: analysis using twitter topic modeling. IEEE International Symposium on Signal Processing and Information Technology (ISSPIT) 2018:558–563. https://doi.org/10.1109/ISSPIT.2018.8642679

Bird D, Ling M, Haynes K (2012) Flooding facebook: the use of social media during the Queensland and Victorian floods. Australian J Emerg Manag 27(1):27–33. https://researchers.mq.edu.au/en/publications/flooding-facebook-the-use-of-social-media-during-the-queensland-a

Blanchard H, Carvin A, Whittaker ME, Fitzgerald M, Harman W, Humphrey B (2010) The case for integrating crisis response with social media. White Paper, American Red Cross

Blei DM, Ng AY, Jordan MI (2003) Latent Dirichlet allocation. J Mach Learn Res 3:993–1022. https://doi.org/10.1162/jmlr.2003.3.4-5.993

Cameron MA, Power R, Robinson B, Yin J (2012) Emergency situation awareness from Twitter for crisis management. In: Proceedings of the 21st international conference companion on World Wide Web. https://doi.org/10.1145/2187980.2188183

Dickerson C (2017, November 13) After hurricane, signs of a mental health crisis haunt Puerto Rico. New York Times. https://www.nytimes.com/2017/11/13/us/puerto-rico-hurricane-maria-mental-health.html?partner=IFTTT

Federal Emergency Management Agency (2014) Emergency response plan. https://www.fema.gov/media-library/assets/documents/89518

Griffiths TL, Steyvers M (2004) Finding scientific topics. Proce Nat Acad Sci 101:5228–5235. https://doi.org/10.1073/pnas.0307752101

Hofmann T (2001) Unsupervised learning by probabilistic latent semantic analysis. Mach Learn J 42(1):177–196. https://doi.org/10.1023/a:1007617005950

Hong L, Davison BD (2010) Empirical study of topic modeling in Twitter. In: Proceedings of the first workshop on social media analytics—SOMA'10. https://doi.org/10.1145/1964858.1964870

Jin X (2020) Exploring crisis communication and information dissemination on social media: Social network analysis of Hurricane Irma tweets. J Int Crisis Risk Commun Research 3(2): 49–80. https://doi.org/10.30658/jicrcr.3.2.3

Jin X, O'Hair HD (2020) Air pollution and health in China. In: O'Hair HD, O'Hair MJ (eds.), Handbook of applied communication research. Wiley-Blackwell, pp 935–954. https://doi.org/10.1002/9781119399926.ch51

Kavanaugh AL, Fox EA, Sheetz SD, Yang S, Li LT, Shoemarker DJ, Natsev A, Xie L (2012) Social media use by government: From the routine to the critical. Gover Inf Quart 29:480–491. https://doi.org/10.1016/j.giq.2012.06.002

Klimberg R, McCullough ED (2016) Fundamentals of predictive analytics with JMP® (Second Edition). SAS Institute Inc

Lachlan KA, Spence PR, Lin X, Del Greco M (2014) Screaming into the wind: examining the volume and content of tweets associated with Hurricane Sandy. Commun Stud 65(5):500–518. https://doi.org/10.1080/10510974.2014.956941

Lachlan KA, Spence PR, Lin X, Najarian K, Del Greco M (2016) Social media and crisis management: CERC, search strategies, and Twitter content. Comput Hum Behav 54:647–652. https://doi.org/10.1016/j.chb.2015.05.027

Lin X, Lachlan KA, Spence PR (2016a) Exploring extreme events on social media: A comparison of user reposting/retweeting behaviors on Twitter and Weibo. Comput Hum Behav 65:576–581. https://doi.org/10.1016/j.chb.2016.04.032

Lin X, Spence PR, Sellnow TL, Lachlan KA (2016b) Crisis communication, learning and responding: Best practices in social media. Comput Hum Behav 65:601–605. https://doi.org/10.1016/j.chb.2016.05.080

Lovejoy K, Waters RD, Saxton GD (2012) Engaging stakeholders through Twitter: how nonprofit organizations are getting more out of 140 characters or less. Public Rel Rev 38:313–318. https://doi.org/10.1016/j.pubrev.2012.01.005

Marken GA (2007) Social media: the hunted can become the hunter. Public Relat Quart 52(4):9–12. https://doi.org/10.1037/e380852004-038

Morris MR, Teeva J, Panovich K (2010) A comparison of information seeking using search engines and social media. In: Proceedings of the 4th International AAAI Conference on Weblogs and Social Media, 10, 23–26. https://doi.org/10.1.1.309.2616

Murphy P (1996) Chaos theory as a model for managing issues and crises. Public Relat Rev 22(2):95–113. https://doi.org/10.1016/s0363-8111(96)90001-6

National Hurricane Center (2017) Hurricane Maria advisory archive. https://www.nhc.noaa.gov/archive/2017/MARIA.shtml?

Navarro M (2017, November 3) Puerto Rico on her mind: how to help a stricken island called home. The New York Times. https://trib.al/6FI4dfx

Paul MJ, Dredze M (2014) Discovering health topics in social media using topic models. PLoS ONE 9(8):e103408. https://doi.org/10.1371/journal.pone.0103408

Resch B, Usländer F, Havas C (2018) Combining machine-learning topic models and spatiotemporal analysis of social media data for disaster footprint and damage assessment. Cartogra Geograp Inf Sci 45(4):362–376. https://doi.org/10.1080/15230406.2017.1356242

Reynolds B (2002) Crisis and emergency risk communication. Centers for Disease Control and Prevention

Reynolds B, Seeger MW (2005) Crisis and Emergency risk communication as an integrative model. J Health Commun 10:43–55. https://doi.org/10.1080/10810730590904571

Rohani V, Shayaa S, Babanejaddehaki G (2016) Topic modeling for social media content: a practical approach. In: 2016 3rd International Conference on Computer and Information Sciences (ICCOINS), 397–402. https://doi.org/10.1109/ICCOINS.2016.7783248

Sadri AM, Hasan S, Ukkusuri SV, Cebrian M (2018) Crisis communication patterns in social media during Hurricane Sandy. Transp Res Rec 2672(1):125–137. https://doi.org/10.1177/0361198118773896

Santiago L, Shoichet CE, Kravarik J (2018, August 28) Puerto Rico's new Hurricane Maria death toll is 46 times higher than the government's previous count. Cable News Network. https://www.cnn.com/2018/08/28/health/puerto-rico-gw-report-excess-deaths/index.html

SAS Institute Inc (2016) JMP® Pro 13. https://www.jmp.com/en_us/software/data-analysis-software.html

Seeger MW, Sellnow TL, Ulmer RR (1998) Communication, organization and crisis. In: Roloff ME (ed.), Communication Yearbook 21, (pp 231–275). Sage

Sin S-CJ, Kim K-S (2013) International students' everyday life information seeking: the informational value of social networking sites. Lib Inf Sci Res 35(2):107–116. https://doi.org/10.1016/j.lisr.2012.11.006

Smith M, Ceni A, Milic-Frayling N, Shneiderman B, Mendes Rodrigues E, Leskovec J, Dunne C (2010) NodeXL: a free and open network overview, discovery and exploration add-in for Excel 2007/2010/2013/2016. https://www.smrfoundation.org

Spence PR, Lachlan KA, Lin X, Del Greco M (2015) Variability in Twitter content across the stages of a natural disaster: Implications for crisis communication. Commun Quar 63(2):171–186. https://doi.org/10.1080/01463373.2015.1012219

Steyvers M, Griffiths T (2007) Probabilistic topic models. In: Landauer T, McNamara D, Dennis S, Kintsch W (eds.), Latent semantic analysis: a road to meaning (pp 427–448). Lawrence Erlbaum Associates Publishers. https://doi.org/10.4324/9780203936399.ch21

Sutton J, Palen L, Shklovski I (2008) Backchannels on the front lines: emergent uses of social media in the 2007 Southern California wildfires. In: Fiedrich F, Van De Walle B (eds.), Proceedings of the 5th International ISCRAM Conference (pp 624–632). Washington, DC

Sutter JD, Santiago L, Shah K (2017, November 21) After CNN investigation, Puerto Rico asks funeral homes to help identify hurricane deaths. Cable News Network. https://www.cnn.com/2017/11/21/health/pesquera-responds-puerto-rico-death-count-invs/index.html

Tang J, Meng Z, Nguyen X, Mei Q, Zhang M (2014) Understanding the limiting factors of topic modeling via posterior contraction analysis. In: Proceedings of the 31st International Conference on International Conference on Machine Learning, 32, 190–198. https://proceedings.mlr.press/v32/tang14.pdf

UNIDOS (2017) Disaster relief & recovery program to support Puerto Rico. https://hispanicfederation.org/unidos/

United Nations International Strategy for Disaster Reduction (2009) Terminology on disaster risk reduction. https://www.unisdr.org/files/7817_UNISDRTerminologyEnglish.pdf

United Nations Office for Disaster Risk Reduction (2015) Sendai framework for disaster risk reduction 2015–2030. https://www.undrr.org/publication/sendai-framework-disaster-risk-reduction-2015-2030

Zinn JO, Taylor-Gooby P (2006) Risk as an interdisciplinary research area. In: Taylor-Gooby P, Zinn JO (eds), Risk in social science. Oxford University Press

Chapter 10
Preparation and Adoption of Risk Sensitive Land Use Plans in the New Federal Context of Nepal

Chandra Laxmi Hada, Rajib Shaw, and Anil Pokhrel

Abstract Nepal is one of the most disaster-risk countries in the world. Increase in frequency and severity—particularly for floods, landslides, and fires and unpredictability of disaster events such as earthquake, is overwhelming Nepal's already limited disaster risk reduction and management (DRRM) capacities and impeding sustainable development progress. Rapid urban growth fueled by migration, high levels of poverty and inequality, unplanned urban developments, climate change, and a culturally pervasive sense of fatalism are the key reasons for increased disaster risk in Nepal. Despite the increased awareness along with preparedness plans, programs, and projects on DRRM, the levels of risk continue to grow. To reduce risk, there are a number of disaster-resilient planning techniques used, among which, the Risk Sensitive Land Use Planning (RSLUP) has been regarded as an evidence-based tool to understand risk, plan and to reduce risk. In Nepal, until so far only eight RSULPs have been prepared, but its use as a planning tool is increasingly getting popular. In the new federal structure, the municipalities have been given full responsibilities for developing the policy legislations, standards for local level development plans and projects. Further, there has been tremendous growth in the generation of hazard, exposure, and vulnerability data fundamental to the preparation of RSLUPs. In this context, it is important to look at an evidence-based land use planning approach as a core element for future development in municipal governments to understand risk while undertaking land use planning, and to integrate RSLUP into the municipal planning process including sectoral plans. This chapter would focus on how to effectively enhance collaborative, participatory, and interactive approach for risk-sensitive land use planning, and fully integrate into a mandatory planning process. The research is to postulate a framework that enables collaborative, participatory, and iterative approach for the implementation of the RSLUP through integration in

C. L. Hada (✉)
SPURS/MIT, Cambridge, USA
e-mail: clhada@yahoo.com

R. Shaw
Keio University, Tokyo, Japan

A. Pokhrel
Plan8 Risk Consulting, Kathmandu, Nepal

R. Djalante et al. (eds.), *Integrated Research on Disaster Risks*, Disaster Risk Reduction,
https://doi.org/10.1007/978-3-030-55563-4_10

municipal development plan. This study will draw on past examples of RSLUP in urban municipalities in Nepal. The unique needs of vulnerable population, including socially excluded groups, will be included in RSLUPs, and activities that address risk. The developed RSLUP framework will serve as a model for other municipalities in Nepal, which is very relevant to the new federal system in the country.

Keywords Nepal · Risk sensitive land use planning · Disaster risk · Urban · Municipality · Restructuring

10.1 Introduction

10.1.1 Overview of Disaster Risk in Urban Municipalities

Nepal is vulnerable to multiple natural hazards—floods, landslides, fires, earthquakes, glacier lake outburst floods, and disease outbreaks. The 2015 Nepal earthquake and its aftershocks caused over US$ 7 billion in damages and losses and killed nearly 9,000 people (Source Ministry of Home Affairs -MoHA, Nepal). The capital city Kathmandu including other cities within Kathmandu valley is the world's most at-risk seismic urban area. The high urban density, haphazard planning, and largely unimplemented building codes have compounded potential earthquake risk in most urban areas. Nepal's current population is estimated at 28.4 million, and it is projected that by 2030 will be 33.3 million and by 2050 will be 36.4 million (World Population Projections 2050). The total population was 26.5 million in 2011, and only 17.1% of the total population lived in 58 municipalities. The Constitution of Nepal 2015 has declared the country a Federal Democratic Republic with 7 provinces, 77 districts, and 753 municipalities (6 metropolitan cities, 11 sub-metropolitan cities, 276 municipalities, and 450 rural municipalities). A total of 293 municipalities out of a total of 753 municipalities are classified as urban municipalities that contributes more than 60% in Nepal (Source: Ministry of Federal Affairs and General Administration—MOFAGA, Nepal).

In addition, Nepal is one of the top ten fastest urbanizing countries in the world with an urbanizing rate of 1.9 per annum. Nepal's urban population grew from 13.9% in 2001 to 17% in 2011 and has now reached to 58.4% in 2017. Several studies indicate that the current pace of urbanization in Nepal will continue but the level of urban growth is still at low level. Urbanization indirectly indicates economic growth, but it also means intensive concentration of population exposed to disaster and then high level of vulnerability. Nepal is a typical example of a country with rapid urbanization coupled with low economic activity but with social development barriers, which culminates to a high level of climate change vulnerability.

The government has taken more steps to advance disaster risk management and improve climate resilience after 2015 earthquake. It enacted the new Disaster Risk Reduction and Management (DRRM) Act in 2017, which promotes DRR focused approach to resilience building. If properly enforced, it can ensure that disaster risk reduction is mainstreamed into development plans and policies to promote sustainable development.

Even though the urban population in Nepal is exponentially increasing, the major chunk of the urban population is still living in a rural setting with limited basic urban amenities and services. The urban municipalities of Nepal are of typical character - more than 50% of the total area in an average is still rural, with agriculture farm land, thatch-roof houses, and narrow roads still common features of the urban municipalities. With the growing urbanization trend, urban disaster risk is also increasing at a rapid pace. Just like other disaster vulnerable countries, Nepal has also recognized this fact and is giving attention to the evolving demand for disaster risk reduction in urban centers of Nepal.

The Local Disaster Risk Management Plan (LDRMP)—2010 issued by the Ministry of Federal Affairs and General Administration (MOFAGA) has been a major guiding document for undertaking risk analysis and development of risk management plan in Nepal. Similarly, Local Adaptation Plan of Action (LAPA) framework prepared by the Ministry of Forest and Environment (MOFE) is being used to prepare local level climate change adaptation actions at the municipal levels. The MOFAGA has begun the process to integrate municipal level DRRM and climate change adaptation planning using a common and coherent tool called Local Disaster and Climate Resilience Plan (LDCRP), which is still in draft version.

Disaster risk in urban areas is increasing due to high population density, unplanned development practices, unsafe construction, and lack of urban planning, poverty, and high inequality in terms of access to public services. As cities continue to urbanize risk level increases particularly in the lower-middle-income countries like Nepal where the pace of urbanization is fastest. Although urban growth rate in Nepal is high, population distribution across cities and towns remains uneven. This can have a variety of effects on the development of urban communities. As Nepal's urban population has exceeded the rural population, the number of urban disasters is increasing and impacts are also on the rise. Reducing disaster risk caused by natural hazards is largely a development issue and needs to be addressed within the context of a wider urban development framework.

10.1.2 Urbanization in the New Federal Structure

Nepal's 2015 constitution set the course for a tectonic shift of power from the federal to the provincial and municipal levels of government. The government is in the process of developing and refining its legal frameworks, comprising of more than a dozen different acts that assign disaster risk reduction and management roles and responsibilities to different Government of Nepal (GON) entities, and resulting in inefficiencies and confusion during and after disaster events. A good step in the process is that under the new federal structure, the recently passed Local Governance Operations Act, 2017 (MoFAGA) shifts the responsibility for disaster risk reduction and management to the municipalities.

The DRRM is among 22 exclusive powers that are now the responsibility of devolved authorities. Schedule 9 of the constitution has spelled out the DRRM functions of all three levels of government—the federal, province, and local levels—with significant decentralization for decision-making, resources management, and service delivery systems. For disaster resilience, the constitution authority to local governments for "Disaster management" falls under Sect. 10.8: on the jurisdiction of local government. However, "Disaster management" also appears on the concurrent list for federal state and local jurisdictions. Low capacity and unfamiliarity with disaster management approaches and principles, particularly in municipalities that have recently graduated to urban municipalities, will likely hamper the ability of local governments to prepare for, reduce risks and respond to disaster events.

10.1.3 Overview of Risk Sensitive Land Use Planning and Its Implications to the New Federal Structure

The 2015 earthquakes and the widespread flooding in 2017 resulted in large-scale loss of life, livelihood, and property. Not only earthquake and floods, but the country has also been facing multiple hazards like windstorm in Tarai and landslides in hilly regions each year. To reduce the risk of such disasters; risk reduction needs to be addressed in land use plan and should be enforced and implemented.

The DRRM Act, 2015 and the National Strategic Action Plan for Disaster Risk Reduction 2018–2030 (Source MoHA) require every local government to prepare their local DRRM Law and action plans. With the support of federal government, the ministry (Ministry of Federal Affairs and General Administration and Ministry of Urban Development), along with the self-initiation of Local government, nearly all 753 local governments are in the process of preparing their local DRRM Law[1] and strategic action plans. With the circular from the federal government, it has made compulsory to local government to prepare such integrated urban development plans.

The newly elected municipal representatives in 2017 and the municipal staff often have limited understanding on DRRM concepts, technical capacities, and finances to undertake disaster risk reduction actions and knowledge to include risk reduction in existing planning and management practices at the local level. As a consequence, disaster risk reduction is often undertaken in a reactive manner following a major disaster. Land use planning however provides an opportunity for proactive risk reduction through location-based and/or structural approaches in the form of comprehensive plans, zoning, and building regulations. If land use planning is formulated based on systematic processes of risk assessment, implemented diligently and monitored continuously, it has the potential to reduce existing, future, and residual risks. In Nepali context RSLUP is prepared as a separate planning document that has been mostly financed by development partners has weak chance of implementation as

[1] A model DRRM Law drafted by the MOFAGA serves as a sample for local municipalities to prepared their local DRRM Law.

mentioned earlier. Overall three risk zones are identified, ranging from high-risk zone (red zone) to low-risk zone (green zone). Different development control strategies have been suggested to each zone.

The Gorkha earthquake in 2015 and the 2017 floods could have largely added to the learning and lessons from the disaster but ongoing isolated planning process related to fundamental development and resilience seems to indicate a completely different picture. The necessity to integrate such planning process is an essential aspect for the resilience urban development.

The impact of natural and man-made hazards can be reduced through adequate planning and incorporating specific DRR measures in land use planning and urban development processes, thus minimizing the human, economic, and environmental damage. By increasing the resilience of communities, DRR also improves the ability of people and infrastructure to recover from disasters. Land use planning can influence reduction of disaster risks through the development of strategic land use plan as well as an assessment of development applications on the basis of the adopted plan.

The RSLUP is a disaster risk reduction tool that helps identify high-risk areas in settlements, land use behaviors, zoning, land use functions, existing building regulation, and incorporate disaster risk management component in urban development plans. RSLUP advocates appropriate planning to ensure reduced vulnerability to the hazards associated with the disaster are taken into account. RSLUP is an integral component of DRR as it ensures locations for proposed development are safe, reducing vulnerabilities to disasters caused by natural hazards. RSLUP is a mechanism that integrates factors related to informed Disaster Risk Reduction (DRR) into land use planning. It is a planning document prepared by development agencies, which identifies the safest areas to prioritize immediate investments in urban development and infrastructure projects. However, experience to date shows that preparation of RSLUP under the auspices of different development organizations and some by municipalities lack implementation due to various reasons. One such reason could be attributed to the preparation of separate municipal plan (viz. periodic plan or Integrated Urban Development) by the municipalities.

10.2 Methodology

The methodology comprises of reviewing past experiences in preparing the eight RSULPs in Nepal along and drawing on global best practices. Similarly, an extensive review and interpretation of policy documents related to planning to make the process for different municipal plans with a special focus on disaster vulnerability reduction provision were carried out. These include DRRM Act, 2017; Local Governance Operations Act, 2017; National Urban Policy, 2007; National DRR Policy, 2017; Land Use Policy, 2015; National Urban Development Strategy 2017, Municipal periodic plan making directives, National Strategic Action Plan for DRR: 2018–2030; National Development Strategy 2017; Building Byelaws, 2006; Urban Planning & Development Act, Building Code for Nepal, 2017; Terms of Reference of Integrated Urban

Development Panning including RSLUP; Methodology of IUDP and RSLUP; and the Sendai Framework for Disaster Risk Reduction and the Urban Agenda, etc.

Interviews with select representatives of Nepali institutions and professionals along with consultations with selected municipal leaders have to be undertaken. For this, the process of RSLUP preparation was studied and interpreted and second, the quantitative method (corelational strategies) is used to understand the people's perception in areas for which RSLUP was prepared. Further, ongoing joint research on RSULP being undertaken by the Lead Author and one of the participating authors will also be used to validate findings. The authors will explore whether RSLUP prepared for urban settings in areas affected by the 2015 Nepal Earthquake have been used for subsequent reconstruction and recovery works. There has been a tremendous growth on the generation of data and information on hazards, micro-zonation assessment, GIS mapping including the use of open source tools and vulnerability information. State of these information will be collected and its relevance in future RSLUPs will be explored.

The sample RSLUPs include the RSLUP for Kathmandu Valley and other seven RSLUPs prepared in different regions and their thorough process was identified and studied. The detailed process of RSLUP viz. Development of Municipal Baseline Information, Multi-hazard Risk and Vulnerability Assessment, and Development of the Proposed Risk Sensitive Land use plan were analyzed. The risk assessment was detailed out and carried out step-by-step review that generates the necessary information for risk sensitive land use planning.

The authors interacted with municipal authorities, communities, private sectors, IT professionals, disaster professionals, planners, and relevant stakeholders. The finding also draws from the Disaster Preparedness Network's conducted on May 30, 2019.

10.3 History of RSLUP in Nepal

The RSLUP takes strategic approaches to reduce the exposure and vulnerabilities to risks through nonstructural approaches that identify the safest locations for prioritizing investments in urban and infrastructure development. While risk-based land use planning can inform urban spatial development plans, comprehensive risk reduction requires social and economic policies and programs that increase the capacity of the urban population to adapt to disaster risks. For example, hazard maps can improve the quality of zoning, but a good zoning map and related regulations may not be enforceable if the people affected cannot afford to retrofit or relocate to avoid the mapped disaster risks. The development of RSLUP has considered four-phased approach viz.

- Development of information and knowledge base needed for evidence-based decision support in planning
- Development of RSLUP and implementing tools and bye-laws

- Participation of stakeholders and concerned in collaborative decision-making; and
- Capacity building and mainstreaming.

The need for preparedness plan, program, and projects to reduce the loss of lives and properties has already been realized and acted upon. There has been an attempt to make different disaster-resilient plans, among which, RSLUP technique is one to help municipal governments in reducing risk. RSLUP has been prepared for Kathmandu Metropolitan City in 2010; RSLUP only for earthquake risks in five municipalities were prepared in 2009; Multi-Hazard RSLUP of Kathmandu Valley (three districts and 22 municipalities) in 2015; Chautara Municipality in 2016; and Changunarayan Municipality in 2017. All RSLUPs were prepared to take the existing conditions to contribute to increased vulnerability of communities, compounding the possible disruption of various functions and destruction of physical assets.

Nepal's historic settlements both within Kathmandu valley and outside resembled elements of planning and sensitivity to disaster risk. Sankhu, for example, is one of the oldest settlements in the Kathmandu Valley, established by the Licchavi Kings. The oldest inscription found in Sankhu is dated 538 AD. Like any other ancient town in the Kathmandu Valley, Sankhu is a compact settlement away from the flood plains of Saalinadi. Aside from the historic necessity for defense and the need for proximity to cultivated farmlands, the compact form of vertically oriented living in Sankhu—or any Newar town for that matter is strongly motivated by concerns to preserve rich farmlands and minimize their use for residential purposes. Similar trend can be observed in Bungamati, Khokana, Lubhu, Kathmandu, Patan, and Kirtipur where settlements have been planned considering risk factors. It is only during recent years, urban settlement has seen a rampant growth and in most cases devoid any planning including for disaster risk. Only seven towns and Kathmandu Valley has experienced the preparation of RSLUPs in Nepal. The following sections provide a summary of RSLUPs prepared.

10.4 RSLUP for Kathmandu Valley

10.4.1 2002 Seismic Risk Assessment in Kathmandu Valley GIZ, EMI, Kathmandu Valley, 2010

The GoN carried out the study on Earthquake Disaster Mitigation in the Kathmandu Valley in 2002 with the support from JICA. The Kathmandu Risk Sensitive Land Use Planning, 2010 is one of two pilot applications of mainstreaming disaster risk reduction in megacities project. It aims to mainstream disaster risk reduction in the development strategies and legislation of Kathmandu Metropolitan City (KMC) through the development of a risk-sensitive land use plan, advocacy campaigns, policy formulation, interinstitutional coordination and strengthening of local institutions. It fully integrates disaster risk reduction within its spatial and physical development strategies including regulatory and nonregulatory planning tools, bye-laws, regulations, and procedures. The program is an initiative of Earthquakes and Megacities Initiative (EMI), German Committee for Disaster Reduction (DKKV), Kathmandu Metropolitan City, Kathmandu Valley Town Planning Committee, and National Society for Earthquake Technology.

10.4.2 RSLUP of Kathmandu Valley UNDP 2015–2016, Kathmandu Valley

The hazard, vulnerability, and risk information are fully integrated into the latest RSLUP of Kathmandu Valley 2015. UNDP's Comprehensive Disaster Risk Management Program supported this initiative. It intended to serve as driving parameters in building the vision, strategies, programs, project, and activities contained therein. It should be the benchmark document that hopes to fill an important gap in directing and controlling sensible development in intended cities. The RSLUP of KV is only approved by Kathmandu Valley Development Authority and sent to ministry council for endorsement. Still, it has not endorsed formally. Other RSLUPs are only in the report format and no further steps have initiated to implement by the implementing agency. RSLUP document is important and should be referred by planners, policymakers, and the government authorities for urban development.

10.4.3 Assessment of Earthquake Disaster Risk for the Kathmandu Valley in Nepal, 2018

The Project for Assessment of Earthquake Disaster Risk for the Kathmandu Valley in Nepal was supported by Japan International Cooperation Agency including consulting companies from Japan in April 2018. Comparing the high risk of a future earthquake in Kathmandu Valley, countermeasures such as the retrofitting of buildings for seismic resistance, land use control, and observance of the National Building Code have not been promoted enough. It becomes a necessary and urgent issue to update the risk assessment for future development plans and policies concerned with disaster risk management. In this circumstance, the project was created with the main components of seismic hazard assessment, seismic risk assessment, and formulation of local disaster and climate resilience plans. The seismic hazard and risk assessment were carried out for the whole Kathmandu Valley and pilot activities targeted on three pilot municipalities: Lalitpur metropolitan city, Bhaktapur municipality, and Budhanilkantha municipality.

Seismic risk assessment was carried out for the purpose of providing basic information for the formulation of disaster risk reduction and management plans of pilot municipalities, as the update of the risk assessment results of 2002 JICA project of The Study on Earthquake Disaster Mitigation in the Kathmandu Valley. It was performed based on the latest situation of buildings, population, etc., and taking into account the new research results on risk assessment method as well as the characteristics of ground motion, building damage, and human casualties caused by the Gorkha Earthquake. The hazard and risk assessment of this project was targeted only for earthquake as the most prioritized disaster. However earthquakes are not the only target disaster in the pilot municipalities considering past disasters and possible disasters, which might occur in the future.

This project, aiming at earthquake disaster risk reduction, which was started immediately after the Gorkha earthquake, has achieved all of its goals for seismic hazard assessment, seismic risk assessment and formulation of BBB plans, LDCRP and SOP for three pilot municipalities, and implementation of CBDRRM activities, which contributed considerably to the commitment of Nepal on the enhancement of DRR and the response to the priority for action and global target of the Sendai Framework for Disaster Risk Reduction (Sendai Framework). It must be understood that the seismic hazard and risk assessment are carried out based on limited data and recognize that there is a long way to go to implement the LDCRP developed by the project. The further actions for mainstreaming seismic disaster risk reduction must be implemented in a planned manner.

The project contributes to the Sendai Framework and recommendations to achieve disaster risk reduction and increase resilience in Kathmandu Valley. And last, a roadmap is proposed for the purpose of concrete disaster risk reduction. It is highly expected that all of the project, aiming at earthquake disaster risk reduction, which was started immediately after the Gorkha earthquake, has achieved all of its goals for seismic hazard assessment, seismic risk assessment and formulation of BBB plans, LDCRP and SOP for three pilot municipalities, and implementation of CBDRRM activities, which contributed considerably to the commitment of Nepal on the enhancement of DRR and the response to the priority for action and global target of the Sendai Framework for Disaster Risk Reduction (Sendai Framework). It must be understood that the seismic hazard and risk assessment are carried out based on limited data and recognize that there is a long way to go to implement the LDCRP developed by the project. The further actions for mainstreaming seismic disaster risk reduction must be implemented in a planned manner.

10.4.4 RSLUP of Changunarayan Municipality

RSLUP for Changunarayan municipality is a product of DCA funded innovation project Promoting new Resilience Approaches in Younger and Smaller Municipalities (PRAYAS). PRAYAS augmented on the urban Disaster Risk Reduction (DRR) project PRAGATI funded by European Union through Civil Protection and Humanitarian Aid (ECHO), which was implemented in Changunarayan municipality of Bhaktapur district.

This approach for Risk Sensitive Land Use Planning is relatively new in the DRR and urban planning sector in Nepal. Disaster Risk Reduction in Megacities—Applications in Metro Manila and Kathmandu is only known significant project that had developed a risk-sensitive land use plan for Kathmandu City. The team has stepped further by initiating RSLUP in the municipality level to strengthen DRR on the local level and capacitate the local government. The adoption of a comprehensive approach comprised of 3 essential components for the RSLUP development built based on its principle: (a) Development of municipal baseline information (b) Multi hazard risk and Vulnerability Assessment (c) Development of Proposed Risk Sensitive Land Use Plan.

RSLUP for Changunarayan municipality was developed based on the approach which constituted of a database with baseline information and subsequent resource mapping and then engaging elected representative and relevant departments of the local government in different consultative meetings and participatory planning works. Communities were involved in every step of the process through KII, FGDs, and interactive workshops for their historical intelligence, current scenario, and perception of the risk evaluating causal factors. The needs of the communities were also instigated in the process with an inclusive approach where the vulnerable people (elderly, women, PwDs, Poor, and insecure households) were part of the study.

10.5 RSLUP of Municipalities Outside of Kathmandu Valley

10.5.1 Batch of Five Municipalities Supported by UNDP/BCPR

The UNDP/BCPR with the support of the Government of Japan initiated an Earthquake Risk Reduction and Recovery Preparedness (ERRRP) program in five high risk south Asian countries: Nepal, Bhutan, Bangladesh, India, and Pakistan. The project in Nepal was a part of the five-country regional program in South Asia that seeks to support regional cooperation through knowledge sharing and development of best practices in the context of the SAARC Comprehensive Framework on Disaster Management agreed in February 2006 (UNDP). The country program for Nepal was implemented through the National Implementing (NIM) project of UNDP-Nepal under project award UNDP/ERRRP-Project Nep/07/010 under a multiyear funding framework for natural disaster reduction from the Government of Japan.

The 5 RSLUPs were prepared in selected five urban municipal areas in the then five different development regions of Nepal. These municipalities are Biratnagar Metropolitan City, Province 1; Hetauda Sub Metropolitan City, Province 3; Pokhara Metropolitan City, Gandaki Province; Birendranagar Municipality, Karnali Province and Dhangadhi Sub Metropolitan City, Sudur Paschim Province.[2] The RSLUP process includes study for assessing the earthquake hazard, risk, and vulnerability in the program municipalities and to develop a framework for earthquake disaster preparedness planning. The program has also aimed to study the land use trends and impact of earthquake hazard in the settlement pattern and land use practice. With these aims, the program has aimed to prepare "Municipal Profiles in terms of Earthquake Risk and Vulnerability" and "Impact of Settlement Pattern, Land Use Practice and Options in High Risk Areas" for its program municipalities.

The report "Impact of Settlement Pattern, Land Use Practice and Options in High Risk Areas" was prepared as a land use profile in terms of earthquake hazard. The report was prepared with a rationale to give an overview of urban development and land use change pattern and risk of development in hazard prone areas. This study assessed the land use change pattern and models the future land use to assess the risk in a scenario of such land use change and earthquake hazard. Based on this scenario study, risk sensitive land use plan was recommended for municipal land use policy formulation and implementation of building regulations and byelaws. The study for the preparation of the land use report undertook the following tasks:

- Review of the existing documents and literatures and best practices
- Review of literatures and documents on regional geology and past earthquake scenarios
- Discussion and interaction with stakeholder municipality, key informants, and local senior citizens
- Historic land use pattern mapping and urban morphology land use dynamic assessment
- Mapping of current land use and trend assessment
- Modeling of land use change and settlement pattern and prediction of future land use
- Assessment of earthquake hazard in land use and settlement pattern
- Recommendations for risk sensitive land use development plan.

This report was expected to serve as a road map for land use planning and formulation of building bye-laws at the municipal levels and provide guidelines to integrate risk sensitive land use plans in municipal level development and periodic plans.

10.5.2 Risk Sensitive Land Use Plan of Chautara Municipality, 2016

With the support of United Nations Development Program/Comprehensive Disaster Risk Management Program (UNDP/CDRMP), the development of Risk Sensitive

[2]Each of the above five urban municipalities represented major urban settlements in the then five development regions.

Land Use Plan (RSLUP) and update Building Bye-laws of Chautara Municipality in Sindhupalchowk District was carried out by the team of GENESIS Consultancy (P) Ltd. and Welink Consultants (P) Ltd. in 2016.

The development of RSLUP in Chautara Municipality has evolved from generation of relevant information, multi-hazard spatial modeling, vulnerability assessment, adaptation of best practices, guided by the "Guideline for Municipal Level Risk Sensitive Land Use Plan"—drafted by the MoFALD. The development of RSLUP in Chautara Municipality was undertaken in a four-phase approach, outlined below:

- Development of information and knowledge base needed for evidence-based decision support in planning
- Development of RSLUP and implementing tools and bye-laws
- Participation of stakeholders and concerned in collaborative decision-making; and
- Development of stakeholder's engagement plan and information, education, communication action plan for implementation of municipal RSLUP.

These plans will be supported by Implementation Action Plan and Institutionalization Guideline, Stakeholder Engagement Plan and IEC Action Plan to enable the municipality and other stakeholders to initiate implementations of the RSLUP and the bye-laws documents.

Many organizations are facilitating efforts to carry out the risk assessments at different level from ward to municipal. Practical Action supported on Pokhara Flood Risk Assessment; Oxfam in Birendranagar; British Red Cross and NRCS in Dhangadhi; Mercy Corps in Dadeldhura and Doti.

10.6 Ongoing Federal and Local Level Initiatives

10.6.1 Resettlement of High-Risk Communities

The Ministry of Urban Development has identified 24 high-disaster risk settlements that require resettlement to safer locations. The resettlement plan follows the GON's Integrated Urban Development Plan concept developed in 2015. The plan intends to resettle scattered high-risk settlement into relatively larger integrated settlements. The Ministry is currently implementing resettlement programs in Rasuwa and Myagdi districts. Other high-risk settlements identified by the Ministry are, in Province 1: Gaurijanj, Jhapa; Chuhandanda, Terathum; Jarayotar, Bhojpur and Khumjung Valley of Solukhumbu. In Province 2: Santapu Matiun and Basantapur, Rautahat and Beltar, Udayapur. In Province 3: Betrawati, Nuwakot; Mude, Sindhupalchowk and Gajuri, Dhading. In Province 4: Chormara, Nawalparasi; Bhachhek Bazaar, Gorkha; Syalpakhe, Syangja, and Humde of Manang. In Province 5; Mainapokhari, Bardiya; Bhalubang, Dang; Supal Bazaar, Pyuthan; Sulichaur, Rolpa. In Karnali Province: Chupra, Dailekh and Gamdadi, Mugu. Similarly, in Sudur Paschim Province the high-risk settlements that require resettlement are Beldandi,

Kanchanpur; Binayak, Accham; Bagargaun, Bhajang; Khodpe Melauli, Baitadi and Gokuleshwor, Darchula.

Integrated settlements to resettle high-risk communities are currently being implemented at Rampur, Okhaldhunga; Bhanga Bhawanipur, Rautahat; Kapilakot of Sindhuli; Balewa of Baglung; Madichaur of Rolpa; Gamtha of Mugu; Chaudhary of Bhajang and in Bajura. A study carried out by National Planning Commission has identified that there are high number of settlements residing in high-risk areas in Bajuara District.

10.6.2 New Town Development Projects by KVDA

The Kathmandu Valley Development Authority (KVDA) is implementing the New Town Development Project covering the 10 million Ropanies of land within Kathmandu Valley. KVDA's mission is to develop "Kathmandu Valley as a Safe, Clean, Organized, Prosperous and Elegant (SCOPE) National Capital Region," so as to foster the global image of Kathmandu Valley as a "livable city with the synergy and harmonization of nature, society and culture."

10.6.3 Integrated Urban Development Plan (IUDP)

In Nepal, municipal planning is largely based on the land use and infrastructure planning in different forms of urban plans from a physical development plan to recent periodic plans. The recently started Integrated Urban Development Plan (IUDP) of 185 municipalities by GON also sought for the integration of DRR concerns in its thematic plan. An integrated urban development plan comprises a system of interlinked actions, which seeks to bring a lasting improvement in the economic, physical, social, and environmental conditions of a city or an area within the city. The key to the process is "integration," meaning that all policies, projects, and proposals are considered in relation to one another. In this regard, the synergies between the elements of the plan should be such that the impact of the plan as a whole adds up to more than the sum of the individual parts if implemented in isolation.

The plan is set out to Long-term Vision and overall Goal, Objectives, and Strategies for Municipalities for 15 years period. The plan also includes the Physical development plan, Social, Cultural, Economic, Financial, and Institutional Development Plan; Environment Sensitive Plan, Urban Transportation Plan, Municipal Transport Master Plan (MTMP), Multi-Sectoral Investment Plan (MSIP), etc., on the basis of vision, objectives, and strategies. This will help to prepare building bye-laws to regulate development in the municipalities integrating Land Use and road network plan and long-term vision of the municipalities. It will also prepare Detail Engineering Design of prioritized 2 subprojects in each municipality amounting not less than 30 Crore for each project.

10.6.4 Smart Cities

GoN has initiated preparing the strategic development master plan for 13 municipalities as smart cities for planned urban development. The report covers the concept of the smart city with necessary features/attributes in the Nepalese context both for national and local level. This will conduct city level assessment by reviewing previous plans (physical development plan, local area plan, profile, etc.) of the study area in the local, regional, and national context. The plan will delineate the area and master plan for the possible development of the smart city.

10.6.5 Classification of Land

Nepal's Land Use Policy, 2015 classifies land into eleven different types. The Department of Survey is in the process of land use mapping for all 77 districts in Nepal. Land use map of 355 rural and urban municipalities in 30 districts has been prepared till date. The report and database of maps has been handed over to respective municipalities in 27 districts. In addition, land use mapping implementation support program is piloted in seven municipalities of seven provinces.[3] The Land Use Policy identifies land pooling as a key priority. Land pooling or land readjustment is a major growth strategy adopted for urban development in Nepal. The land pooling (LP) projects help minimize premature conversion of farmlands and control urban sprawl. It is an effective tool to provide housing, infrastructure, and services in an efficient manner. The many land pooling projects such as Ichangunarayan LP Project, Bagamati Nagar LP Project, Dibyaswori—Thimi LP Project, Kamerotar LP Project, Byasi-Bhaktapur industrial state LP project, Nayabazar LP Project, Chamati LP Project, and Manaohara LP Project are already implemented over the time.

10.6.6 Interaction with Mayors and Deputy Mayors

The telephonic and in-person discussion was carried out with each municipal mayors and or deputy mayors of those municipalities where RSLUPs were developed. An assessment carried out by International Office of Migration in 2019 showed that the concept of "risk assessment" and Risk Sensitive Land Use Planning (RSLUP) is like déjà vu to all municipalities. The risk assessments have not been undertaken even in high earthquake and flood risk municipalities in Nepal. The hazard assessment for some wards have been carried out and completed, but results are not available within the municipality for subsequent update or use for sectoral assessments. The disaggregated data on gender, caste, persons with disabilities (PWDs), elderly, and

[3]As presented by Mr. Sumeer Koirala, Department of Survey, GON on 30 May 2019 during the Risk Sensitive Land Use Planning Workshop.

children exist in all municipalities. However, there is limited practice and knowledge in integrating them into development planning including DRRM. As per the municipal engineers, they had made request to UNDP and other International Organizations to provide technical and financial support in the preparation of RSLUP.

10.7 Implementation of RSLUP

The development of the RSLUP received the backing and support of officials from the municipalities, as well as from other ministries and agencies of the Government of Nepal (GoN). The RSLUP builds on previous and existing land use plans, land use maps, and land use related programs of the government, as well as policies, initiatives, and studies in disaster risk management (DRM) that effect in different municipalities. The RSLUP was prepared to align with the higher-level physical framework plans but never endorsed it and took the appropriate steps for its adoption and implementation through various government agencies. However, the Planners and the implementing agencies are weak to engage the wider stakeholder's participation and collaboration into the process of preparation, adoption, implementation, and enforcement to strengthen the legal and institutional frameworks.

Eight RSLUPs are prepared to date but the very process of RSLUP, which involves public participation as a ritual in the process of preparation, has been the center of the debate. RSLUPs made to date have hardly made it to the public sector's development decisions. Such limitation of RSLUP prepared, among others, is conspicuously prominent in case of all-important city bylaws, which largely fails to integrate the risk reduction strategies. This study thus aims to contribute to the current gaps in the planning process of RSLUP in terms of real participation of stakeholders in building urban disaster resilience.

The RSLUP of KV is only approved by Kathmandu Valley Development Authority and sent to ministry council for endorsement. Still, it has not endorsed formally. Other RSLUPs are only in the report format and no further steps have initiated to implement by the implementing agency. RSLUP document is important and should be referred by planners, policymakers, and the government authorities for urban development. But none of these RSLUPs are implemented in single municipality and shortcomings are mainly the public participation and lack disseminating the information to the community. The building bye-laws of each municipality should be prepared based on the RSLUP document. Currently, Department of Urban Development and Building Construction (DUDBC) is preparing very generic building bye-laws without analyzing the hazard and risk information. This is the implementable document at the municipality level. The physical boundary of the municipality (Chautara and other) was changed or increased after the development of RSLUP. Municipal authorities are confused and very weak on making decisions to move forward.

The absence of local representatives at the municipal level weakens the institutional system due to frequent transfer of the government officials from one to another. The knowledge base is transferred along with the officials. So, it is difficult to keep on advocating the importance of RSLUP. In short, there is no complete process and implementation mechanism tool for RSLUP.

RSLUP is an initiation of broader planning scenario and envisioning of the future urban form to provide the basic right of any citizen, i.e., safer place to live and prosper. RSLUP's recommendations and guidelines should be implemented through a legal document "building bye-laws" by the municipal authority. The building bye-law must be implementable, acceptable, and inclusive to translate the future safer growth vision of the municipal region. RSLUP's recommendations and guidelines can be implemented through various mechanisms and tools related to urban development, land management, financial incentivizing such as land readjustment (land pooling), transfer of development rights, infill development, building pooling, etc. At the same time, institutional capacity building of the municipal authority is required to implement the RSLUP and its implementing tools efficiently and effectively. The integration of RSLUP to strengthen urban governance is the utmost importance.

10.8 Application of RSLUP in Selected Municipality and Key Learning

Initially, the development of RSLUPs received support from officials of the municipalities, as well as from other ministries and agencies of the Government of Nepal (GoN). The RSLUPs were prepared to align with higher-level physical framework plans but were never endorsed and for this reason there were no subsequent steps for its adoption and implementation by various government agencies. The implementing agencies are weak to engage the wider stakeholder's participation and collaborate during the process of preparation, adoption, implementation, and enforcement to strengthen the legal and institutional frameworks. Till date, eight RSLUPs have been prepared for urban settings in Nepal. Past experience has shown that the process of RSLUP and their implementation status is very much limited and has hardly made it to the public sector's development decisions. Such limitation of RSLUP prepared, among others, is conspicuously prominent in the case of city bye-laws that largely fails to integrate the risk reduction strategies.

No doubt, risk-sensitive land use planning has played an effective role in reducing disaster risks in Philippines and Bangladesh. Risk-sensitive land use planning seeks to adopt the best land use options and translate the assessment of risks into the appropriate location of land uses, functions, and facilities into land use regulations and policies. It is a potentially powerful mitigation tool as it seeks to mitigate the risks and vulnerability from several geological and other hazards, generally through physical measures such as strengthening structures to withstand earthquakes, limiting development in flood and landslide-prone areas, placing development away from geologically unstable areas.

The utility of RSLUP has immense possibility in an effective disaster risk reduction in Nepalese context, in the post-2015 earthquake. Methodologically integrating risk reduction strategies in land use planning will contribute in reducing the urban disaster risk via systematic development and application of DRR policies, strategies, and practices to enhance resilience throughout the society, by limiting or avoiding the hazard exposure, within the broad context of sustainable development, which in other word is Risk Sensitive Land Use Planning (RSLUP). Planning hierarchy is conceived for this Municipality as a linkage between regulatory framework of suitability zones defined by color zones, land use zones as defined under the national land use policy and the implementation units at the land parcels levels.

DUDBC/MoUD is in the process of preparation of RSLUP of Api municipality and other municipalities in far western part of Nepal. There is a major gap of any clear direction on how the municipalities would implement RSLUP, develop building byelaws, and issuing building permits. Interpretation of the overall planning framework is too complex a process for municipalities to take right decisions.

Risk-sensitive land use planning has played an effective role in generating an evidence-based approach for reducing disaster risks in high-risk countries such as in the Philippines and Bangladesh. RSLUP seeks to adopt the best land use options and translate the assessment of risks into the appropriate location of land uses, functions, and facilities and into land use regulations and policies. RSULPs process, if done correctly serves as a tool for disaster risk awareness and more importantly a tool to reduce actual risk while also enabling monitoring risk over time.

Chapter 11
Counting Down to Day Zero: Exploring Community-Based Water Management Strategies in Western Cape Province Drought, South Africa (2017/2018)

Clayton Hazvinei Vhumbunu

Abstract This research explored the various community-based water conservation and water management strategies employed in the face of the Western Cape Province Drought that occurred in the season period of 2017/2018. Specifically, the study analyzed the effectiveness of the water conservation strategies and assessed the implementation complexities of these strategies, as well as proffered interventions that can be adopted to ensure that water resources are optimally used as the countdown to Day Zero, set for 2020. In the process, findings of the study and analysis presented salient lessons to other countries in Africa, Asia, Latin America, North America, Middle East, and Europe that may be faced with similar natural disasters that threaten water supply and availability. The key concepts of *Community-based Water Management (CBWM)* and *Community-Based Disaster Risk Management (CBDRM) constituted the conceptual framework of analysis.* In terms of methodology, this study made use of questionnaires that were distributed to 29 purposively sampled respondents comprising officials in charge of water management and disaster management in the Western Cape Provincial Government, Western Cape district and local municipalities, and academics and/or researchers specializing in the four universities in Western Cape. From the gathered data, it was established that although there were notable challenges of resource constraints, conflicts of interests, infrastructure decay, information gaps, resistance to change, retrogressive community attitudes, policy inconsistences; community-based water management strategies assisted to complemented municipality-initiated efforts to reduce the risk of drought and mitigate the socio-economic impact of the natural disaster, thereby pushing further the envisaged Day Zero. The chapter recommends the strengthening of community-based water management strategies through the involvement of government and non-state actors as well as prioritization and alignment of community-based water management strategies with municipal, provincial, national and global frameworks and plans on disaster management and climate change adaptation to facilitate more robust coordination drought risk mitigation, responses, and preparedness.

C. H. Vhumbunu (✉)
School of Social Sciences at the University of KwaZulu-Natal (UKZN), 238 Mazisi Kunene Road, Durban, South Africa
e-mail: Vhumbunuc@ukzn.ac.za; cvhumbunu@gmail.com

R. Djalante et al. (eds.), *Integrated Research on Disaster Risks*, Disaster Risk Reduction,
https://doi.org/10.1007/978-3-030-55563-4_11

Keywords Community-based water management strategies · Disaster management · Disaster risk reduction · Day zero · Western cape province

11.1 Introduction

The severe drought of 2017/2018 experienced in Western Cape Province, South Africa's fourth largest province situated in the South-Western coast of the country, which is third most populous with over 6 million inhabitants, brought tremendous damage to the province's economic activities while disrupting the lives and livelihoods of the people mostly through water shortages. The drought was so severe to the extent that the Western Cape Provincial Government had to declare a provincial state of disaster in accordance with Sect. 41(1) of the Disaster Management Act, 2002 (Act 57 of 2002). For this reason, 'Day Zero' was introduced in the provincial capital, City of Cape Town which hosts 4 million residents, in order to focus the attention of all residents on water consumption management, drought mitigation, and drought risk reduction. Day Zero, initially puts on 12 April, 2018, later moved to the end of April, May, July, and August, 2018, and then finally to 2019 and 2020; was an envisaged day when the City would run out of municipal water. While the Western Cape Drought of 2017/2018 remains very significant and instructive in presenting future lessons on how collaborative community-based strategies can be fundamental in mitigating and managing the risk of natural disasters, no in-depth applied research has been conducted to empirically explore the community-based water management strategies and practices that were adopted by the City of Cape Town and the rest of the Western Cape Province with a view to reduce and mitigate the drought risks and effects. In this chapter, the community-based water conservation and water management strategies and practices implemented in Western Cape Province within the framework of climate change adaption and disaster risk reduction, so as to push further the imminence of Day Zero in the province given the recurrence of droughts in Western Cape. Insights emerging from the study are critical in providing lessons for other drought-stricken communities, especially in developing countries in Africa, Asia, Latin America, and across the globe as countries implement their National and Regional Climate Change Adaptation Strategies as well as pursue the key priorities and targets of the Sendai Framework for Disaster Risk Reduction.

11.2 Contextualizing the Western Cape Drought Within the International Disaster Risk Reduction Governance Framework

The efforts to mitigate, manage, and reduce the risks of the Western Cape drought of 2017/2018 through collaborative community-based practices and strategies, coordinated by the Provincial Government in collaboration with the City of Cape Town and the other five district municipalities, namely Cape Winelands District Municipality, Central Karoo District Municipality, Garden Route District Municipality, Overberg District Municipality, and West Coast District Municipality together with all the 24 local municipalities, can only be well fully explored and understood within the proper sub-national, national, regional, and global contexts.

It is a fact that South Africa faces increasingly high levels of natural disaster risks, which have been intensifying in recent years largely as a result of climate change and dynamics of global weather systems, hence the frequency of extreme hydrometeorological hazards. As a result, droughts, veld and urban fires, floods, tropical storms, and tropical cyclones have all been common across different parts of South Africa and the wider region. In March and April 2019 alone, the country has been affected by the effects of Cyclone Idai and Cyclone Kenneth mostly in KwaZulu-Natal (KZN) coastal areas, while Gauteng and Cape Town experienced severe floods in January 2019.

Faced with such exposure to frequent and recurrent weather hazards, climate-related disasters, and other natural disaster risks, which aggravate the vulnerability of citizens, the need for a regular review of comprehensive disaster preparedness, disaster risk management, and disaster risk reduction plans is essential, while the alignment of such interventions with global disaster risk reduction framework is equally indispensable as it is obligatory. At a local level, South Africa's Municipal Systems Act 32 of 2000 obliges every municipality to have a Disaster Management Plan or a disaster management strategic action as part of its Integrated Development Plan (IDP) and most of the municipalities have these in place. At the national level, the White Paper on Disaster Management of 1999 provides policy, the legislative and institutional framework for disaster risk management, while the Disaster Management Act of 2002, supported by the National Disaster Management Framework (NDMF) gazetted in April 2005 and later by the National Disaster Management Guidelines of May 2008, assists to provide an integrated, uniform, and coordinated approach to disaster prevention, disaster preparedness, disaster management, disaster mitigation, disaster risk reduction, and post-disaster recovery and rehabilitation. Thus the legislation provides for the establishment of Disaster Management Centers (DMCs) and Disaster Risk Management Advisory Forums (DRMAFs) at national, provincial, and local government levels, with the Inter-Governmental Relations Framework Act of 2005 providing the legal basis for cooperation and collaboration between and among the three spheres of government in disaster management.

The South African policy and legal framework disaster management and disaster risk reduction is aligned to the Southern African Development Community (SADC) Regional Disaster Preparedness and Response Mechanism, the SADC Regional Disaster Risk Reduction and Resilience Programme, the SADC Regional Indicative Strategic Development Plan, and the Africa Regional Strategy for Disaster Risk Reduction (ARSDRR). However, with climate-related disasters now by far the most frequent natural disasters globally, it is imperative to view disaster management efforts within the framework of climate change policy frameworks, including the Western Cape Drought of 2017/2018.

The Draft National Climate Change Adaptation Strategy (NCCAS) of South Africa, awaiting finalization and adoption, outline the national vision for addressing climate change and the country's transition to a climate-resilient economy while the National Development Plan (NDP) also mentions the importance of measures and policies key in disaster management and environmental sustainability. This is also aligned not only to the Common Market for Eastern and Southern Africa (COMESA)-East African Community (EAC)-SADC Tripartite Programme on Climate Change Adaptation and Mitigation, and the African Common Position on Climate Change which all aim at addressing the impacts of climate change through effective adaptation and mitigation interventions in order not only to build social and economic resilience in communities but also to the global disaster management frameworks and climate change adaptation regimes such as the Sendai Framework for Disaster Risk Reduction (2015–2030) which is a key policy framework on disaster risk reduction as well as the 2030 Agenda for Sustainable Development that stresses the need to reduce disaster risks.

It is against this background that the key research questions underpinning this chapter specifically sought to understand the various challenges, complexities, and dynamics of drought mitigation, water conservation, and water management adopted and utilized in Western Cape in the wake of Day Zero in order to mitigate and reduce the effects of the drought disaster on the social and economic livelihoods of communities in the province. The significance of such an inquiry is that it is relevant to the Sendai Framework for Disaster Risk Reduction (2015–2030)'s priority of "understanding disaster risk(s)" and the goal of "preventing new and reducing existing disaster risks through the implementation of integrated and inclusive…measures that prevent and reduce hazard exposure and vulnerability to disaster…" (United Nations 2015: 12). The lessons that emerged from this research on Western Cape communities, disaster management institutions, municipalities, and other stakeholders are key in pursuit of Sendai Framework for Disaster Risk Reduction (2015–2030)'s priorities of "strengthening disaster risk governance to management risk" and "investing in disaster risk reduction for resilience" as well as "enhancing disaster preparedness for effective response" all in pursuit of the global targets to reduce the number of people affected, and disruption posed to basic services globally, by disasters (United Nations 2015: 14). Further, the attainment of SDGs Goal 2 on ending hunger, achieving food security, and promoting sustainable agriculture; Goal 6 on ensuring availability and sustainable management of water and sanitation; Goal 11 on making cities and human settlements inclusive, safe, resilient, and sustainable; Goal 14 on conserving and sustainably using oceans, seas, and marine resources; and Goal 15 promoting sustainable use of terrestrial ecosystems are all linked to the research aims and objectives of the study as insights emerging from the study provide lessons for other drought-stricken communities, especially in developing countries in Africa, Asia, Latin America, and across the globe as countries as pursuing the key priorities and targets of both the Sendai Framework and SDGs as well as other related frameworks such as the New Urban Agenda.

11.3 Conceptual Framework

The two key concepts of *Community-based Water Management (CBWM* and *Community-Based Disaster Risk Management (CBDRM)* constitute the conceptual lenses and frames of analysis for this research. Whereas water resource management pertains to the optimum planning, developing, distributing, and managing of resources, *Community-based Water Management* relates to the inclusive local level involvement of communities in the management of their water resources in a participatory approach to ensure water supply security, reliability, and accessibility by all members of the community in terms of quality and quantity against competing demands, greater extremes, uncertainties, and water scarcity (Kobayashi et al 2014; World Bank 2017). Thus, the goal of CBWM is to ensure that all community residents manage to secure adequate water supply for drinking, sanitation, household use, energy generation, agriculture, water-based recreation, and other uses (World Bank 2017).

While traditionally CBWM has been popularized as a model common in rural communities since the government and other state institutions have very minimum control of water supply and management, the practice has now spread to cities and towns due to the diversification of water sources as well as disasters such as droughts. Authors such as Tantoh and Simatele (2017) have argued that while Community-based Water Management, itself a form of community-based natural resource management, may be viewed as an effective resource management strategy, in some instances local communities may lack the ability to manage water resources in an inclusive and sustainable manner thereby resulting in intra-community struggles over control of water resources. This has become imperative in light of drought recurrence across the globe, with the World Bank (2017) noting that water scarcity affects over 40% of the global population.

Bruns (2008), *as cited in* Day (2009: 52), *has argued that community-based water resource management is advantageous in that local water users often possess indigenous knowledge on water resources, water needs, and historical change that has occurred related to water use. This assists in the prudent implementation of community-based water resource management strategies. To Bruns* (2008), *the fact that water users recognize the importance of water in their subsistence-based livelihoods facilitates relationship-building between and among water users. Another merit of community-based water resource management is that communities can easily monitor their agreed water usage arrangements every day and oftentimes they have historical mechanisms for participatory conflict management and resolution evolve and adapt to global challenges (Bruns in* Day 2009*: 52). However,* Day (2009)*also acknowledges that the concept of community-based water resource management can be made difficult to implement given the non-homogenous nature of communities as societal members disparities in wealth, knowledge and power often leads to conflicts over interests, needs, priorities, views, and perspectives that may be very difficult to reconcile. It is not uncommon community-based water resource management strategies to be frustrated by contestations over access to water resources, or have their* *own traditional community-based water resource management systems being threatened by undue external interference and influence from political elites, Government institutions, Non-Governmental Organizations (NGOs), et cetera.*

Community-based Water Management is conceptually linked and related to the concept of Community-Based Disaster Risk Management (CBDRM). CBDRM has been used interchangeably with related concepts such as community-driven disaster risk reduction (CBDRR), community-based disaster management (CBDM), community-driven disaster risk management (CDDRR), community-based disaster preparedness (CBDP), community-managed disaster risk reduction (CMDRR) (see Lassa et al 2018). *It has to be noted that, stricto sensu, these concepts may differ, in the same manner,* disaster risk management, disaster risk reduction, and disaster preparedness both conceptually and operationally. *CBDRM promotes the involvement of potentially affected communities in disaster risk management at the local level by building their capacities to assess their vulnerability to natural disasters and develop strategies necessary to mitigate the impact of these disasters* (United Nations International Strategy for Disaster Risk Reduction 2009). *At the core of these concepts is the involvement of communities in making decisions and implementing disaster risk management strategies, actions, and initiatives.*

One of the most fundamental components of CBDRM is community contingency planning (CCP) which provides a holistic picture of disaster risk reduction by adopting an approach that integrates natural disaster elements of risk, hazard, and vulnerability (Surjan et al. 2016). *To* Yodmani (2001), *CBDRM has objectives of reducing vulnerabilities, increasing capacities of disaster-struck communities to cope, minimizing loss and damage to people, property, and environment, and facilitating post-disaster recovery. As such, one has to understand that while CBDRM is conceptually 'local', operationally it involves intergovernmental collaboration and linkages for better delivery. Of late there has been more emphasis on the summoning of indigenous knowledge systems to complement non-indigenous knowledge as part of community-based strategies to mitigate the risk and impact natural disaster* (see, for instance, Dube and Munsaka 2018; Trogrlić et al. 2019; Balay-As et al. 2018; Haque 2019; Mugambiwa 2018; Iloka 2016).

Thus, within the conceptual matrix, Community-based Water Management is conceptually located within the ambit of the broader concept of *Community-Based Disaster Risk Management. The two make up a conceptual framework of elements that are keys in minimizing vulnerabilities and disaster risks in communities and dealing with adverse impacts of disasters. The assumption, and reality, as argued by* Enshassi and Shakalaih (2016), *is that communities are no longer passive actors with regards to disaster management. Rather, they now participate to complement efforts toward disaster risk management and disaster risk reduction. Effectively, in times of droughts,* Community-based Water Management becomes a function and sub-set of *Community-Based Disaster Risk Management. Of course, both CBWM and CBDRM will be guided by* water governance structures in instances of drought as these determine the management and usage of water, that is, the 'who gets what, when and how?' aspects of water resources in any community. Rogers and Hall (2003) aver that water governance involves mediating behavior through values, policies, laws, regulations, institutions, and actions pertaining to the allocation and regulation of water use and management.

11.4 Methodological Approach

This study made use of questionnaires that were distributed to the selected key informants/respondents between the period of June 2018 and April 2019. Purposive sampling was used to identify and select questionnaire respondents. The questionnaire respondents comprised officials in charge of water management and disaster management in the Western Cape Provincial Government, officials from the Community Development Services Departments or the Infrastructure, Planning and Development Department of Western Cape district municipalities and local municipalities; and academics and/or researchers specializing in local government management and/or public administration in Western Cape.

In selecting the district municipalities and local municipalities of focus, consideration was made that the Western Cape Province has thirty (30) municipalities distributed as follows:

- One (1) Metropolitan Municipality (City of Cape Town)
- Five (5) district municipalities (namely, Cape Winelands District Municipality, Central Karoo District Municipality, Garden Route District Municipality, Overberg District Municipality, and West Coast District Municipality), and
- Twenty-four (24) local municipalities as shown on the map in Fig. 11.1.

From the $n = 5$ district municipalities, $n = 15$[1] were randomly selected to constitute the sample for the study, using the representative formulae of $n = 3$ local municipalities per each district municipality on the basis of ensuring geographical balance. In addition, $n = 5$ respondents were selected to represent the five district municipalities of Cape Winelands District Municipality, Central Karoo District Municipality, Garden Route District Municipality, Overberg District Municipality, and West Coast District Municipality. As for academics and researchers, $n = 1$ academic was identified from each of the four universities in Western Cape, namely the University of Western Cape, the University of Stellenbosch, the University of Cape Town, and Cape Peninsula University of Technology (CPUT) using a purposive sampling technique whereby respondents specializing in local government management and/or public administration were selected respond to the questionnaires. In addition, there were $n = 2$ respondents from both the Western Cape Disaster Management Centre and the National Disaster Management Centre. Out of the total $n = 29$ questionnaires

[1]The 15 local municipalities selected were Breede Valley Local Municipality, Drakenstein Local Municipality, Langeberg Local Municipality, Beaufort West Local Municipality, Laingsburg Local Municipality, Prince Albert Local Municipality, Bitou Local Municipality, George Local Municipality, Hessequa Local Municipality, Cape Agulhas Local Municipality, Overstrand Local Municipality, Swellendam Local Municipality, Bergrivier Local Municipality, Cederberg Local Municipality, and Matzikama Local Municipality.

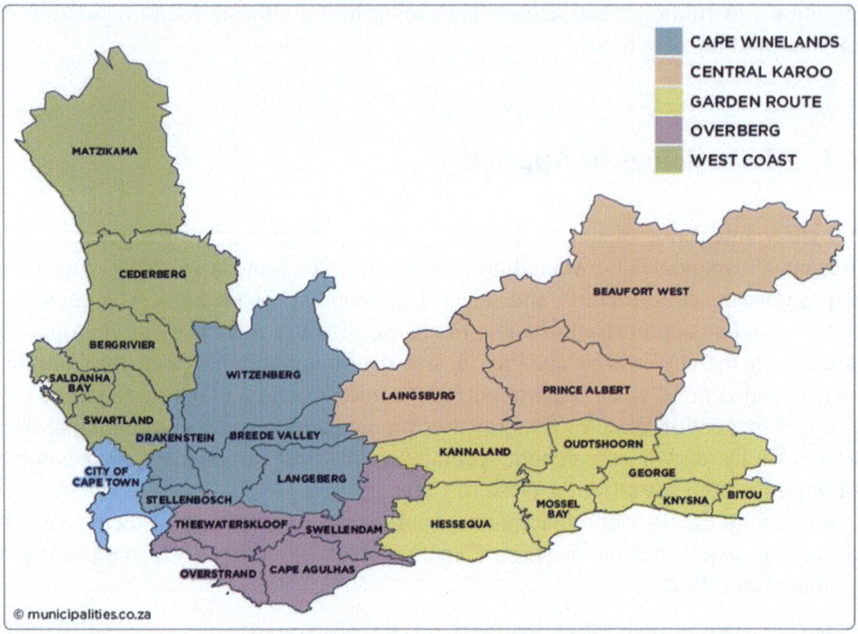

Fig. 11.1 Map of Western Cape Province. *Source* The Local Government Handbook: South Africa (Yes Media 2019)

distributed, $n = 24$ were returned, thereby constituting a response rate of 82,75% as depicted in the Table 11.1.

11.5 Results

From the data collected in the form of questionnaires responses secured from a total of 24 ($n = 24$) respondents, there were five key themes that emerged. The themes were related to the key questions asked to respondents with regard to the community-based water conservation and water management strategies and practices adopted by the City of Cape Town and the rest of the Western Cape Province with a view to reduce and mitigate the drought risks and effects as well as to push further the imminence of Day Zero in the province given the recurrence of droughts in Western Cape.

The themes revolved around the efficacy of the drought-instigated community-based water management strategies in Western Cape; challenges in implementing community-based water management strategies, and stakeholder perceptions on the effectiveness of community-based water management strategies, and possible suggestions and recommendations for sustainable water management and future drought mitigation in Western Cape Province. The results of the study are presented, analyzed, and discussed in the thematic sub-sections below.

11.5.1 Drought-Instigated Community-Based Water Management Strategies in Western Cape

Most of the responses (95%) were of the awareness of all the community-based water management strategies that were coordinated by the municipalities across the Western Cape Province. They converged that the municipalities in the Western Cape Province managed to promote community-based water management strategies to ensure the efficient use and reasonable access to water resources by citizens across the drought-stricken province.

In most cases, however, it is the municipalities that introduced water services infrastructure and the communities would devise strategies on how to manage these at a local level to allow for sustainability, equity in water use, and order. An official from the Western Cape Provincial Government noted;

> In the cities such as the City of Cape Town, our emphasis was on a variety of initiatives to manage the scarce water resources. Our intention was try by all means to push as far as possible the much dreaded Day Zero by reducing increasing demand and increasing water supply. In a drought situation this is not an easy task. We paid more attention to urban water demand management (through water-wise approaches, tough and restrictive water tariffs) and promoting the use of alternative water supplies (drilling of boreholes, managing communal water points and encouraging water re-use). These two strategies helped a lot because we managed to reduce water usage by half without creating a serious crisis of shortages. Desalination plants have also been set up. For example OSMOSUN Solar Powered Desalination Plant in Hessequa Local Municipality constructed with co-funding from France in December 2018, and it produces 150 kiloliters of water daily. Other desalination plants are Monwabisi Desalination Plant which produces 7 million liters per day as well as Strandfontein Desalination Plant, Quality Filtration Systems Desalination Plant and V & A Waterfront Desalination Plant. But these are a bit expensive to set up and operate though. On its part, the Central Government assisted with bulk water supply and storage.

The comments from the Western Cape Provincial Government resonate with most municipalities. However, each municipality has adopted its own strategy and approach to suit prevailing conditions and magnitude of water crisis depending on demand and supply constraints. In the City of Cape Town employed Day Zero Water Restrictions as a strategy. These were also adopted in several municipalities within their urban areas. What residents would do was to put in place mechanisms to ensure cooperation from residents themselves. A City of Cape Town official hinted;

> When we initially put Day Zero at 22 April 2018, and later 20 May 2018 water consumption per person was put at 87 L per day, targeting an overall city water consumption of 500 million liters per day. By then we had plans to abstract more groundwater from three aquifers at Cape Flats, Table Mountain and Atlantis together with the Water Re-use Projects at Zandvliet, Cape Flats, Macassar, Potsdam and Ahtlone in addition to spring water and river water projects at Newlands, Oranjezicht, and Lourensriver as part of our Cape Town City

> Water Augmentation initiatives. We also had plans to operationalize at least two of our four desalination plants and also proposed to impose a Drought Levy on residents. Because of the initial non-cooperative attitude despite the imposition of fines exceeding R5.000, we had to install water management devices (WMDs) as residents exceeded restriction levels. We also deployed pressure management technology to reduce water consumption. Residents staying around water points would organize themselves to protect these water sources and also put in place rules on quantities that one has to get per each chance. We saw this happening at the Newlands Spring along Kildare/Springs Ways where Riyaz Rawoot (a resident around Newlands) managed to put PVC [polyvinyl chloride] pipes to allow for more people to access water.

Indeed, The City of Cape Town continuously adjusted the Levels of water usage targets from Level 5 (in October 2018) to Level 3 restrictions (from December 2018) for residential, non-residential and commercial users in the city (City of Cape Town 2019a, b). These had different measures relating to water use for household purposes, swimming pools, personal and commercial car washes, park sprays, etc. The lower the restriction levels would mean lower tariffs (City of Cape Town 2019a, b). The community-based initiatives to reduce wastage of water resources and ensure adherence to the water usage restrictions may have contributed to the success of the overall efforts to push further Day Zero.

Different community engagement initiatives had to be employed to conscientize the public on the wisdom of embracing the various community-based water management strategies. Most municipalities made use of the various media platforms to disseminate such information. One respondent from Cape Winelands District Municipality confirmed;

> Community radio stations played a very important role. Strategies to save water only work when all residents work together, and they have to be convinced to do so. There were water management messages spread on a number of community radio stations in Western Cape such as *Bush Radio FM, Radio Namakwaland, Voice of the Cape, UCT Radio, Radio West Coast FM, Valley FM* and *Eden FM.* These and many more help people to appreciate the desperation of our situation and know the dangers of not cooperating to push Day Zero further. I think it worked in averting the prospects of Day Zero, despite a lot of criticism that we got from the general public especially on social medial platforms. What communities would do was to organize local area meetings through councilors that would monitor the implementation of the various water management measures.

The comments from the Cape Winelands District Municipality official were also re-echoed by an official from the City of Cape Town. He responded;

> The power of the media was felt. Countless times in town, locations and resting places we could hear information targeting residents to cooperate with the Metropolitan Municipality in implementing measures and strategies to manage water use. *KFM, Good Hope FM, Cape Talk, Radio Voice of the Cape, Umhlobo Wenene FM,* and others broadcasted messages and interviews to that effect which complemented SABC Radios, SABC TV and eNCA. The same was carried in newspapers such as Cape Times, Cape Argus, Daily Voice and Vukani newspapers. We had regular Water Reports that we would publish to update our residents. In March this year we also launched, together with UCT, the *Cape Town Drought Response Learning Initiatives Film* that reflected on the drought and told a vivid story of the drought. It was shown across town and the province.

An official from Drakenstein Local Municipality stated that the municipality introduced a number of water serving strategies which were complemented by community-based water management strategies. The official retorted;

> Drakenstein Local Municipality introduced water use restrictions which prohibited watering or irrigation of gardens, flowers, parks and sports fields and open spaces within certain times, control of water use in ablution facilities, registration of boreholes and waterwells in compliance with the Department of Water and Sanitation, washing and rinsing of vehicles on personal and commercial basis, prohibition of the use of ornamental water fountains, restrictions on filling up swimming pools, and prohibition of establishing new landscaping or sports fields. Residents associations managed to keep encouraging residents to adhere to water restrictions whilst watching misuse and wastages of water which assisted to entrench a water conservation culture in communities.

From the questionnaires, the apparent finding was that there were more grassroots initiated and bottom-up community-based water management strategies adopted and implemented in rural areas of Western Cape as compared to urban areas to manage the risks of the 2017/2018 drought. These included the management of water assets in communities, arrangements for the maintenance boreholes and hand pumps, This was evident in all the responses (100%) received from West Coast District Municipality, Garden Route District Municipality, and Overberg District Municipality where there are most of the rural areas of Western Cape are located. An official from West Coast District Municipality stated;

> When working with the rural communities of Ebenezer and Rietpoort, we were impressed by their own initiatives towards managing water at village level. In Ebenezer, which is situated about 40 km from Vredendal, community members agreed to make financial contributions that would be used to pay community security personnel to protect their boreholes as well as pay for transport expenses for the community representative to report any breakdown and maintenance requirements all the time to our Community Service and Technical Services Departments which are responsible for the maintenance and management of water drinking supply systems in both rural and urban areas within the municipality.

The same was re-iterated by Garden Route District Municipality respondent. In rural communities of Garden Route District Municipality, just as was the case in Genadendal and Suurbraak rural areas of Overberg District Municipality, a number of community-based water management strategies were operational as indicated by the respondent;

> Community-based water management strategies were prevalent in the rural area communities of Haarlem, Kranshoek, Friemersheim, Zoar and Slangrivier during the drought. They had their community wells that they would maintain collectively as villages. With the water table going down everyday, the communities would dig further together organized under their traditional leadership. They also agreed on the water usage mechanisms and restrictions. The same also happened for water wells used for gardening and irrigation. This brought some sense of ownership and responsibility.

The community-based water management strategies appeared to be more focused on addressing issues of protecting water sources from vandalism, wear and tear while also ensuring there is equity in water usage. There were cases of community-managed borehole systems in Beaufort West Local Municipality employed to allow for equity and sustained access to water within communities. One official from Beaufort West Local Municipality confided;

> In Beaufort West, the townships of Murraysburg, Kwa-Mandlenkosi, Merweville and Nelspoort were hard hit by drought which exposed over 50,000 residents to water stress. Our local dam, Leeu-Gamka Dam, along the Leeuw River completely dried up. People resorted to the 32 boreholes we could manage to deliver since we did not have much funds. Communities managed to get together and establish teams of youths and elders who would control these water points to avoid vandalism, ensure equitable water excess and maintain order when people queue to fill up their buckets. It worked well because there were very few negative reports. There were cases were communities would use their own resources to fence and secure these boreholes.

Thus, from the respondents, the main community-based water management strategies adopted related to the use and protection of water sources by community members in order to ensure equity, fairness, and sustainability in water usage against a background of looming Day Zero. What can be observed is that the community-based water management strategies were more organic, community-initiated, and bottom-up in the Western Cape Province rural areas than they were in urban areas were the Municipalities initiated most of the strategies and residents would complement these through community strategies for sustainability.

11.5.2 Challenges in Implementing Community-Based Water Management Strategies

From the responses, the community-based water management strategies implemented in Western Cape Province during the drought of 2017/2018 varied from one community to the other.

However, one of the most common challenges was conflicts relating to the use of water points or access to water sources. In Beaufort West Local Municipality, there were conflicts relating to clash over prioritization as some residents fetched water from boreholes for their animals while accusations of water theft were common. An official from Beaufort West Local Municipality confirmed;

> There was a reported cases of water theft and conflicts over water use in villages and communities mostly outside the main towns, especially those involved in agriculture, keeping goats, lamb and cattle. Whilst a number of farmers came from as far as Mpumalanga Province to deliver animal feed bales, there were arising conflicts on allocation of the animal feed. Some farmers and residents clashed as non-farmers felt it was unfair for the farmers to take huge quantities of water for their animals at the expense of human beings. We should understand communities are not homogenous entities, interests and priorities differ. Given this, conflicts are plenty.

The same challenges were also reported to have been experienced in different local municipalities that make up Cape Winelands District Municipality, Garden Route District Municipality, Central Karoo District Municipality, West Coast District

Municipality, and Overberg District Municipality. In Cape Winelands District Municipality, for example, there was stiff competition for water resources between community residents and farmers mostly involved in horticulture. An official from Cape Winelands District Municipality lamented;

> We were inundated with reports and cases of conflict over access and use of water involving residents and farmers that are into grape farming and citrus farming. Some farmers resorted to the dripping irrigation system but other could not. Out Drought Monitoring Committee met a lot of conflicts cases handled by Water User Associations in the district municipality. For example when the Misverstand Dam was almost drying up, and water was released from the Berg River Dam, farmers clashed with riverine communities when they disobeyed the Government's instruction not to pump water for irrigation. Likewise, farmers themselves sometimes had disagreements as some violated set rules for abstracting scarce water from the canal networks, pipelines and irrigation schemes in Stellenbosch Farms, Touws River, Robertson, Paarl, Ceres, Wellington, etc. A number of conflicts emanated from failure to adjust to the reality of rationed water whereas at times it was just a manifestation of retrogressive community attitudes driven by just being selfish and inconsiderate.

It may be understandable why disagreements ensued across communities over the implementation of community-based water management strategies. There was tension emanating from the heavy losses incurred by the farmers as their crops succumbed to water stress. There were estimates that by July 2018, Western Cape Province recorded an economic loss of R5.9 billion in agriculture, shedding off 30 000 jobs in the process while its agricultural exports were reduced to 20% (Chambers 2018).

In a trend that may be explained by the existence of limited community-rooted conflict resolution mechanisms or dispute management systems, most respondents from predominantly urban municipalities (90%) reported conflicts in the implementation of community-based water management strategies. A respondent from the City of Cape Town stated;

> Conflicts erupted most at water collection points in townships. Residents could simply not agree on relatively smaller issues relating to observing norms on queuing for water, having too many water containers per person and theft of water in other people's properties. For instance, almost every day there would be fist fights at Newlands Spring [along Kildare/Springs Ways]. And SAPS [South Africa Police Services] would be called to restore order and normalcy. This is despite the fact that residents had their own residents community representatives mandated with the same function.

There were also challenges relating to deficiencies of technical capacity, organizational capacity as well as financial capacity on the part of communities to sustain their community-based water management strategies. In most of the responses (75%), failure of communities to maintain their own water supply infrastructure and successfully mobilize resources for the repair of such was a serious concern. A respondent from the Western Cape Provincial Government avers;

> Most community-initiatives were around installing hand pumps, boreholes, wells, irrigation canals, water ponds and community tanks. In some cases the municipalities would install these and leave them for communities to use and manage by themselves. First, most communities, especially those in rural areas such as *Friemersheim, Ebenezer, Haarlem, Genadendal,*

Mamre, Kranshoek, Rietpoort, Pniel, Suurbraak, Saron, Zoar and Slangrivier had no financial capacity to crowd fund the repair and maintenance of their water infrastructure whenever they broke down. In such instances they would wait for long periods of time after reporting the cases to their municipalities whilst the municipalities "raise resources" for repair. In other cases, very minor faults on the boreholes or hand pumps would leave communities with no water supply just because there were no skilled residents to attend to them.

The above and other challenges such as lack of adequate support from municipalities and community NGOs, absence of effective norms for dispute resolution, increasing pressure, and demand for water, non-existence of written rules on water use and access to water supply points, among others were highlighted in most responses (over 80%).

11.5.3 Stakeholder Perceptions on the Effectiveness of Community-Based Water Management Strategies

The perceptions of stakeholders covered by the research on the effectiveness of community-based water management strategies were divergent. However, the majority (70%) were of the view that the adopted community-based water management strategies within their areas of jurisdiction were effective for various reasons.

The main reason why those who agreed that the adopted community-based water management strategies were effective was that these managed to allow for efficiency, equity, and sustainability in water usage. The predominantly urban municipalities were of the view that the reduction in water usage in 2017/2018 is a reliable indicator of the effectiveness of community-based water management strategies. A respondent from the University of Stellenbosch stated;

> There hasn't been much rains since 2017 except occasional storms and dam levels across the Province have been low with a gradual increase recorded. However, water use habits changed. Community based water management strategies in urban areas helped to ensure compliance with restrictions on water use whilst those in rural areas were effective to some extent in that despite the huge demand for water especially for irrigated farming, there was no case of Day Zero.

The argument submitted by an academic from the University of Stellenbosch augurs well with the factual reality on the ground. However, there was a due acknowledgment that the success of community-based water management strategies should not be considered in isolation as they were complemented by the efforts made at all spheres of Government, that is National, Provincial and Local Government spheres. This was confirmed by a respondent from the City of Cape Town;

> Community based water management strategies were, and are still, helpful. Out municipality has around 5 million residents. But we managed to reduce water use by 50% between 2016 and 2018. Water demand reduced exponentially, and as we speak, four municipalities are applying Level 1 Restrictions, two municipalities are applying Level 2 Restrictions, two municipalities are applying Level 2B Restrictions, eight municipalities are applying Level 3 Restrictions, five municipalities are applying Level 4 Restrictions, and two municipalities are applying Level 5 Restrictions.

The ability of all communities to work together with municipalities to ensure that the Day Zero was pushed further since 2017 was also raised by many respondents (65%) as a sign of the success of community-based water management strategies wherever they were applied. A researcher from the Cape Peninsula University of Technology (CPUT) was emphatic;

> It is not a secret that the major water sources in the six dams that make up the Western Cape Integrated Water Supply System declined. In November 2017, all these dams were somewhere around 38% capacity levels which resulted in the adoption of the Day Zero Concept. Day Zero was initially set for 22 April 2018. It was thought from Day Zero, dams would fall to below 13.5%, and all taps would run dry and residents would collect a water rationed amount of 25 L per day. By January 2018, the dam levels had fallen to around 26%. The fact that we managed to push back Day Zero from March, 2018, 12 April 2018, 22 April 2018, 16 April 2018, 11 May 2018, and now 2020 is a good sign of triumph and success. Water consumption was also cut from a daily high of 580 million liters to less than that. To me this was mainly due to community based water management strategies. Without these, all municipality efforts would have gone to waste.

Consistent with views from the CPUT respondent, the six Western Cape dams that comprise the Western Cape Integrated Water Supply System, that is, Berg River, Steenbras Lower, Steenbras Upper, Theewaterskloof, Voelvlei, and Wemmershoek continued to decline in water levels since early 2017. However, if water consumption levels had continued as before, Day Zero was inevitable. Community-based water management strategies assisted to reduce water demands as communities agreed on norms and rules to reduce water usage. On the other, these strategies assisted to increase water supply sources, in both rural and urban areas mostly by the drilling of community boreholes and water wells. A researcher from the University of Stellenbosch concurred;

> Though I can't really quantify the level of success, I am of the view that community based water management strategies were two-pronged: water demand reduction, and water supply augmentation. Communities in rural areas and urban areas managed well their water supplies. Were it not for this, municipalities would have crumbled with pressure, and they couldn't have managed keep the water supply levels without avoiding the looming Day Zero. It was a collaborative and complementary effort. Because municipalities used their legal instruments to manage water too. We know the City of Cape Town was strict with its Water Amendment By-Law of 2018 to enforce water control, regulation and management measures.

If one traces the dam levels in Western Cape Province from 2016 up to the present day, it may confirm the perceptions from the stakeholders that water management strategies contributed immensely to the postponement of Day Zero. Perhaps, with better early warning systems, the water conservation strategies would have been commenced earlier and water usage restrictions applied more gradually (Table 11.2).

While a few respondents (10%) were not sure whether community-based water management strategies were effective, there were, however, some respondents (20%) who were of the view that community-based water management strategies would have achieved better results if they were well-coordinated by municipalities and well-resourced by NGOs, Government, and other community-based organizations. It is highly likely that their views were based on the limitations of the community-based water management strategies. A response from a University of Cape Town researcher as follows:

> Community based water management strategies rarely achieve their objectives by themselves given the obvious capacity issues that affect most community initiatives, especially in rural areas. Are communities in Western Cape capacitated enough, be it financially, resource-wise, technically, and organizationally? If municipalities, Department of Water and Sanitation, NGOs, etc. were to fill in these gaps and assist communities in implementing community based water management strategies, no doubt the results would have been far much better and more visible.

In all, community-based water management strategies were judged to have been effective in the overall in as far as managing and reducing the risks of drought effects and impacts on the socio-economic and political lives of the people of Western Cape Province is concerned.

11.6 Conclusions and Recommendations

From the responses gathered in the research, it is clear that all in all, community-based water management strategies attempted with notable success to facilitate not only the responsible use of water resources but also presenting a strong case of how natural disaster risks can be well-managed from a grassroots level, itself a paradigm shift from the traditional practice of waiting for Government and NGOs to manage and reduce the risk of natural disasters.

Challenges were indeed faced in implementing community-based water management strategies across most of the $n = 22$ municipalities covered in the research. These related to conflicts among community members; the lack of financial resources, organizational skills, and technical skills needed to mobilize resources for repair and maintenance of water supply infrastructure; the absence of effective rules for conflict resolution and dispute settlement; the lack of complementary assistance from Government and NGO actors. It was also established that community-based water management strategies implemented in the rural areas, mostly Rietpoort, Genadendal, Ebenezer, Haarlem, Saron, Mamre, Zoar, Kranshoek, Slangrivier, Pniel, Friemersheim, and Suurbraak suffered most from lack of finances and technical skills needed for the success of most community-based water management strategies.

Within the context of natural disaster management and natural disaster risk reduction, this Chapter recommends that drought mitigation, water conservation, and water management should prioritize local people as central. For Western Cape Province, there are already projects that by 2050, the rainfall in the province will likely decrease by about 30% from current levels. The South African National Disaster Management Framework (NDMF) of 2005 and the Western Cape Provincial Disaster Management Framework should drought risk mitigation strategies in general, and community-based water management strategies in particular should be streamlined and embraced as a key part of community-based disaster management. It thus recommended that the Disaster Management Centers at National, Provincial and Municipal levels collaborate to strengthen community-based water management strategies and institutionalize them so that drought risk mitigation, preparedness, responses, rehabilitation, and recovery are enhanced. The same institutions should partner with government and NGO actors partner to invest in community-based water management structures to reduce the risk of drought in Western Cape and build community resilience in light of imminent droughts since already there have be0en predictions that Western Cape will experience moderate to strong warming over the next 100 years. It would also help to further educate the communities on how best they can implement their community-based water management strategies so that they utilize the knowledge for better drought risk assessment, drought risk mitigation, and drought risk reduction. All these will be in consistent with the four key priorities of the Sendai Framework for Disaster Risk Reduction.

With the South African Draft National Climate Change Adaptation Strategy published on 19 May 2019 for public comments, it is recommended that when adopted, priority should be given to channel resources toward innovative and proactive community-based measures to sustainably plan, respond, and adapt to the effects of climate change, especially in vulnerable provinces such as Western Cape, which will assist in building climate-resilient societies. With specific regard to droughts, municipalities need to thrive toward constructive alignment of their respective Disaster Management Plans, Water Strategies, and Climate Change Policies with the National Climate Change Adaptation Strategy, Provincial and National Disaster Management Framework, and global frameworks on disaster management such as the Sendai Framework to allow for sustainability and facilitate smooth global partnerships in disaster management down to the community levels.

Table 11.1 Questionnaire respondents, questionnaires distributed, and questionnaires returned

Institution/Questionnaire respondents	Questionnaires distributed	Questionnaires returned
Western Cape Provincial Government	1	1
Western Cape Disaster Management Centre	1	1
National Disaster Management Centre	1	1
City of Cape Town	1	1
Cape Winelands District Municipality	1	1
Garden Route District Municipality	1	0
Central Karoo District Municipality	1	1
West Coast District Municipality	1	1
Overberg District Municipality	1	0
Breede Valley Local Municipality	1	1
Drakenstein Local Municipality	1	1
Langeberg Local Municipality	1	1
Beaufort West Local Municipality	1	0
Laingsburg Local Municipality	1	1
Albert Local Municipality	1	1
Prince Albert Local Municipality	1	1
Bitou Local Municipality	1	1
George Local Municipality	1	0
Hessequa Local Municipality	1	1
Cape Agulhas Local Municipality	1	1
Overstrand Local Municipality	1	1
Swellendam Local Municipality	1	0
Bergrivier Local Municipality	1	1
Matzikama Local Municipality	1	1
Cederberg Local Municipality	1	1
University of Western Cape	1	1
University of Stellenbosch	1	1
University of Cape Town	1	1
Cape Peninsula University of Technology (CPUT)	1	1
Total	**29**	**24**

Source Author's construction based on data gathered during fieldwork

Table 11.2 Water storage levels in major dams comprising Western Cape Province Water Supply System (WCWSS) (from 2016 to 2019)

Dam	Capacity (ML)	Percentage storage			
		2016	2017	2018	20 May 2019
Berg River	130 010	27.2	31.7	39.1	69.1
Steenbras Lower	33 517	36.2	25.5	33.7	38.5
Steenbras Upper	31 767	55.7	55.7	56.2	73.8
Theewaterskloof	480 188	30.8	14.6	12.1	35.9
Voelvlei	164 095	21.3	16.4	14.4	54.6
Wemmershoek	58 644	47.3	36.1	47.5	42.0
Total stored	898 221	275 581	185 755	189 375	412 815
Percentage storage		30.7	20.7	21.1	46.0

Source City of Cape Town (2019a, b)

References

Balay-As M, Marlowe J, Gaillard JC (2018) Deconstructing the binary between indigenous and scientific knowledge in disaster risk reduction: approaches to high impact weather hazards. Int J Disaster Risk Reduct 30:18–24

Bruns B (2008) Community priorities for water rights: some conjectures on assumptions, principles and programmes. In: Van Koppen B, Giordano M, Butterworth J (eds) Community based water law and water resource management reform in developing countries, CABI, pp 28–45

Chambers D (2018) Western Cape agriculture adds up the cost of the continuing drought. Business Daily. Available https://www.businesslive.co.za/bd/national/2018-07-19-western-cape-agriculture-adds-up-the-cost-of-the-continuing-drought/. Accessed 23 May 2019

City of Cape Town (2019a). City of Cape Town: dam levels report. Available https://resource.capetown.gov.za/documentcentre/Documents/City%20research%20reports%20and%20review/damlevels.pdf. Accessed 21 May 2019

City of Cape Town (2019b) Residential water restrictions explained. Available https://www.capetown.gov.za/Family%20and%20home/residential-utility-services/residential-water-and-sanitation-services/Residential-water-restrictions-explained. Accessed 22 May 2019

Day SJ (2009) Community-based water resources management. Waterlines 28(1):47–62

Dube E, Munsaka E (2018) The contribution of indigenous knowledge to disaster risk reduction activities in Zimbabwe: a big call to practitioners. Jàmbá J Disaster Risk Stud 10(1):1–8

Enshassi A, Shakalaih S (2016) Community based disaster risk reduction: a case study from Palestine. In: Proceedings of 2016 2nd international conference on disaster management and civil engineering (ICDMCE'2016), Kyoto, Japan, pp 15–21

Haque M (2019) Indigenous knowledge and practices in disaster management: experiences of the coastal people of Bangladesh. In: Disaster risk reduction. Singapore, Palgrave Macmillan, pp 59–72

Iloka N (2016) Indigenous knowledge for disaster risk reduction: an African Perspective. J Disaster Risk Stud 8(1):272

Kobayashi K, Ari IRD, Escobar IC, Schaefer A (2014) Community based water management and social capital. IWA Publishing, London

Lassa JA, Yos Bo, Yulius N, Silvia F, Alex O, Herman L (2018) Twenty years of community-based disaster risk reduction experience from a dryland village in Indonesia. Jàmbá J Disaster Risk Stud 10(1):1–10

Mugambiwa SS (2018) Adaptation measures to sustain indigenous practices and the use of indigenous knowledge systems to adapt to climate change in Mutoko rural district of Zimbabwe. Jàmbá J Disaster Risk Stud 10(1):1–9

Rogers P, Hall AW (2003) Effective water governance. Stockholm, Global Water Partnership. Global Water Partnership Technical Committee (TEC) Background Papers No. 7, Sweden, Elanders Novum

Surjan A, Shimpei K, Juha IU (2016) Risk and vulnerability. In: Sustainable development and disaster risk reduction. Springer, Tokyo, pp 37–55

Tantoh HB, Simatele D (2017) Community-based water resource management in North-West Cameroon: the role of potable water supply in community development. S Afr Geogr J 99(2):166–183

Trogrlić RŠ, Grant BW, Melanie J, Duncan, Marc JC van den H, Adebayo JA, Faidess DM, Mwafulirwa J (2019) Characterising local knowledge across the flood risk management cycle: a case study of Southern Malawi. *Sustainability* 11(6):1681

United Nations (2015) Sendai framework for disaster risk reduction (2015–2030). United Nations Office for Disaster Risk Reduction (UNISDR). United Nations, New York

United Nations International Strategy for Disaster Risk Reduction (2009) UNISDR terminology on disaster risk reduction. Geneva, Switzerland. Available at https://www.unisdr.org/files/7817_UNISDRTerminologyEnglish.pdf. Accessed 20 April 2019

World Bank (2017) Water resource management. Available at https://www.worldbank.org/en/topic/waterresourcesmanagement. Accessed 22 March 2019

Yes Media (2019) The local government handbook South Africa 2019: a complete guide to municipalities in South Africa, 9th edn. Cape Town, Yes Media. Available at https://municipalities.co.za/provinces/view/9/western-cape. Accessed 14 March 2019

Yodmani S (2001) Disaster risk management and vulnerability reduction: Protecting the poor. Paper presented at The Asia and Pacific Forum on Poverty organized by the Asian Development Bank, in Manila, Philippines. Available at https://www.adpc.net/V2007/IKM/ONLINE%20DOCUMENTS/downloads/PovertyPaper.pdf. Accessed 2 April 2019

Part III
Emerging Topics in IRDR: Post-disaster Recovery, Public Health, Role of Young Scientists, Gender

Chapter 12
Climate Change Adaptation Strategies in Primary Health Care

Hastoro Dwinantoaji, Hasti Widyasamratri, Mila Karmilah, and Sakiko Kanbara

Abstract Global climate change has already had observable effects on the environment. Floods in Semarang, Indonesia are caused by a combination of coastal floods and flood tides due to climate change. Key health impacts of climate change are considered from the perspective of extreme weather events, infectious diseases, and drought-related problems. The aim of the study was to explore the adaptation strategies the Kemijen communities take to reduce the negative effects of floods on human health. An exploratory case study approach was carried out from April to June 2018 in Kemijen, Semarang. Data were obtained through field survey, face-to-face in-depth interview, observations, and literature studies. In total, 102 health cadres were chosen through total sampling. Twenty-three percent of health cadres reported the period in which they were suffering from floods already lasted for more than 8 years. Types of illnesses due to floods in their area were diarrhea, dengue fever, skin diseases, and leptospirosis. Community program led by Municipal Health Office, *Puskesmas* Karangdoro and health cadres have been implemented to monitor dengue fever cases. Public kitchen was provided through *Dasa Wisma.* Medical supplies and assistance needed to recover from flood-related diseases were provided through *Posyandu.* We concluded that primary health care (PHC) implemented by health cadres plays an important role in preparing for extreme events, monitoring and responding to infectious disease outbreaks due to changing patterns of vector- and waterborne diseases by providing extra support for communities. Insights into the integration of PHC adaptation strategies and strengthening the role of health cadres

H. Dwinantoaji (✉)
Alumni of Cooperative Doctoral Program for Disaster Nursing, Graduate School of Nursing, University of Kochi, Kochi, Japan
e-mail: hastoro27@gmail.com

Lecturer and Researcher at Department of Mental Health and Community Nursing, Faculty of Nursing, Universitas Andalas, Padang, Indonesia

H. Widyasamratri · M. Karmilah
Department of Urban and Regional Planning, Faculty of Engineering Sultan, Agung Islamic University, Semarang, Indonesia

S. Kanbara
Cooperative Doctoral Program for Disaster Nursing, Graduate School of Nursing, University of Kochi, Kochi, Japan

R. Djalante et al. (eds.), *Integrated Research on Disaster Risks*, Disaster Risk Reduction,
https://doi.org/10.1007/978-3-030-55563-4_12

in flood risk management can help to meet the goal of the Sendai Framework for Disaster Risk Reduction (SFDRR) that calls for the participation of people disproportionately affected by disasters and highlights the need to strengthen the link between health security and human resources management and disaster risk reduction (DRR).

Keywords Adaptation strategies · Climate change · Health cadres · Primary health care (PHC)

12.1 Health and the Sendai Framework for Disaster Risk Reduction (SFDRR) 2015–2030

Disaster impacts are strongly influenced by physical, social, economic, and environmental factors (Aitsi-Selmi and Murray 2015). Reducing disaster risk, hence, requires concerted action across a wide range of sectors, institutions, and disciplines. The SFDRR calls for an all-off society engagement and partnership to reduce disaster risk, with inclusive and accessible participation of people disproportionately affected by disasters. It also identifies the need to strengthen the link between health security and human resources management and disaster risk reduction (DRR) (Aitsi-Selmi et al. 2015). The SFDRR 2015–2030 highlights concerns on human health and well-being that are common to DRR, climate change, and sustainable development (Gaillard and Mercer 2013). There are more than 30 explicit references to health, referring to the implementation of an all-hazards approach to managing disaster risk, including links to epidemics and pandemics, several references to the International Health Regulations (World Health Organization 2005) and to rehabilitation as part of disaster recovery. Further, primary health care (PHC) focuses on basic services to improve health status, which in turn builds community resilience and provides the foundation for responding to emergencies (Murray et al. 2015). Policies and strategies focusing on PHC can contribute to decreasing vulnerability and preparing households, communities, and health systems for emergencies. By focusing on the health risks of the community, our study explored strategies the Kemijen communities take to reduce disaster morbidity and mortality at the urban communities by averting the major threats to floods. Our study initiates to break through disaster planning and response by involving and utilizing health cadres at the community level and cooperative interaction with the *Puskesmas* and Municipal Health Office that concerned about health security and human resources management and DRR in the urban area.

12.2 Climate Change and Health in Indonesia

On the other hand, there is a high level of consensus that climate change is occurring now (Bray 2010). One of the potential consequences of climate change is sea level rise (SLR) that may threaten the coastal regions of the world and impact human life. Due to climate change, the global mean sea level has been observed to rise about 15–20 cm in the past century (Pachauri et al. 2014; Nicholls and Cazenave 2010; Moore et al. 2010), and it is predicted that this rise will continue in the twenty-first century (Moore et al. 2010).

The health impacts of climate change can occur through a number of direct and indirect causal pathways, and the severity is in part determined by the adaptive capacity of the population (McMichael and Lindgren 2011). Those groups at risk include poorer countries and communities, those geographically vulnerable to extreme weather events, and those highly dependent on agriculture for their livelihood.

Indonesia is already experiencing prolonged droughts, increases in temperature, and increased frequency of extreme weather events and heavy rainfall, all of which are expected to intensify with climate change (Network 2009). As an archipelago, the country is exceptionally vulnerable to SLR and the inundation of coastal areas. Global estimates of the Intergovernmental Panel on Climate Change (IPCC) predict that the sea level in Indonesia will rise 100 cm due to increased sea water temperature from 1.3 degree Celsius to 4.6 degree Celsius in the year 2100 (McMichael and Lindgren 2011; Zhu et al. 2010). It will clearly deepen the stress of coastal areas, causing floods, erosion of the shores, and destruction of coastal structure (Harwitasari and Ast 2011; Zhu et al. 2010) and will increase health problems that it is largely attributable to a decrease in water quality and seawater intrusion into the freshwater sources.

Kemijen urban village is located in the northeast part of the East Semarang subdistrict on the northern coast of the island of Java, Indonesia which is one of the slum areas in Semarang city. In Kemijen, the worst floods occur as tidal floods; these are caused by a combination of coastal floods and flood tides (Harwitasari and Ast 2011; Marfai et al. 2008); this is then further reinforced by possible waves and SLR (Marfai et al. 2008). The tidal floods have a major impact on the coastal areas in Kemijen, Semarang. Floods can negatively affect drinking water, sanitation, access to health care, and safe food. Floods happened has the potential to be in human insecurity for inhabitants. Key health impacts of climate change are considered from the perspective of extreme weather events, infectious diseases, and drought-related problems.

WHO had demonstrated the important role of community health volunteers (CHVs) in achieving Goal 3 related to health indicators in the Sustainable Development Goals (SDGs). The CHVs in Indonesia are known as health cadres, were built on the national women's Family Welfare Movement (PKK) in the 1970s (World Health Organization 2012). They were trained to conduct health and nutrition promotion activities in each village. Health cadres are selected by their communities based on a set of characteristics, including educational achievement, how integrated they are within the community, personality, dedication to service, and willingness to commit to the time requirements to support services at the integrated health services post (*Posyandu*). The aim of study was to explore the adaptation strategies the Kemijen's health cadres take to reduce the negative effects of floods on human health. Additionally, we overviewed the evidence for human-induced climate change, the predicted health impacts, and the role of health cadres in flood risk management and PHC provisions as the climate change adaptation strategies.

12.3 Methodology

An exploratory case study approach was carried out from April to June 2018 in Kemijen, Semarang. Data were obtained through surveys, field observations, and literature studies. In total, 102 health cadres were chosen through total sampling. The primary data was collected through surveys, face-to-face in-depth interviews, and field observations. Meanwhile, the secondary data was retrieved from the literature and data from local institutional actors. A field survey, face-to-face in-depth interview, and observations were conducted to investigate the experience of health cadres in Kemijen regarding floods, their negative environmental health due to floods and their strategies for adaptation. The characteristics of respondents, floods, and the negative environmental health due to floods were presented by using graphs and charts. Furthermore, we show the pathways by which climate change can impact on health and potential PHC adaptive strategies.

12.4 Results: Urban Vulnerability in the Context of Climate Change

Climate change is also predicted to cause more flooding in some regions due to more frequent heavy rainfall events. Coastal communities are especially vulnerable to coastal surges exacerbated by a combination of rising sea levels and more intense storms (Nicholls and Cazenave 2010; Harwitasari and Ast 2011; Marfai et al. 2008). Recent climate observations indicate that the dry season in Semarang is becoming longer with more intense rainfall events occurring during the rainy season (Pachauri et al. 2014). This changing rainfall pattern contributes to existing problems with flooding and drought. Many poor urban communities are situated in parts of Semarang city that are at high risk of climate change impacts from floods and coastal inundation. Currently, almost all of Semarang's districts experience severe disruption of unclean water, pollution of sanitation system, infectious diseases, inadequate access to health care, food availability, and safety from flooding (Pachauri et al. 2014; Network 2009; Harwitasari and Ast 2011; Zhu et al. 2010). Residences on the coast are particularly at risk from SLR and land subsidence. Human activities such as infrastructure development and groundwater exploitation became the driving factors of the land subsidence, especially in the North Semarang (Harwitasari and Ast 2011). The soil structure in this area tends to be sandy so it is very easy to subside (Marfai et al. 2008). Hence, there are some concerns that this might stimulate migration from coastal areas to other parts of the city, creating additional social and demographic pressures. Furthermore, rural people migrating into these settings may face heightened environmental and health insecurity.

In total, 102 health cadres were used in this study with all representing a different household. The respondents came from different areas within the Kemijen urban village. It is relevant to dig further into the specific characteristics of the respondents and their households. In particular, the factors of family type, education level, occupation, and monthly family income are important within this research. Most respondents were women aged 37–69 years (Fig. 12.1) living within the nuclear family type (Fig. 12.2) and housewife (Fig. 12.3). For the majority of the respondents (72%), the highest educational level they reached was senior high school (Fig. 12.4). Only 3% of the respondents gained a higher educational level. Moreover, monthly family income level of almost all respondents was below the Regencies Minimum Wage in

Fig. 12.1 Age group

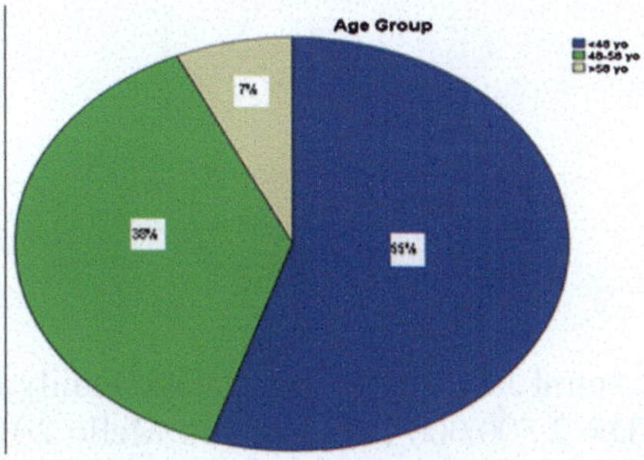

Fig. 12.2 Family type

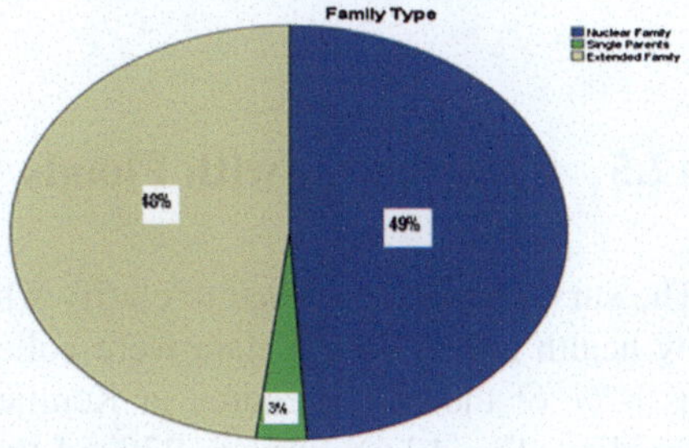

Fig. 12.3 Occupation

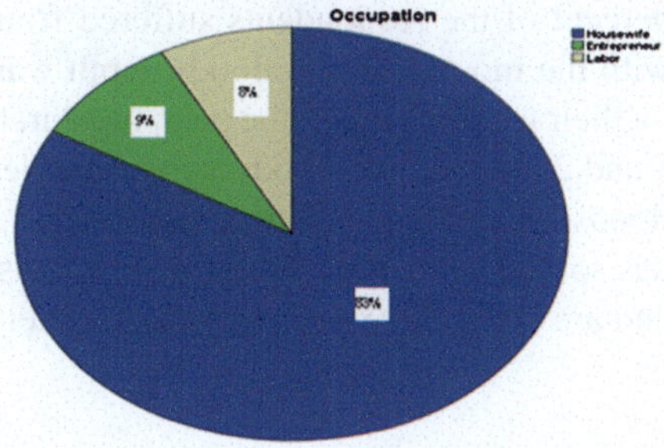

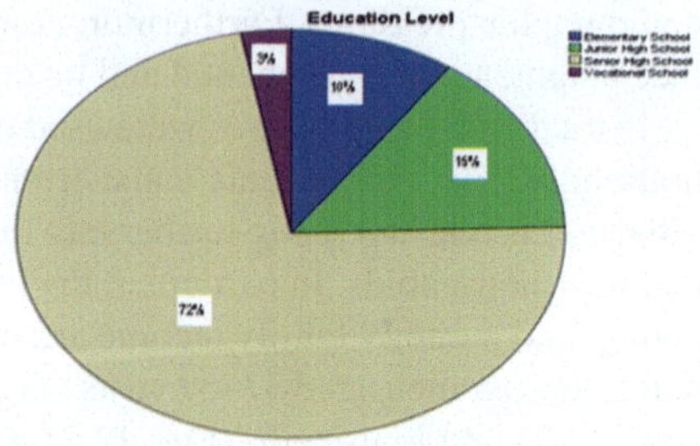

Fig. 12.4 Education level

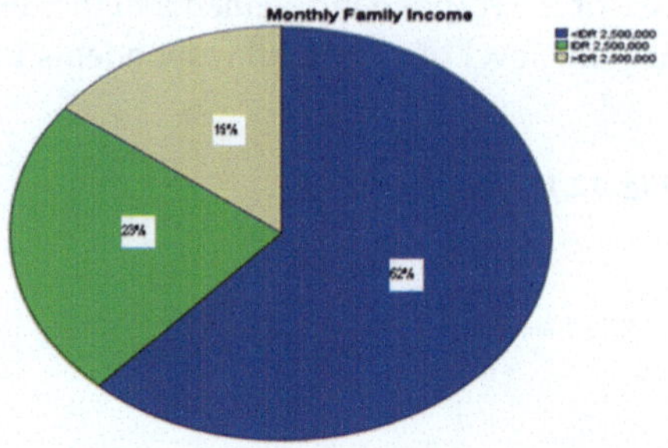

Fig. 12.5 Monthly family income

Central Java 2019. The average family income level in Semarang city lies at about IDR 2,500,000 (Comola and Mello 2011). Only 15% of the respondents said that they have a family income of more than IDR 2,500,000 (Fig. 12.5). These low family income levels could be a problem when people need to adapt to the increasing floods.

12.5 Experiences with Floods

The survey initially wished to clarify whether the floods were actually experienced by health cadres. These data were collected at a specific time moment. As many as 87% of the health cadres in Kemijen reported that they were suffering from flooding (Fig. 12.6). Further, 23% of them indicated that the period in which they were suffering from floods already lasted for more than 8 years (Fig. 12.7). Fifty-two percent of the respondents suffered from flooding between 1 and 24 h (Fig. 12.8) with the most common floods depth was between 0.51 and 1 m (Fig. 12.9). Based on their experiences in the past, the duration of floods appeared to be between about 1 and 24 h since the floods arrived and left along with the tides. Even sometimes the floods are much worse than average. Some of them reported that the polder system has solved all of their flooding problems. However, others still experience flooding and are not so sure that the polder system has fixed everything.

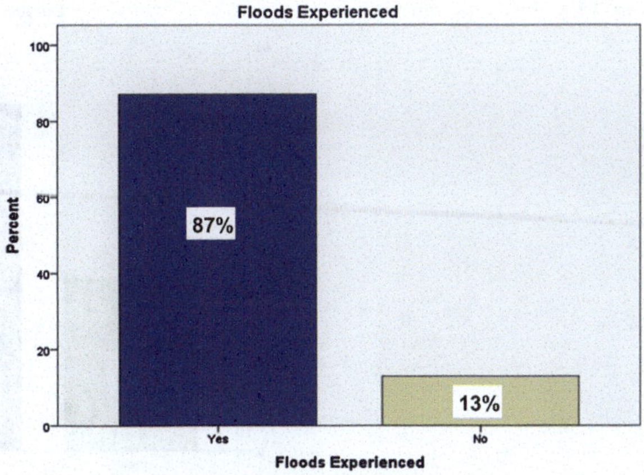

Fig. 12.6 Floods experienced

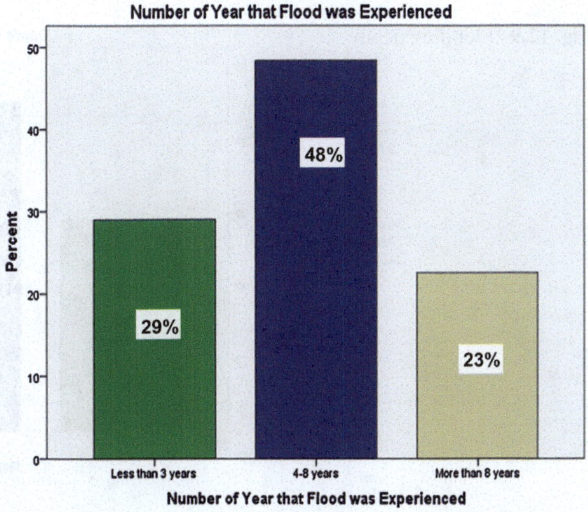

Fig. 12.7 Flooding period

12.6 Water Scarcity, Waterborne Diseases, and Water-Related Vector

In many parts of the world, water scarcity results in inadequate access to safe drinking water and this can lead to the spread of infectious diseases via fecal contamination of drinking water (waterborne diseases) (Zakar et al. 2012). The poor quality of drinking water coupled with a lack of basic sanitation is an important factor in the spread of diarrhea, which is the second most common contributor to the disease burden in developing countries (Mor and Griffiths 2011). In many countries, water scarcity compels the local population to use wastewater for irrigation purposes, but this serves to disperse microbes into the environment.

Fig. 12.8 Flooding duration

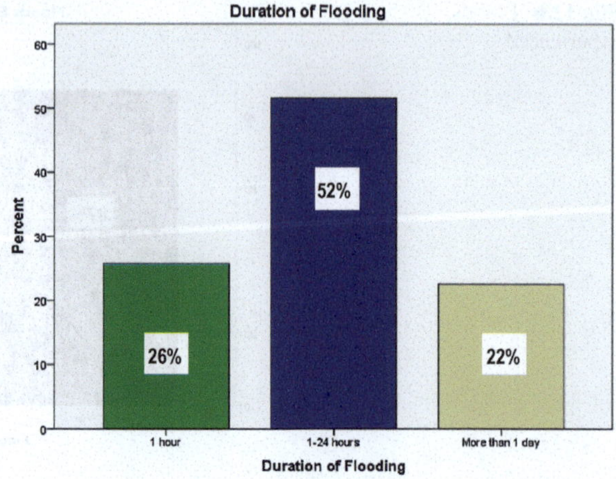

Fig. 12.9 Flooding depth

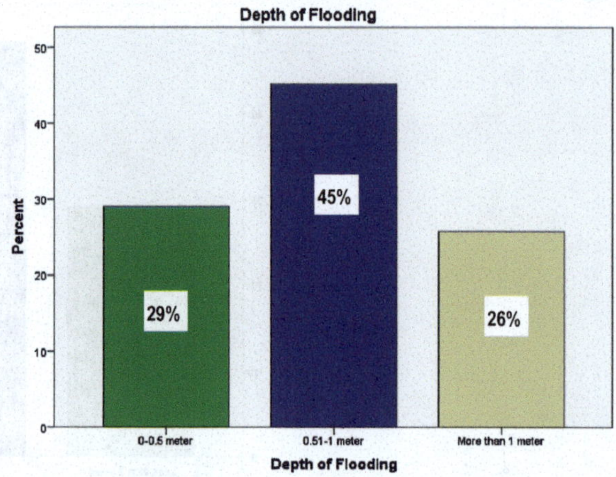

Additionally, water scarcity is also a barrier to basic sanitation and hygiene and can lead to the spread of skin diseases that can otherwise be prevented by promoting improving hygiene. Stagnant water also provides essential habitat for the intermediate hosts of important parasites. Human infection can result following an exposure to water in which these hosts and parasites live (water-based diseases) or through the bite of insects that live or breed in water (water-related diseases) (Zakar et al. 2012; Mor and Griffiths 2011).

Figure 12.10 shows the case of diseases that most commonly occurred in Kemijen. Health cadres reported there were diarrhea (29 of 116 cases), dengue fever (23 of 116 cases), skin diseases (26 of 116 cases), leptospirosis (13 of 116 cases), malnutrition (5 of 116 cases), acute respiratory infections (8 of 116 cases), and typhus (12 of 116 cases). The floods-related health issues most frequently reported by health cadres

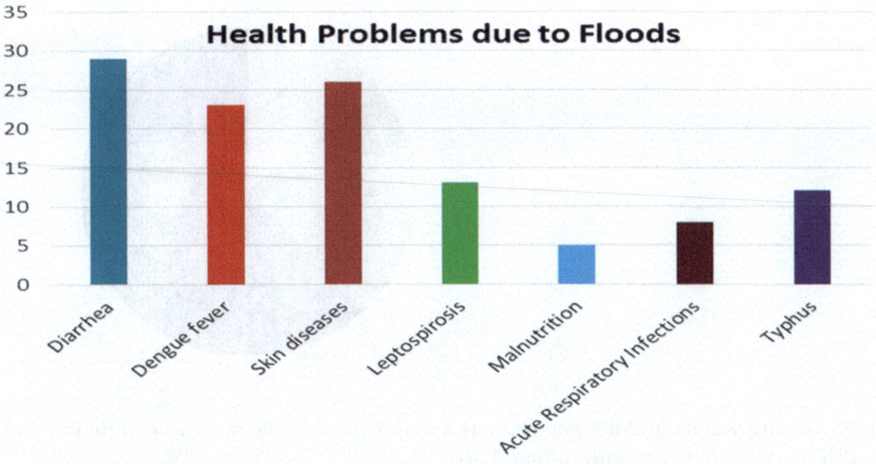

Fig. 12.10 Health problems due to floods

were diarrhea, dengue fever, and skin diseases among children. Water that is contaminated by wastewater can lead to diarrheal diseases, skin infections, and leptospirosis. There are some households that use water from shallow wells for cooking, bathing, and washing, which still indicate a risk of direct contact with pathogens. Consequently, contaminated water creates health problems even in their daily life. In addition, stagnant water and slow running water are located close to households throughout Kemijen and children were repeatedly seen playing and bathing in the water where solid waste, sewage, apparent bacterial growth, and living larva were evident. Accumulated solid waste in urban areas is a domain for insects, parasites, and rummaging animals. These life forms increase the prevalence of mosquito-borne diseases, rodent-borne diseases, and waterborne diseases. It is clear that flooding and sanitary issues have mutually reinforcing effects.

Many infectious diseases are sensitive to climatic conditions. Temperature, rainfall, and humidity variously affect the replication, maturation, and viability of the pathogen, the vector organism (where applicable), and the range and abundance of any reservoir or intermediate animal species (McMichael et al. 2007). In Indonesia, one disease thought to be particularly susceptible to climate change is dengue fever, which is difficult to treat and can be fatal. Pattern outbreak of dengue virus infection is also influenced by climate and humidity, where the hot temperatures (28–32 degree Celsius) and high humidity is an intermediate vector of dengue virus in order to stay alive (Karina et al. 2015). Kemijen Village is included Dengue Hemorrhagic Fever (DHF) endemic as this area is an area often affected by high tides. DHF is a disease caused by the dengue virus which is transmitted to humans by the bite of infected *Aedes aegypti* mosquitoes throughout the tropical and subtropical world (Wahyono et al. 2017). It is one kind of disease which becomes the public health problem and endemics in Semarang. Stagnant water or slow-flowing water attracts mosquitoes that use the water as breeding grounds. The larva that was observable in

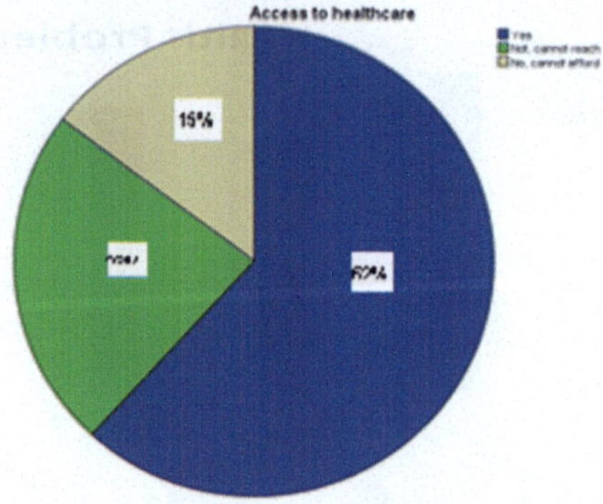

Fig. 12.11 Access to healthcare

slow-flowing waters and dengue fever is a major issue in the area according to local health workers and community members.

Further, leptospirosis cases were also found in Kemijen. Leptospirosis is one of the rodent-borne neglected diseases but health problem in day. Transmission of Leptospirosis occurs by contact with water or humid soil contaminated with urine from rodents infected with *Leptospira Sp* (Yuniarto and Ramadhani 2010). The leptospirosis cases tended to increase in the rainy season, particularly in times of floods. Flooding can cause the rats to come out of hiding and wandering in a residential neighborhood. The floodwaters can be the medium of transmission of *Leptospira Sp*, and can be a place to live *Leptospira Sp*. Flooding can expand *Leptospira Sp* transmission so that it can increase the risk of exposure to human (Febrian and Solikhah 2011).

To be able to recover from the abovementioned illnesses, it is important for people to have access to sufficient health care, also in times of floods. Over half of health cadres are covered by health insurance. However, some of them indicated that they are not covered by health insurance. Consequently, 15% of the respondents have no access to health care because they cannot afford it (Fig. 12.11). More than 20% of the respondents cannot reach healthcare facilities when floods occurred but this does not seem to be the biggest issue. This can also have more indirect negative social consequences regarding their monthly family income and family type.

12.7 Primary Health Care Adaptation Strategies

Many poor urban communities are situated in parts of cities that are at high risk of climate change impacts from floods and coastal inundation. Further, much of the climate change-related displacement will occur in developing regions where public health resources are inadequate (McMichael and Lindgren 2011; McMichael et al. 2007). Given the broad range of potential health impacts attributable to climate change, regional responses from the health professional will be required. Disaster preparedness will be determined by the vulnerability of communities and might include preparation for heat waves, coastal inundation, or rapid urbanization. Hence, rural people migrating into these settings may face heightened environmental and health insecurity. In other regions, early monitoring for vector-borne diseases or waterborne illnesses may be an important role for PHC providers. Primary care may also have a role in the mitigation of climate change through encouraging environmentally sustainable healthcare services (McMichael et al. 2007; Bell and Towle 2011). Figure 12.12 shows the pathways by which climate change can impact on health and potential PHC adaptive strategies.

Based on the face-to-face in-depth interview with health cadres and local institutional actors and empirical literature (Jayanimitta et al. 2018; Setyono and Yuniartanti 2016), it is known that disaster preparedness groups in Kemijen urban village (KSBs) on floods and dengue fevers take on the responsibilities and control from their grass-roots activities. Additionally, the Banger Polder Management Board created a mission and vision of "dry feet for all" with the auspicious goal of ridding the residents of Kemijen from coastal inundation and rainwater floods. The establishment of the Banger Polder Management Board has an impact on the formation of local adaptive capacity. Residents are involved in the Banger Polder Board and in the management of the Banger Polder System. This is done to increase knowledge and empower the community so that people are aware to protect their environment.

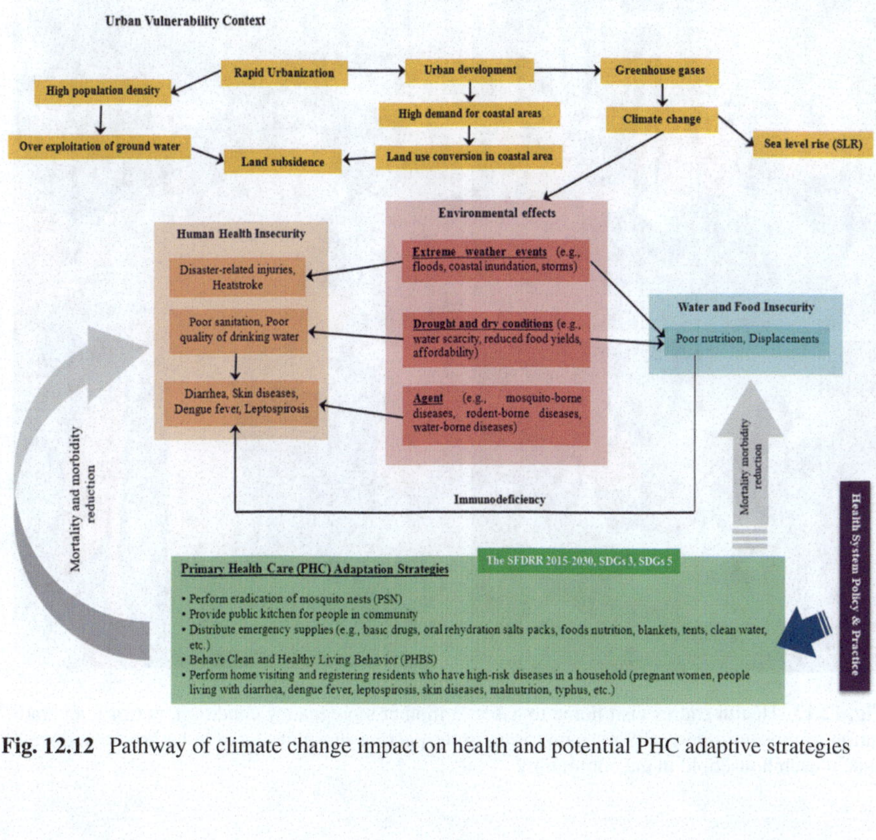

Fig. 12.12 Pathway of climate change impact on health and potential PHC adaptive strategies

In this case, they approached the local women's organizations in Kemijen, including Family Welfare Movement (PKK), *Dasa Wisma* and *Posyandu*. Every month, these women's organizations held a regular meeting in order to increase knowledge and educate women in the neighborhood. The board explained the reason they choose the local women's organizations as follows: (1) generally, women have a responsibility to keep their family and children; (2) women also have a responsibility to manage their family's finance; (3) women hold a responsibility to keep the cleanness of their neighborhood (e.g., the person usually throw garbage into the river is a woman); (4) women usually work to gather neighborhood's data voluntarily; and (5) empower women to participate in maintaining the environment in their neighborhood.

Health cadres were built on the PKK. They are selected by their communities based on a set of characteristics, including educational achievement, how integrated they are within the community, personality, dedication to service, and willingness to commit to the time requirements to support services at the integrated health services post (*Posyandu*). Health cadres perform an eradication of mosquito nests (PSN) in particular monitoring on inspecting mosquito larva in the bathroom and water tank by visiting every household in the community (Fig. 12.13). This activity includes the making of monthly documentation and disseminating their documentation to

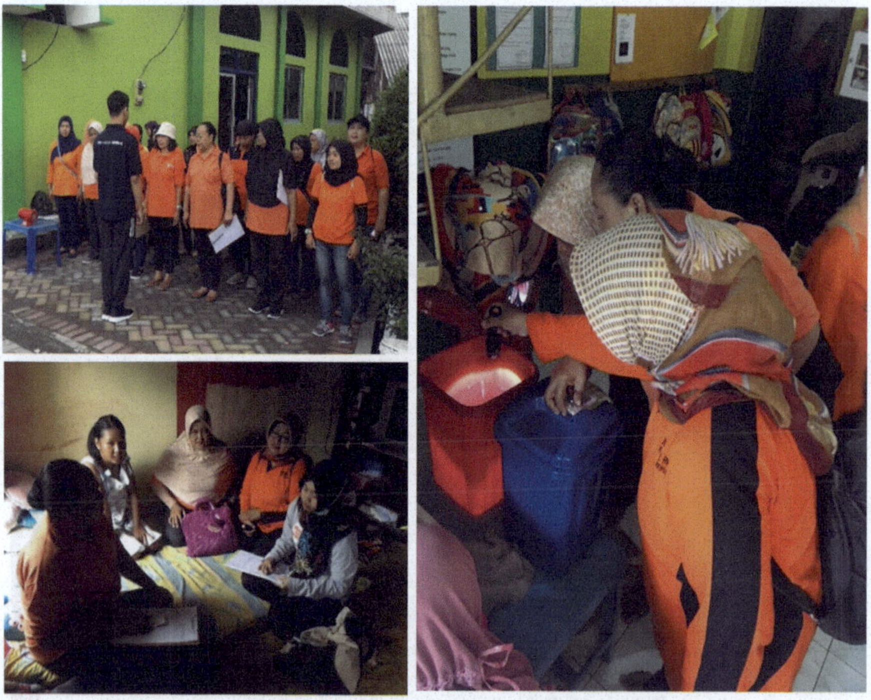

Fig. 12.13 Health cadres visit home to assess a mother's pregnancy condition, perform an eradication of mosquito nests (PSN): monitoring on inspecting mosquito larva in bathroom and water tank at each household in the community

the neighborhood meeting and submitting the report to their respective subdistrict and *Puskesmas*. Awareness to ensure the cleanness of water resources has also been increasing. This is so because there is a culture among the community that people especially women feel embarrassed if not behaving a clean and healthy lifestyle. According to the Municipality Health Office, every household is expected to behave Clean and Healthy Living Behavior (PHBS) for reducing the incident rate of dengue fever by applying 3 M Plus. This term is used to indicate some activities, namely cleaning water tank weekly, covering water tank, recycling used household equipment, and protecting them from mosquito attack. Ideally, the score of households with Larvae Free Index (LFI) should be over 95% (Subdirektorat Pengendalian Arbovirus 2011), however, the existing LFI score in Kemijen is only 70–80% on average.

Furthermore, *Dasa Wisma* and *Posyandu* have an important role in flood events. A public kitchen is established during emergencies through *Dasa Wisma*, distribute basic drugs, and provide medical assistance needed to recover from flood-related diseases is provided through *Posyandu* (Fig. 12.14). Even though key informants focused on the practical values of *Dasa Wisma* and *Posyandu*, these practices also seem to have other purposes regarding flooding. An example is education about hygiene. Through *Dasa Wisma*, some of the female inhabitants claimed to learn about cleaning after flooding and not to throw garbage into the river. Through *Posyandu* some health cadres claimed also to learn about hygiene in order to prevent diseases.

Fig. 12.14 Provide public kitchen and medical assistance service, distribute basic drugs, and behave Clean and Healthy Living Behavior (PHBS) in the community

Since *Posyandu* is more practice-oriented, education is mainly given through *Dasa Wisma*.

12.8 Conclusion

The important barrier for a system shift toward a more inclusive approach to DRR and climate change adaptation comprising PHC adaptation measures is the lack of rewards system and certification, and inadequate the mechanism of protection and incentive for health cadres. Whereas PHC implemented by health cadres plays an important role in preparing for extreme events, monitoring, and responding to infectious disease outbreaks due to changing patterns of vector- and waterborne diseases by providing extra support for communities. In the long-term, therefore, policy-makers need to consider the health cadres and the other healthcare providers at the community who involved in DRR issues so we are well prepared for the predicted health impacts of climate change and begin to implement long-term adaptive health policies. Insights into the integration of PHC adaptation strategies and strengthening the role of health cadres in flood risk management can help to meet the goal of the SFDRR 2015–2030 that calls for the participation of people disproportionately affected by disasters and highlights the need to strengthen the link between health security and human resources management and DRR.

Acknowledgments I would like to acknowledge Prof. Sakiko Kanbara, Professor at Faculty of Nursing University of Kochi, Principal Supervisor and Advisor from Japan for guiding my dissertation research, and Dr. Mila Karmilah and Dr. Hasti Widyasamratri, Faculty Member at Department of Urban and Regional Planning, Faculty of Engineering Sultan Agung Islamic University from Semarang in Indonesia, for their warm guidance and support and opportunity given to coordinate this research project. In addition, I sincerely thank all health cadres in Kemijen, *Puskesmas* Karangdoro, and local institutional actors in the East Semarang subdistrict for their efforts in providing PHC and monitoring health of people in flood events.

References

Aitsi-Selmi A, Murray V (2015) The Sendai framework: disaster risk reduction through a health lens

Aitsi-Selmi A, Egawa S, Sasaki H, Wannous C, Murray V (2015) The Sendai framework for disaster risk reduction: Renewing the global commitment to people's resilience, health, and well-being. Int J Disaster Risk Sci 6(2):164–176

Bell E, Towle N. Climate change and primary health care: towards better regional health impact and risk assessments. In National Rural Health Conference. 2011:1–13.

Bray D (2010) The scientific consensus of climate change revisited. Environ Sci Policy 13(5):340–350

Comola M, De Mello L (2011) How does decentralized minimum wage setting affect employment and informality? The case of Indonesia. Rev Income Wealth 57:S79-99

Febrian F, Solikhah S. Analisis Spasial Kejadian Penyakit Leptospirosis di Kabupaten Sleman Propinsi Daerah Istimewa Yogyakarta Tahun 2011. Kesehatan Masyarakat: Jurnal Fakultas Kesehatan Masyarakat Universitas Ahmad Dahlan. 2013;7(1).

Gaillard JC, Mercer J (2013) From knowledge to action: Bridging gaps in disaster risk reduction. Prog Hum Geogr 37(1):93–114

Harwitasari D, Van Ast JA (2011) Climate change adaptation in practice: People's responses to tidal flooding in Semarang Indonesia. J Flood Risk Manag 4(3):216–233

Jayanimitta ME, Puspasari DA, Widyahantari R, Kristina D, Ratnaningtyas T, Setionurjaya A, Anindita YA. Beyond Vulnerability Assessment: Impact of Developments toward Local Adaptive Capacity in Kemijen City Village, Semarang City. In IOP Conference Series: Earth and Environmental Science 2018;123(1):012007. IOP Publishing.

Karina A, Sari SY, Sumardi U, Setiawati EP (2015) Incidence of dengue hemorrhagic fever related to annual rainfall, population density, larval free index and prevention program in Bandung 2008 to 2011. Althea Med J 2(2):262–267

Marfai MA, King L, Sartohadi J, Sudrajat S, Budiani SR, Yulianto F (2008) The impact of tidal flooding on a coastal community in Semarang Indonesia. Environmentalist 28(3):237–248

McMichael AJ, Lindgren E (2011) Climate change: present and future risks to health, and necessary responses. J Intern Med 270(5):401–413

McMichael T, Blashki G, Karoly DJ (2007) Climate change and primary health care. Aust Fam Physician 36(12):986

Moore JC, Jevrejeva S, Grinsted A (2010) Efficacy of geoengineering to limit 21st century sea-level rise. Proc Natl Acad Sci 107(36):15699–15703

Mor SM, Griffiths JK (2011) Water-related diseases in the developing world. Encyclopedia of environmental health. Elsevier, New York, pp 741–753

Murray V, Aitsi-Selmi A, Blanchard K (2015) The role of public health within the United Nations post-2015 framework for disaster risk reduction. Int J Disaster Risk Sci 6(1):28–37

Network AC (2009) Responding to the urban climate challenge. Boulder, Colorado, USA

Nicholls RJ, Cazenave A (2010) Sea-level rise and its impact on coastal zones. Science 328(5985):1517–1520

Pachauri RK, Allen MR, Barros VR, Broome J, Cramer W, Christ R, Church JA, Clarke L, Dahe Q, Dasgupta P, Dubash NK (2014) Climate change 2014: synthesis report. In: Contribution of Working Groups I, II and III to the fifth assessment report of the Intergovernmental Panel on Climate Change. IPCC

Setyono JS, Yuniartanti RK (2016) The challenges of disaster governance in an Indonesian multi-hazards city: a case of Semarang. Central Java. Procedia-Social and Behavioral Sciences. 227:347–353

Subdirektorat Pengendalian Arbovirus (2011) Informasi umum demam berdarah dengue. Dit PPBB Ditjen PP dan PL KemKes RI, Jakarta

Wahyono TY, Nealon J, Beucher S, Prayitno A, Moureau A, Nawawi S, Thabrany H, Nadjib M (2017) Indonesian dengue burden estimates: review of evidence by an expert panel. Epidemiol Infect 145(11):2324–2329

World Health Organization (2008) International health regulations (2005). World Health Organization

World Health Organization. Sasakawa health prize: stories from South-East Asia. WHO Regional Office for South-East Asia

Yuniarto B, Ramadhani T (2010) Kajian Epidemiologi Kejadian Leptospirosis Di Kota Semarang Dan Kabupaten Demak Tahun 2008. Jurnal Litbang Pengendalian Penyakit Bersumber Binatang Banjarnegara, BALABA, pp 7–11

Zakar MZ, Zakar R, Fischer F. Climate change-induced water scarcity: a threat to human health. South Asian Stud (1026-678X) 27(2)

Zhu X, Linham MM, Nicholls RJ (2010) Technologies for climate change adaptation-Coastal erosion and flooding. Danmarks Tekniske Universitet, Risø Nationallaboratoriet for Bæredygtig Energi

Chapter 13
Capacity-Building Strategy for Creating Disaster- and Climate-Risk-Sensitive Development Plans—A Case Study of Multi-Stakeholder Engagement in Sri Lanka

A. M. Aslam Saja, S. M. Lafir Sahid, and M. Sutharshanan

Abstract The emphasis on disaster management sector had been on standalone disaster risk reduction or response preparedness plans for the past decades. There was a lack of an integrated approach to disaster risk reduction and development interventions at all levels. In the context of disaster-development nexus, two standalone approaches had evolved in the past: Community-Based Disaster Risk Management (CBDRM) without proper integration mechanisms within development projects and community development through Village Development Plans (VDP) without due consideration of risk reduction aspects. Both isolated approaches have reduced not only the aid effectiveness, but proven to be not enhancing the resilience of communities. In this context, this chapter discusses a new risk-sensitive approach to development planning piloted by DIPECHO (Disaster Preparedness project of European Commission Humanitarian Office) partners in Sri Lanka by drawing from lessons from past experiences in implementing standalone CBDRM projects and community development projects. Whilst this project addressed four priority actions of the Sendai Framework for disaster risk reduction, the links were made to the local development approach at the community level to build resilient communities. The Sri Lanka Community Resilience Framework (CRF) has been the driving concept behind the risk-sensitive approach in development planning. The state officials at the sub-national (district) levels were trained as trainers on CRF and risk-sensitive development planning process. The community-level state staff officials and local government officers were then trained by district state officials through cascade training programs. This chapter details the capacity-building strategy, process, and tools for creating risk-sensitive development plans as a locally led sustainable approach. One of the key lessons from the risk-sensitive development approach identified the

---

A. M. A. Saja (✉)
South Eastern University of Sri Lanka, Oluvil, Sri Lanka
e-mail: saja.aslam@seu.ac.lk

S. M. L. Sahid
DIPECHO Consortium Project, Plan International, Yangoon, Myanmar

M. Sutharshanan
ACTED, Colombo, Sri Lanka

R. Djalante et al. (eds.), *Integrated Research on Disaster Risks*, Disaster Risk Reduction,
https://doi.org/10.1007/978-3-030-55563-4_13

absence of efforts to integrate climate change adaptation measures. The lack of an integrated and holistic approach highlighted the importance of further strengthening the **L**inkages in **D**isaster **R**eduction, **D**evelopment, and **C**limate **C**hange **A**daptation (***LDRDCCA***), aligning with three post-2015 global agendas.

Keywords Climate risk · Community resilience · Disaster mainstreaming · Disaster risk · Risk-informed development

Abbreviations

CBDRM	Community-Based Disaster Risk Management
CRF	Community Resilience Framework
DMC	Disaster Management Centre in Sri Lanka
DRM	Disaster Risk Management
DRR	Disaster Risk Reduction—both these terms were used interchangeably in the literature
DIPECHO	Disaster Preparedness project funded by the Humanitarian office of European Commisison
DS	Divisional Secretariate which is the next level of administration to the GN. A collection of DS divisions form the district and Sri Lanka is divided into 25 administrative districts
GN	Grama Niladari, which is the lowest state administrative level at the community level in Sri Lanka
RS	Risk-Sensitive
VDP or CDP	Village or City Development Plans

13.1 Introduction

The Disaster Management Policy places a strong emphasis on saving lives, protecting livelihoods and strengthening recovery from disasters with an approach to build back better, and enabling safe living whilst preventing and reducing vulnerability. National and local Disaster Risk Reduction (DRR) strategies and plans are key for effective implementation of the Sendai Framework and for risk-sensitive development planning and implementation from grass-root level to national level (UNDRR 2019). To address this, the Ministry of Disaster Management and Disaster Management Centre (DMC) has designed an approach for building disaster-resilient communities with a vision of "Safer Sri Lanka" by bridging the gap between Disaster Risk Reduction (DRR) and long-term development objectives.

The resilience framework is an attempt to protect development gains in the long term and to reduce dramatic declines in development caused by disasters. This highlights the overlapping nature of disaster mitigation, preparedness, relief, and recovery work, and thus necessitates the incorporation of those activities into development. This has further ensured that the voices of communities at risk are represented at local and national levels to bring the field realities to policy discussions and better reflect the needs and capacities on the ground. Based on this, the DMC appointed a technical working committee to look at the options to harmonise the community-based DRR efforts in the country within a broader framework of building resilient villages, and build consensus amongst DRR practitioners on common indicators in order to be able to monitor the increased resilience of communities. The DIPECHO 8 consortium, as part of the initiative, facilitated by the Sri Lanka Comprehensive Disaster Management Programme (SLCDMP) and the Disaster Management Centre (DMC), has agreed to provide its financial and technical support.

The Community Resilience Framework (CRF), which was drafted with the support of the technical working group was placed for wider consultation. The pilot testing of operationalising CRF in 5 districts had commenced (Kilinochchi, Mullaithivu, Puttalam, Rathnapura and Kurunegala) and this process aimed for validation and finalisation of CRF. In view of the pilot implementation of CRF, it was agreed to develop capacities required on resilience planning and development. Hence, the initial training program on the implementation of CRF trained 25 front-level development representatives from Kilinochchi and Mullaithivu. Trainers who attended the training of trainers program on CRF, further facilitated the implementation of CRF at divisional level in both districts through a cascading training approach devised by DIPECHO project. In addition, they have been capacitated as facilitators of CRF and as a result, a resource pool for the region was now established with the trained trainers, who have also gone through the complete process of trainer assessment.

The objective of this chapter is to present the capacity-building strategy created and implemented successfully with the engagement of disaster management and development stakeholders for developing risk-sensitive development plans in Sri Lanka. This chapter also briefs a historical perspective of Disaster Risk Management (DRM) in Sri Lanka, particularly Community-Based DRM (CBDRM) and the challenges in achieving the intended impacts in the communities by implementing CBDRM strategies, and the need for the risk-sensitive development planning process.

13.2 Literature Review

The linkages between disaster impact and development investment are well-established. For example, disasters delay development programs, decrease economic potential, and reduce human capital (Bendimerad 2003). The regional and national development plans should also identify disaster risks and integrate disaster risk reduction measures in addition to the contingency disaster preparedness plans, so that communities can learn and take actions to reduce the impact of emerging disaster risks (Wilkinson et al. 2016). Due to rapid urbanisation and increase of cities across the globe, the exposure to disasters and therefore disaster risk has been increasing, that demands for more effective risk-sensitive development planning. However, the development stakeholders are encountered with difficulties in effective integration of disaster risk reduction measures in their development work due to short-term focus and limited resources available for addressing immediate and ad hoc community needs (Schipper and Pelling 2006). Further, the Sendai Framework for Disaster Risk Reduction (SFDRR) sets a target (Target F) to sufficiently increase international cooperation to developing nations to launch effective national policies that enhance local public financing for risk-sensitive sustainable development (UNDRR 2019).

Risk-sensitive development is a decision-making process based on disaster-related risk information that enables understanding of multiple concurrent threats and complex risks to and arising from development investment decisions and implementation based on those information and decisions (Sarah Opitz-Stapleton 2019). Risk-informed development and risk-sensitive development are often used interchangeably in the literature. Some frameworks and tools were developed to mainstream disaster risk over the past decade. For example, Benson et al. (2007) proposed 14 guidance notes highlighting critical factors to mainstream disaster risk in development programing. Mitchell (2003) proposed an operational framework with four key thematic areas that include politics and legislation, policy, knowledge, and practice with over 20 indicators and benchmarks. The experiences from Southeast Asian countries in risk-sensitive development planning have highlighted many challenges including the systematic integration of disaster risk information into development programing and design, implementation, and maintenance of individual projects across all relevant sectors (UNESCAP 2018). However, the research on risk-sensitive development tools, matrices, and guidance is still at the early stages, since the growing recognition and need for risk integrated development was realised in the post-2015 global agenda than ever before.

For sustainable development to be realised in line with 2030 United Nations Sustainable Development Goals, risk sensitivity in the development planning and initiatives are inevitable that allows for development to become a driving vehicle to reduce disaster risks, avoid creating further risks and help to build a resilient community (Sarah Opitz-Stapleton 2019). There has been some experience and lessons on developing and implementing risk-sensitive land use plans and infrastructure in the past (McAllister 2015; Papazoglou et al. 2000). However, risk-sensitive development at different levels—from community to national, as a holistic approach has a long way to go in a way that systematically integrates disaster and climate risks from the development conceptualisation, planning, designing, and implementation. Further, formulation of risk-sensitive development approach was promoted in the Community Resilience Framework (CRF) to align with the local priorities of operationalising Sendai Framework for Disaster Risk Reduction (2015–2030) (Saja Aslam et al. 2016).

13.3 Community Resilience in the Context of Disaster Risk Management

Whilst Sri Lanka has historically experienced the impacts of multiple hydro-meteorological disasters such as floods, drought, landslides as well as geological disasters such as tsunamis and earth tremors, more than 90% of the disaster victims in Sri Lanka result from hydro-meteorological disasters (SLCDMP 2014). Concurrently, the frequency and severity of natural disasters attributed to rapid urbanisation, industrialisation and unplanned development have escalated, causing serious loss and increased vulnerability in terms of human, socio-economic, cultural, infrastructure and environment.

13.3.1 Disaster Risk Management—A Brief Historical Overview in Sri Lanka

Sri Lanka historically embraced an ecosystem-based approach to Disaster Risk Reduction (DRR) within a tank and village culture. This system was disturbed over the time period and the resilient capacity of communities gradually eroded. Since independence, the government has established systems to provide welfare assistance and from the 1990s it formalised provision of relief and shelter assistance through the Ministry of Social Services, and Community-Based Disaster Management programs (CBDRM) were also launched.

Historically, Sri Lankan communities have had their own systems and strategies of managing disaster risk dating back many centuries. This history of Disaster Risk Reduction (DRR) goes back to the village tank (reservoir) system, which was based on an ecosystem management approach and supported the country's agrarian civilisation (DMC 2015). Ancient community governance systems covering agriculture, irrigation and social institutions were abandoned in later years, but some of these reappeared following their recognition as valuable tools for social governance (DMC 2015). Although the term was not formally used, community participation was practiced at various levels in post-colonial times. During the past two decades, various agencies including the Disaster Management Centre (DMC) piloted programmes towards community participation in risk reduction.

The 2004 Tsunami and the subsequent Disaster Management Act significantly altered the existing disaster management system in Sri Lanka. The changes included the establishing necessary institutional and administrative foundations for disaster management, developing disaster management plans at different levels (national to local), establishing coordination for risk reduction, initiating mainstreaming Disaster Risk Reduction (DRR) into other development sectors, and knowledge building programs at various levels towards creating a safer nation. New disaster management policy assigned collective responsibility, equality, diversity and inclusion towards building a safer nation for all with increased resilience to disasters and identified Community-Based Disaster Risk Management (CBDRM) as the tool for risk reduction at the local level (DMC 2015).

13.3.2 Community-Based Disaster Risk Management (CBDRM)

Community involvement is the corner stone to identify vulnerability scales and to improve their resilience. At-risk communities are actively engaged in the identification, analysis, treatment, monitoring and evaluation of disaster risks in order to reduce their vulnerabilities and enhance their coping capacities (DMC 2015). The involvement of highly vulnerable social groups such as women, children, the elderly and the disabled is considered paramount in CBDRM.

CBDRM implies the participation of communities including most vulnerable groups in the identification of risk and risk management proposals including mitigation of potential hazards, enhancing coping capacity, preparedness planning and response. Whilst the emphasis by the Disaster Management Centre has been on the development of preparedness and response plans at all levels, some NGOs have engaged in piloting and implementing preparedness, response and risk reduction and development activities at the community level within the past 20 years. Whilst some organisations have included DRM into development planning at the lowest government administrative levels through inclusive approaches, others have confined their activities only to create village/city-level development plans without due consideration of risk reduction aspects.

13.3.3 Key Lessons and Challenges in CBDRM Approach in Sri Lanka

Whilst the current practice of CBDRM is well accepted as a tool and the process for building safer communities, challenges arose in (1) practical implementation of CBDRM and (2) affordability of mainstreaming into government community development programs. Therefore, current practices of CBDRM needed re-examination and a common CBDRM with the contextual approach has to be considered, which also mainstreams risk into the national and regional development plans and programs.

The National Disaster Management Plan (2013–2017) and the Sri Lanka National Disaster Management Policy emphasised the importance of community engagement in disaster management. Further, CBDRM is prioritised in the National Policy and Plan on Disaster Risk Reduction. However, CBDRM was identified as a standalone intervention where clarity on its connectivity in influencing scaling up development planning is not adequately addressed. Hence, it becomes of paramount importance to look for opportunities on how development planning can be influenced through the CBDRM approach. CBDRM implementation in Sri Lanka had many limitations and challenges (DMC 2015):

(a) The number of casualities is decreasing, the impact on the economy and livelihoods of the communities affected by disasters is escalating. The CBDRM, whilst it functions as a tool to develop community preparedness capacities, has so far failed to look into development needs in general.
(b) Community participation and local leadership in making decisions and sharing outcomes jointly has to arise through an organically grown system. The project-oriented CBDRM approach typically consists of preconceived activities without adequate room for changes or adaptation based on local risk perspectives.
(c) Lack of or limited government and multi-stakeholder participation. The disaster management activities/processes are seen as limited to the Disaster Management Centre, not the business of other departments involved in development planning and implementation.
(d) Since disaster management is not a subject devolved to the local governance system, local authorities do not allocate adequate financial and physical resources required for the implementation of DRR activities. Further, most local authorities fail to consider DRM activities and programs as their priorities, and often expecting external support through international agencies.

13.3.4 Concept of Community Resilience and Risk-Sensitive Development

The community or village is the first affected level in a disaster and must use coping and survival strategies to face the situation before outside help arrives. Communities facing simultaneous or repeated shocks are better supported when humanitarian action also addresses the underlying vulnerabilities and builds capacities to better cope (DMC 2015). The community-based approach complements the defects of the top-down approach in development planning and disaster management which fails to address local needs, ignores the potential of indigenous resources and capacities, and may even increase people's vulnerabilities.

CBDRM covers a broad range of interventions to reduce disaster risks designed by people in at-risk localities and are based on their urgent needs and capacities. However, with the paradigm shift from emergency management to disaster risk reduction, there is more stress on proactive pre-disaster interventions (such as prevention, mitigation and preparedness). These sectors are not in the control of the community themselves and often beyond their scope. Similarly, most disasters by nature are transboundary and need higher level interventions in addressing risk factors. Thus, a more comprehensive community-centred approach is required to address the limitations of the CBDRM framework, which enables risk-sensitive development including the principles of CBDRM. Risk assessments and reduction strategies must be integrated into public investment policies and planning including community-based interventions. Policymakers and other relevant key stakeholders should encourage the development and financing of plans in a coordinated and coherent manner across sectors recognising community inputs (DMC 2015).

The following three concepts discussed in this chapter converged to mainstream disaster risks in the development process:

Integrated Community-Based Disaster Risk Management (ICBDRM): The traditional approach to CBDRM was augmented to ICBDRM by an all-inclusive approach from the lens of disaster risk management. From the disaster management point of view, ICBDRM was promoted amongst the disaster management stakeholders at the practitioner and policymaker levels.

Risk-Sensitive Development Planning (RSDP): The advocacy for greater risk sensitivity in the development planning was promoted amongst the development practitioners and policymakers through the concept of risk-sensitive development. This concept was advocated through the lexicon familiar amongst development stakeholders to mainstream disaster risk reduction and it was found to be effective than talking about ICBDRM to the development partners.

Community Resilience Framework (CRF): The concept of community resilience was the converging point for disaster and development practitioners. CRF was the driving theme behind the ICBDRM and RSDP, which provided a common meeting point as an overarching framework for discussion amongst the key stakeholders.

13.4 Method

The main goal of the Community Resilience Framework (CRF) is to help Sri Lankan communities achieve resilience through the understanding of underlying risk factors. It intends to provide an insight into community-based resilience for any stakeholder working on DRR. For policymakers, it is a document that will assist in designing strategies to overcome obstacles imposed due to natural or man-made hazards.

The key objective of CRF is to guide and facilitate local authority (including divisional secretariat and local government) officials and other stakeholders in disaster management to work towards resilience building at local to divisional level (DMC 2015). This entails guiding development planning committees to assess risks and accordingly plan disaster preparedness so as to reduce community vulnerability.

In this process, the disaster risk assessment is led by the community and facilitated by local authority development planning groups together with external stakeholders. The representatives from the most vulnerable community will also be a part of the risk assessment whilst the inclusion of women, children, persons with disabilities and the aged will be ensured. Community Risk Assessment leads to the formulation of development plans at the community level with adequate emphasis on risk reduction measures as well as exploring the required resources. Annual Divisional Development Plans and other sources facilitate the budgetary requirement, which leads to the successful implementation of these plans. The implementation takes place with consideration given to mainstreaming the Disaster Risk Reduction into development.

Specific objectives of the Community Resilience Framework (DMC 2015) are to.

- Direct decision-makers to implement development plans with adequate incorporation of risk reduction measures and to mobilise resources towards resilience
- Provide direction to individuals, communities, organisations and practitioners who engage in building community resilience
- Provide a common implementing guideline for all actors who support and ensure that community resilience is strengthened, and ensure community-level risk reduction initiatives are sustainable.

13.4.1 Process of Operationalising Community Resilience Framework (CRF)—Sri Lanka

First Phase: Field testing the Integrated CBDRM (ICBDRM) model in two districts for customisation and validation. This included the following key components:

(a) A prescribed ICBDRM model implemented through the Community volunteers.
(b) Activities were fully implemented by the agencies (Oxfam and Practical action in this pilot project) and with coordination of District Disaster Management Units.
(c) Limited to district-level implementation.
(d) Village Disaster Management Plans were prepared in government formats.
(e) Mostly looking at preparedness as a focus.

In this phase, the implementing agencies also tried to understand the possible coordination with the government in advocating the Integrated CBDRM (ICBDRM) as a way forward and advocated in accepting the ICBDRM model as a tool also for community preparedness.

Second Phase: Since the inception of the Disaster Management Road Map for Sri Lanka in 2005, various Disaster Management stakeholders including government and non-government agencies have implemented CBDRM interventions across the country. The policy shifted to risk-based planning aligning with the Sendai Framework for DRR adopted in March 2015 in the Sri Lanka Comprehensive Disaster Management Plan [SLCDMP (2014–18)] and Education Sector Development Framework and Programme [ESDFP (2012–2016)].

To have a greater impact at the community and national level, the Government of Sri Lanka was advocated to agree on a unified version of community-based disaster management model to harmonise CBDRM methodology and approaches across the country at all levels from national to communities. More importantly, unified vision or common characteristics of CBDRM were defined and reinforced through an affordable and sustainable model where DRM stakeholders both at the national and local level can play an instrumental role in operationalising the model. It is quite evident that standalone project model does not comply with the requirement for replicability and sustainability. A complementary or a common CBDRM model that is affordable by the local DRM stakeholders needed to have evolved and validated by the stakeholders of Disaster Management and other policymakers from sectoral ministries.

The project aimed to work with and support the Ministry of Disaster Management (MDM) and Ministry of Education (MoE) to take the lead in implementing the ICBDRM model that can promote technically, operationally and financially affordable best practices from field-tested experiences towards building resilient communities.

In order to ensure that the communities are safer to disasters, it is fundamental that all its members, including vulnerable groups such as women, children, people with disabilities, can better cope and avert disasters and risks are reduced. Thus the project promotes an inclusive approach to disaster risk management, recognising that different people face different risks and barriers, so responses are tailored and empowering. The approach, including community mobilisation, capacity building, and inclusive risk assessment, planning and mitigation ensures the full and meaningful participation of all groups and individuals in identifying and reducing risks; promotes equality of rights and opportunities for them in encountering potential risks; appreciates and responds to their diverse characteristics, capacities and vulnerabilities; contributes to resilience for everyone by removing barriers that keep excluded people further away from engagement.

Third Phase: This is the final phase of operationalising CRF, where the contribution to the national framework for risk-sensitive development planning was scaled up.

(a) Advocated for Community Resilience Framework as a common model for community-based risk management in Sri Lanka.
(b) Trained the human resource capacity of the government as a sustainable approach.
(c) District planning mechanisms and the common format were replicated in other divisions.
(d) Simplified tools and processes for risk-sensitive development planning as a win–win, where the tools were simplified and updated to suit the time and technology of the government system (Fig. 13.1).

13.5 Results and Discussion

Implementation of Community Resilience Framework (CRF) training for the key divisional and district-level development state officials has highlighted further revision to refine the training manual on Community Resilience Framework (CRF) developed by the technical working committee on DRR under the DMC. The need for fine-tuning the existing training guidelines along with the addition of training aids and

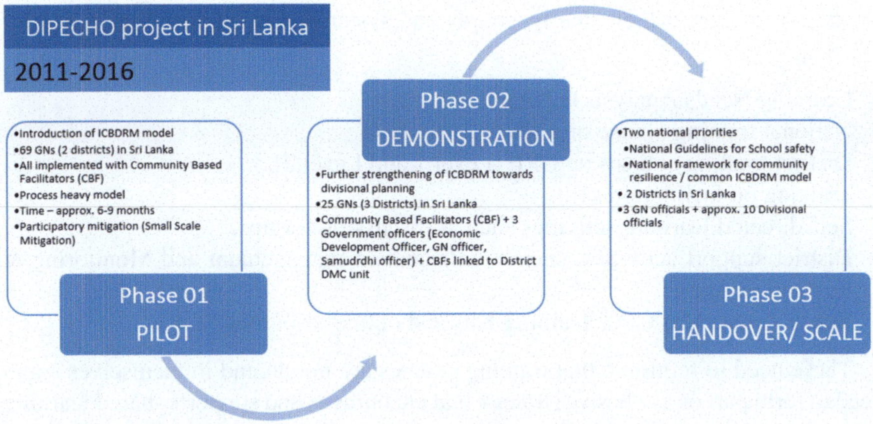

Fig. 13.1 Three phased approach for operationalising CRF. Source (Authors)

tools was identified during the cascade training programs implemented in districts by the trained trainers of CRF.

In addition to the refining and finalisation of the CRF training manual along with the new training aids, an appropriate training strategy needs to be conceived so that all key stakeholders implementing training programs on CRF could adapt a uniform approach. A uniform training strategy is to make the learning service accessible to all key staff members involved in development planning and implementation of development projects. This will be achieved through the establishment of learning service lead by DMC which is able to provide appropriate, contextualised and affordable learning support in CRF. Hence training, supporting, assessing and creating a pool of assessed/accredited local trainers in CRF at district level helped in achieving the roll-out of CRF to divisional and GN level. This requires an approach to implement CRF that facilitates a wide range of learning options. The programme can be developed and delivered through an appropriate mix of the following learning methods, including but not limited to.

- Direct group learning such as workshops, briefings and seminars.
- Self-directed group learning such as discussion forums, action learning sets and communities of practice.
- One-to-one learning such as coaching and mentoring.
- Self-managed learning such as interactive learning CD, videos, publications/resources.

13.5.1 *Training Strategy of CRF Trainers and Facilitators*

The key concept for the training strategy on Community Resilience Framework (CRF) is to build on and develop initiatives that are already in place, but within a cohesive strategy which clearly, simply and systematically links all the activities together.

- Learning Needs Analysis (LNA),
- National training course on CRF,
- District training courses on CRF [Level 1 and Level 2],
- Training on facilitation skills,
- Self-directed learning activities such as discussion forums,
- District support activities, such as information management and Monitoring & Evaluation, and
- Resources such as CDs, Training Kits and online resources.

These need to reinforce that training courses are not an end in themselves—and need to form part of a cohesive strategy that encourages and supports shared learning and a positive learning culture, and maintains quality of learning. This could be developed through a staff workshop initially at the divisional level by the Divisional Secretary (DS) under the guidance of District Secretary (Government Agent), and supported with appropriate systems, resources and ongoing communication by the Assistant Director of District Disaster Management Centre (DDMC).

The key strategy for the implementation of Community Resilience Framework training program, therefore, is to build on and develop these initiatives which are already in place, but within a cohesive strategy which clearly, simply and systematically links all the activities together: Consistently the feedback from the past workshops conducted on CRF under DIPECHO project at district and divisional level, highlights that the practical, experiential approach to training is received extremely positively, despite the 'paradigm of formal training courses with didactic presentations by content experts to passive learners'. Hence, approach to training of this kind—using activities, case studies, experiential learning—is influencing the culture of learning and were well received by the development practitioners who have already a degree level formal qualification.

Another key element of the training on CRF is in Training of Trainers Program for the district and divisional level state officials involved in planning such as Assistant Directors of Planning (ADP), Divisional Secretaries (DS). The assessment process for trainers is seen to be very positive and successful. The piloted strategy ensured the effectiveness and impact of this on the programme through the assessment process, since they were trained in cascade training within their own district and divisions effectively, and to be a resource person for the future CRF training programs.

13.5.2 Community Resilience Framework Training Process

In order to enhance the knowledge and skills on Community Resilience Framework (CRF), a number of state officials were selected for training as CRF trainers. This selection could be done at the district level. However, in order to ensure that all divisions were given equal opportunity, the selection of state officials was done at the

Divisional Secretariat level. Key state officials at the higher-authority level included: Divisional Secretary of the division, Assistant Divisional Secretary of the division, Director of Planning at district administration, Assistant Director of Planning of the division, Assistant Commissioner of respective Local Government, and Assistant Director of District Disaster Management Centre of the respective district (Fig. 13.2).

In addition, the following mid-level staff from divisional administrative offices were also selected: National Disaster Relief Service Officer, Development Officer

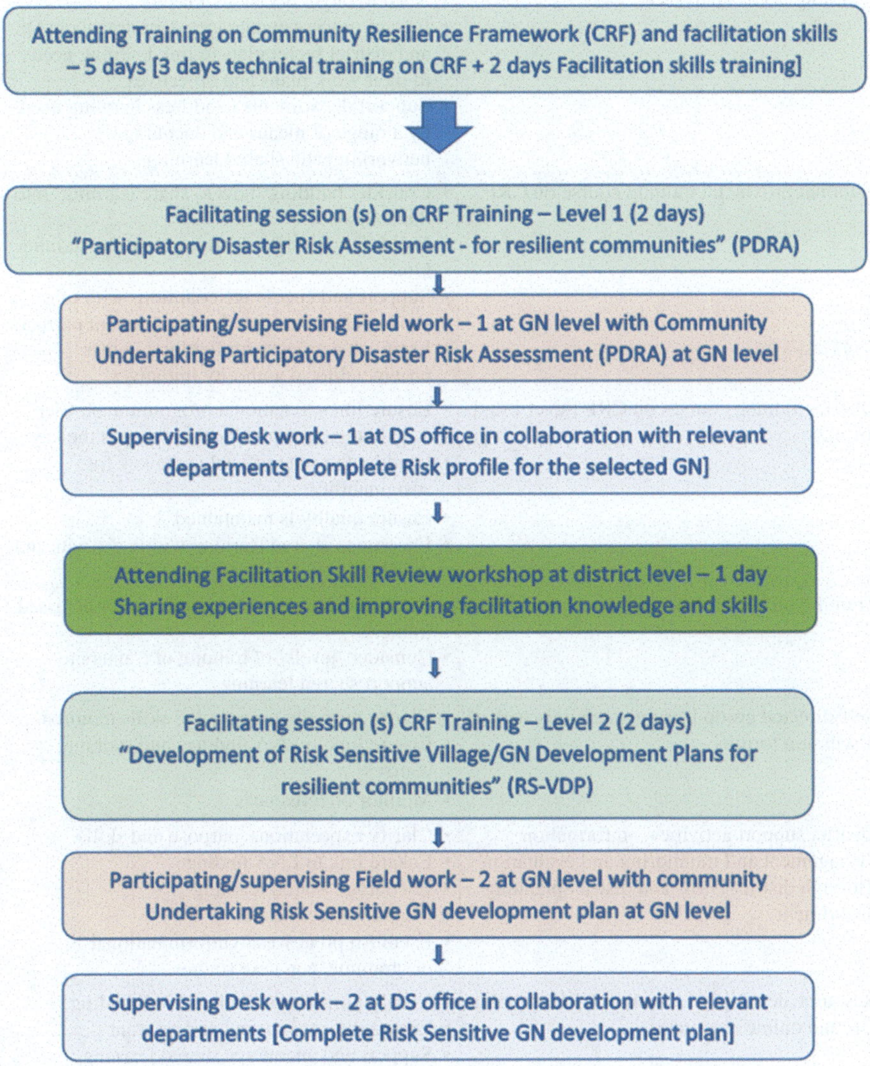

Fig. 13.2 CRF Training of Trainer flowchart. Source (Authors)

involved in DS division development planning, Social Service Officer, and Disaster Management Center Staff members (Table 13.1).

However, the same category of other staff members who were involved in disaster management and development work can also be included, based on the learning needs

Table 13.1 Key elements of CRF training strategy and their descriptions

Key elements	Purpose and their description
Learning needs analysis (LNA)	• Need to target decisions-makers • Ensure process encourages departments in DS and district secretariats to link learning needs to their work plans and objectives • Support departments to address learning needs by a range of means and encourage networking and shared learning
National/provincial training course on CRF	• Consider building 'how to share learning' into the courses • Build 'how to share learning' into the training kits • Support and encourage coalitions with other stakeholders particularly development partners • Establish a process for linking to other provincial/local authority initiatives
District training courses on CRF [level 1 and level 2]	• Ensure links to national programme on CRF • Local languages and local training is the key • Lead by the state official is the key for sustainability • Ensure quality is maintained • Encourage shared learning within districts and divisions
Training on facilitation skills	• Establish clear terms of reference for assessed trainers • Consider 'levels' of training of trainers to support shared learning
Self-directed group learning activities such as discussion forums	• Clarify purpose, process and skills required • Ensure link to LNA findings and training courses • Monitor effectiveness
District support activities—information management and monitoring and evaluation Through district DMC and district planning department	• Clarify expectations, purpose and skills • Ensure link to LNA finding • Encourage shared learning • Monitor effectiveness • Establish processes to inform national programme, e.g. case studies
Resource development such as CRF, training kits and online resources	• Clarify purpose and link to LNA finding • Ensure they are current and relevant • Support and encourage shared learning • Consider different levels of need • Assess need for translations

analysis and the requirements at the respective DS divisional level. The complete process of training CRF trainers is shown in the flowchart Fig. 13.2. Each CRF trainer should be encouraged to complete a post-training reflective journal capturing lessons learnt and challenges and participation in the post-training review and evaluation. Hence, the complete process includes the following steps and products.

In order to become "CRF Trainers", the following activities should be completed by each CRF trainer.

I. Three days Community Resilience Framework Training attendance as a participant.
II. Attending two days of Facilitation Skills Training as a participant.
III. Facilitating at least one session in the first or second cascade training.
IV. Evidence of involvement in the fieldwork during and/or after the first and second cascade training programs.
V. Completing a post-training reflective journal, capturing lessons learnt and challenges
VI. Participation in the post-training review and evaluation.

Based on the training strategy and lessons from the CRF implementation by DIPECHO, a set of steps are proposed to train a number of government staff members from the Divisional Secretariate Divisions [DSD] and Grama Niladari Divisions [GND] who have been selected for implementation of CRF. Government staff members who were trained by Master Trainers undertook the development of "Participatory Disaster Risk Assessment" and "Risk Sensitive Village/City Development Plans" once they completed two levels of CRF training.

At the end of the process, Risk Sensitive—Village Development Plan (RS-VDP) was produced for each of the village/city (GND) which is the lowest administrative level in Sri Lanka. The outcome of this process is twofold. One side, it is aimed at building the capacity of the key government staff members who can train their subordinates, on the other side, trained government staff members responsible for the village/city facilitate the participatory risk assessment and planning were headed by the respective communities. The government staff members completing the process of attending the two levels of training programs and subsequently implementing fieldwork were certified as CRF Facilitators. This includes key staff members who are responsible for the village/city divisions so that after attending the training they facilitate the participatory risk assessment and planning in their respective lower level administrative division.

It is recommended to select the following key staff members for the cascade training programs in districts: Grama Niladari (GN) who is the village administrator of selected GN divisions, Economic Development Officer (EDO) of the selected GN divisions, Samurdi Development Officer (SDO) of the selected GN divisions and any other state official responsible for the respective village or city division. Preparation of PDRA and Risk—Sensitive Village/City Development Plan was done by the communities under the facilitation and collaboration of trained CRF Facilitators. This process comprised:

I. Two days of Community Resilience Framework Training—I (PDRA) attendance as a participant.
II. Facilitating Participatory Disaster Risk Assessment (PDRA) at the respective division.
III. Two days of Community Resilience Framework Training—II attendance as a participant.
IV. Facilitating Risk Sensitive—Village Development Plan (RS-VDP) at the respective division.

The following key lessons from implementing cascading capacity-building programs were identified: the complete process is said to be "lead by, facilitated by, undertaken by" the responsible state officials and their departments, the combination of development and DM state officials lead by the departments of District/Divisional Development Planning and District DMC have formed a greater drive, a sense of responsibility is created to act according to the action plan drawn at the end of the training programs, the format of CRF training into two sectional training programs and the time gap allowed between is found to be an effective approach and could be the best way of delivering CRF training modules at the district/divisional level (Fig. 13.3).

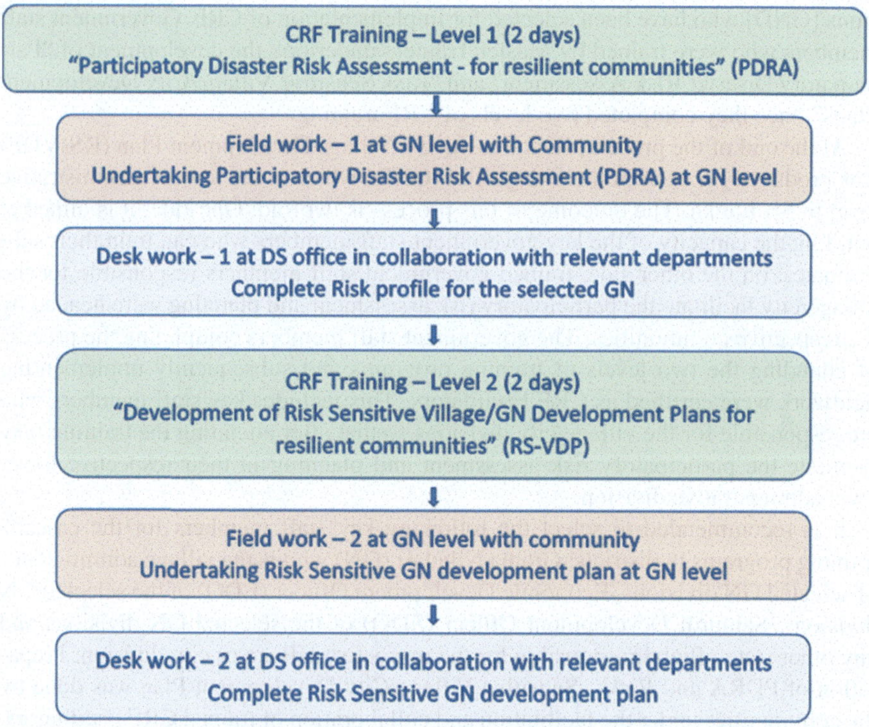

Fig. 13.3 CRF cascading training strategy. Source (Authors)

However certain challenges have also existed. Some of the key challenges in this process were as follows: the process is very time-intensive from the community side as well as for CRF Facilitators, absence of a monitoring system imposes a key challenge to produce the final outcome. Master Trainers have encountered difficulties in creating a good flow between modules such as linking the previous module to the next, capturing key lessons in each module and providing few additional technical inputs/know-how, and Master Trainers found it difficult to structure some modules in logical steps and adapting the content of some modules to the target audience, which has caused some confusion amongst trainees, missing some key tools during the fieldwork. Further, the mainstreaming climate change adaption strategies had been lacking in the risk-sensitive development planning and implementation in this pilot project, which was found to be an important gap in operationalising the community resilience framework as an overarching driving framework to build resilient and sustainable communities.

13.6 Concluding Remarks and Recommendations

The growing impact due to disasters demand risk-sensitivity in development planning more than ever before. An appropriate resilience investment is key to implement the Sendai Framework for disaster risk reduction priorities. However, over the past decades, neither the standalone approach to reduce risks at the community level, through CBDRM approaches, nor development plans implemented without due concern to risk reduction integration were effective and brought expected impact. The lessons in implementing CBDRM approaches and development initiatives without proper mainstreaming of disaster risks have lead the concept of risk-sensitive development planning as a concept to drive the community resilience framework in Sri Lanka.

The sustainable approach to roll-out risk-sensitive development planning and implementation is to build the capacity of the state stakeholders in disaster management and development planning departments from the community to regional level. This manuscript discussed the cascading capacity-building approach implemented by the DIPECHO project in the Northern Province of Sri Lanka as a pilot project to operationalise Community Resilience Framework (CRF). Initially, the divisional staff attached to development planning and disaster management were trained as trainers, who then trained the community-level state staff members who are the direct facilitators of risk-sensitive development planning and implementation. This process was successful, since the state departments felt the ownership from the beginning of the project cycle, compared to externally driven, non-governmental organisation supported and facilitated method of implementing a project that had been the traditional approach in the targeted communities.

This study has drawn following key recommendations from the implementation of CRF at district/divisional level through cascading training approach: capitalising on the positive lessons and improve the process and approach for the implementation of CRF at district/division/village levels, developing new monitoring tools to ensure sustainability of the process after the exit of the pilot project, creating a "Community of Practice" (CoP) at district level under District DMC by establishing a resource pool of CRF Master Trainers/Facilitators, fine-tuning/complementing the trainer guide for the purpose of training the CRF Facilitators at district/division level, and updating the current model from the lessons drawn from the replication and advocate it for further uptake from the national to village/city levels. Finally, the gaps in the integration of climate risks into the risk-sensitive development planning need to be addressed by further reinforcing the Linkages between Disaster Reduction, Development, and Climate Change Adaptation (LDRDCCA).

Acknowledgements The authors wish to thank Oxfam and ACTED Sri Lanka for allowing materials and reports produced as part of the 8th DIPECHO (Disaster Preparedness project funded by the European Commission Humanitarian Office) project in Sri Lanka to be reflected in this paper. The contents of this publication and views expressed in this publication are the sole responsibility of the authors and can in no way be taken to reflect the views of project stakeholders including ECHO, Sri Lanka Disaster Management Centre, and the current affiliations of authors.

References

Bendimerad F (2003) Disaster risk reduction and sustainable development. In: World Bank Seminar on The Role of Local Governments in Reducing the Risk of Disasters, Held in Istanbul, Turkey (vol. 28, pp. 57–75). Citeseer

Benson C, Twigg J, Rossetto T (2007) Tools for mainstreaming disaster risk reduction: guidance notes for development organisations. ProVention Consortium

DMC (2015) Community resilience framework. Ministry of disaster management/disaster management center (DMC), Sri Lanka. https://www.preventionweb.net/files/54485_communityresiliene cfwsl.pdf (Accessed 20 Februrary 2020)

McAllister T (2015) Research needs for developing a risk-informed methodology for community resilience. J Struct Eng 142(8):C4015008

Mitchell T (2003) An operational framework for mainstreaming disaster risk reduction. University College of London

Papazoglou IA, Bonanos G, Briassoulis H (2000) Risk informed decision making in land use planning. J Risk Res 3(1):69–92

Saja Aslam A, Lafir Sahid S, Sutharshanan M (2016) A localized approach in risk-sensitive development planning to build resilient communities: an experience from Northern Sri Lanka. In: Proceedings of the 7th International Conference on Sustainable Built Enviornment 2016 (vol. 7, pp. 110–120, vol. Symposium C). ICSBE 2016

Sarah Opitz-Stapleton RN, Kellett J, Calderone M, Quevedo A, Peters K, Mayhew L (2019) Risk-informed development From crisis to resilience. UNDP, ODI

Schipper L, Pelling M (2006) Disaster risk, climate change and international development: Scope for, and challenges to, integration. Disasters 30(1):19–38. https://doi.org/10.1111/j.1467-9523.2006.00304.x

UNDRR (2019) Global assessment report on disaster risk reduction. In U. N. O. f. D. R. Reduction (ed). Geneva, Switzerland: United Nations Office for Disaster Risk Reduction (UNDRR)
UNESCAP (2018) Risk-informed development: using disaster risk information for resilience (vol. Conference Report).
Wilkinson E, Lovell E, Carby B, Barclay J, Robertson R (2016) The dilemmas of risk-sensitive development on a small volcanic island. Resources 5(2):21

Chapter 14
Leveraging Youth Engagement in Disaster Risk Reduction Through Science, Engineering, Technology, and Innovation in Indonesia

Fajar Shidiq, Mizan B. F. Bisri, Nuraini Rahma Hanifa, Risye Dwiyani, Irina Rafliana, and Ardito Kodijat

Abstract This chapter provides an overview of Indonesian youth and young professionals' (YYP) engagement in Disaster Risk Reduction (DRR) and explores how the current national framework enables the environment for the YYP to best contribute to reducing disaster risks. DRR activities involving YYP have been growing for years. However, many of them are scattered as there is no official DRR platform exists focusing on youth and young professionals. Thus, the overall potentials of YYP for DRR have not been well utilized as a part of the nation's resilience building. The first part of the proposed chapter will update the current academic discourse on youth engagement in DRR and general characteristics of Indonesian YYP engagement in DRR. The second part of the chapter will discuss the interlinkages between Indonesian policy on disaster management and youth development as well as discussion on the potential role of youth. The third part of the chapter will exhibit insight from U-INSPIRE, a platform of youth and young professionals working on innovation,

F. Shidiq (✉) · M. B. F. Bisri
IRDR Young Scientist Fellow, Beijing, China
e-mail: sdq.fajar@gmail.com

F. Shidiq · M. B. F. Bisri · N. R. Hanifa · R. Dwiyani · I. Rafliana
U-INSPIRE Indonesia, Jakarta, Indonesia

F. Shidiq
Climate Change Adaptation and Disaster Research Center, Indonesia Defense University (PSB API UNHAN), Sentul, Indonesia

M. B. F. Bisri
United Nations-University-Institute for Advanced Study of Sustainability (UNU-IAS), Tokyo, Japan

N. R. Hanifa
Center for Earthquake Science and Technology, Research Center for Disaster Mitigation (CEST-RCDM, ITB), Bandung Institute of Technology, Bandung, Indonesia

I. Rafliana
Indonesian Institute of Sciences (LIPI), Jakarta, Indonesia

A. Kodijat
UNESCO Jakarta Office, Jakarta, Indonesia

R. Djalante et al. (eds.), *Integrated Research on Disaster Risks*, Disaster Risk Reduction,
https://doi.org/10.1007/978-3-030-55563-4_14

science, and technology for resiliency that supports DRR policy and action at the local, national, and international level in line with Sendai Framework. Given the ability of U-INSPIRE to mobilize Science, Engineering, Technology, and Innovation (SETI) actions from YYP at all phases within the disaster management cycle in Indonesia since its establishment, as well as diffusion and uptake from YYP of other countries to establish their own U-INSPIRE, potentially this approach can become a model on leveraging YYP engagement in DRR.

Keywords Youth · Young scientists/professionals · SETI · DRR · Indonesia

14.1 Background: Literature Review on Youth and Disaster Risk Reduction

The first concrete effort to connect youth and DRR was made in 2000, where the United Nations International Strategy for Disaster Reduction (UNISDR) initiated the campaign "Disaster Prevention, Education and Youth" (UNISDR 2000). From there, generally, the role of youth has been limited to risk-knowledge dissemination and communication through formal and informal communication networks (Mitchell et al. 2008; Haynes and Tanner 2013; Fernandez and Shaw 2014). At that time, in principle, youth is being grouped altogether with children as the target of DRR education and training interventions, as clearly expressed in Article 3-ii-h of the Hyogo Framework of Action (HFA) 2005–2015: "… promote the inclusion of DRR knowledge in relevant sections of school curricula at all levels and the use of other formal and informal channels to reach youth and children with information" (UNISDR 2005). There has been a limited space for youth to engage in international framework DRR discourse as well as in the DRR policy-making at the local level, in the decade before and after the adoption of HFA 2005–2015 as well as during the Second UN World Conference on DRR in Kobe, Japan in 2005.

Nevertheless, the recognition for youth's contribution to the implementation of DRR at their respective community and globally was gaining momentum during the Third UN World Conference on DRR in Sendai, Japan, in March 2015. The youth was recognized as an essential and strategic partner in the DRR framework development and implementation during the world conference as well as eventually in the conference output, the Sendai Framework for Disaster Risk Reduction 2015–2030 (SFDRR) (Cumsikey et al. 2015). The SFDRR 2015–2030 includes a substantial change: "the recognition for children and youth as agents of change which require member states to give the space and modalities for contribution to disaster risk reduction, in accordance with legislation, national practice, and educational curricula" (Article 36, a, ii), for achieving the seven targets of SFDRR (UNISDR 2015).

Prior to the SFDRR 2015–2030, some countries have provided space for youth to formally engage in DRR policy-making (Fernandez and Shaw, 2014) and it is widely expected that the adoption of SFDRR can further strengthen the position of youth. In addition, Cumiskey et al. (2015) have provided statistical evidence that the number of youth involved in DRR activities in various regions keeps on increasing.

While a number of stakeholders engaged in the formulation of the SFDRR have recognized the benefits of building the capacity of youth in their voluntary commitments (Cumsikey et al. 2015), there is a need of locating the framework for youth in the DRR engagement and a concerted institutional setup at the national level of each country to fully engage their youth. Many good practices on youth engagement in DRR are rather limited to specific case studies, anecdotal evidence, and the SFDRR itself did not mention a specific role for youth (and children) in DRR (Chatterjee et al. 2015).

However, there are several challenges to youth engagement in DRR. First, is the definition of youth itself. In general, the UN generally defines "youth" as the age group between 15 and 24, which rooted in UN's General Assembly Resolution A/36/215 and 50/81 whereas it mainly defined as such for statistical purpose and adopted by most of the UN agencies, e.g., UNICEF, WHO, UNFPA, UNESCO (United Nations 2013). The UN also acknowledges other definitions employed by Member States and other UN entities, e.g. UN-Habitat recognizes youth up to 32 years of age. Furthermore, in terms of participation from the children and youth, as one of the nine major groups of civil society, the United Nations Department of Economic and Social Affairs (UNDESA) mandated the United Nations Major Group for Children and Youth (UNMGCY) for facilitating the participation of children and youth up and until the age of 30 (Cumsikey et al. 2015). It is the UNMGCY that also promotes the engagement of children and youth in the negotiation process on the Third UN World Conference on Disaster Risk Reduction (WCDRR), which resulted in agreement on the SFDRR 2015–2030.

Second, especially in Asia, lack of structured and institutionalized opportunities made the role of youth very limited in the decision-making process, including DRR. Fernandez and Shaw (2014), by describing the case of the Philippines, showcased that a supporting environment by adults and mainstream development actors is essential to fully make use of DRR knowledge and skills of youth. Furthermore, it is understood that in the past few decades, young people have been gradually moving away from engagement in institutionalized structure toward greater involvement in caused-oriented political activism (UNDESA 2015), e.g. including on DRR, climate change or environmental cause. Third, the niche of youth engagement in DRR is also a competing one where it may fall under the part of disaster education and school safety as well as community-based DRR (CBDRR), hence the term youth-centered DRR is still relatively new as a stand-alone concept, both in Indonesian and Asian context. Therefore, real and integrated inclusion and follow up is yet to be seen at the country and local level, and must be built upon lessons learned of past youth engagement in disaster risk reduction—including in Indonesia.

14.2 Objectives

This chapter provides an overview of Indonesian youth engagement in DRR and explores how the current national framework delivers enabling environment for the youth to best contribute to reducing risks. DRR activities involving youth have been growing recently. It reviews various policies and programs of the Indonesian government related to youth and DRR and further proposes a conceptual framework of effective partnership with Indonesian youth in DRR. Furthermore, this study outlines the requirement and essential checklists for effective youth engagement in fulfilling the targets of the SFDRR 2015–2030, and particularly guiding framework of Article 36a-ii on children and youth engagement in DRR, both in the Indonesian context and its applicability to other countries. This case study highlights one of the largest groups in our society, i.e., youth, who is often overlooked in the development cycle, especially in Indonesia and many other developing countries.

14.3 Methodology: Data Collection and Analysis

The case study focuses on literally any youth-led or youth-related movements or organizations, which are based in Indonesia that may contribute to DRR. We discuss some prominent and publicly recognized organizations and events from local to the national level, led by youth, working with the youth, or targeting youth as the beneficiaries. As mentioned in the previous section, there is no official statistics on the engagement of youth on DRR activities in Indonesia. Accordingly, we tried to fill in the gap on the topic by pooling data of youth engagement in DRR through four steps: (1) internet-based search on youth-related organizations with activities related to DRR; (2) internet-based search on youth-related programs on DRR; (3) utilization of database of Indonesian applicants to the Children and Youth Forum (C&Y Forum)[1] during the Third United Nations World Conference on Disaster Risk Reduction (WCDRR) on March 18–22, 2015, and (4) member database of U-INSPIRE Indonesia, a Youth and Young Professional Platform in Science, Engineering, Technology, and Innovation for DRR. The writers acknowledge the limitation of the data collections on this chapter as there is no official statistical database related to the youth engagement on DRR in Indonesia. However, we believe that by collecting and analyzing the data through steps taken above, it helps representing a better picture of current youth involvement in DRR in Indonesia.

As the first and second steps, data and information collection were done through extensive online data mining by using Google search engine from October 2016 to January 2017. For the first approach, the authors used the combination of keywords such as "youth", "Indonesia", and "disaster" or "disaster risk reduction", both in English and Indonesia, resulting in 38 youth organizations. For the second approach, on the DRR-related events, the authors used the keywords such as "youth", "disaster", "events", or "activities" and the combinations of them both in Indonesian and English. What differs "events" from "organizations" is that "events" do not necessarily a part of a sustained program of an organization. It could be a one-time or even a spontaneous event recorded on the Internet in the form of full report or news, which indicates youth movements in DRR. Events that are held in several locations in

[1] https://www.wcdrr.org/conference/events/1025 (last accessed November 15, 2016).

Indonesia under the same name and moment are counted as one event. As a result, we found 38 DRR-themed events held in Indonesia that involve Indonesian youth from 2010–2015.

In the third step, the C&Y Forum was selected because its registration has a large number of applicants from Indonesian youth (446), where the authors can tap relevant information, e.g., age, applicants' province of residence and hometown, gender, affiliation to youth organizations, potential skills, and contribution to the C&Y Forum, DRR experiences (professional, academic, and other) and their plan to use knowledge and skills gained from the C&Y Forum. The forum was an important official side event of the WCDRR, as the only side event specifically for youth and directly organized by UNMGCY and directly supported by the United Nations International Strategy for Disaster Risk Reduction (UNISDR). Even in the actual forum Indonesian youth were the second-largest delegation with 190 youth (19% from the total participants) just behind the host country—Japan (23%) (Cumsikey et al. 2015).

In the fourth step, member database of U-INPSIRE is selected because: U-INSPIRE is the first official platform for YYP on DRR in Indonesia, it has a wide range of membership, it gains momentum and receives positive response by government and other DRR stakeholders after the Sixth Session of Global Platform for Disaster Risk Reduction (GPDRR) on October 2019 in Geneva, where Indonesian government acknowledged U-INSPIRE as an official platform for YYP on SETI for DRR in Indonesia. As of October 27, 2019, the U-INSPIRE has 188 members and has partnered with 11 government institutions, 7 international agencies, 18 NGOs/CBOs, 4 private sectors, and 3 media channels.

14.4 Indonesia: Home to Disasters, Home of the Youth

As one of the countries located along the ring of fire, where about 90% of the world's earthquakes occurred, Indonesia faces frequent earthquakes and volcanic eruptions (Brown 2015). This is due to the geographical features of Indonesia which is located on the convergent plate boundaries (Simkin et al. 2006). Despite the high number of earthquake occurrences in Indonesia, frequencies of hydrometeorological disasters are even higher due to the rapid development and deforestation. The number of climate-related hazards has been increasing in the past 26 years with flood having the highest percentage of disaster type (31.2%), followed by tornado (20%), landslide, peat fire, and drought (BNPB 2016). Regarding the occurrence and impacts between geophysical disasters and climate-related disasters, Djalante (2016) concluded that geological disasters have been extremely deadly, while climate-related disasters occurred more often and caused more damage.

Those disasters continuously threatening 261,890,900 population of Indonesia, of which 63.36 million population (25%) are youth (BPS 2018). Demographically, youth in Indonesia is classified as the population between 16 and 30 years of age (BPS 2017). Based on the Indonesian Youth Statistics, the youth sex ratio is 102.36, which means in every 100-female youth there are 102 male youth (BPS 2017). The number of youth as compared to the total population who reside in urban areas (25.22%) and in rural areas (23.19%) are quite similar. On average, Indonesian youth have attended 10.21 years of education, at least grade 1 of high school and it is beyond the state's minimum education level of junior high school (9 years of education). The penetration of the Internet to youth is considerably high with more than half of youth have the access to the Internet. Smartphones become a highly used tool to access the Internet with a percentage of 93.42%, followed by laptop 22.34%, computer 16.9%, and tablet 4.34%.

Overall, 43.78% of Indonesian youth have graduated from high school, 31.99% completed their junior high school, 18.51% just graduated from elementary school, and 4.67% did not complete elementary school education. The rate of school participation shows increasing trends where the rate in 2014, 23.53%, was a jump from 19.05% in 2012 and 21.16% in 2013. The illiteracy rate of Indonesian youth is at 0.64%, with most of them reside in rural areas.

In the context of disasters, youth is often classified with children, as vulnerable groups. Even so, past researches have shown that youth can contribute in sharing information of disaster preparedness with their community (Ronan et al. 2008), brought creative and resourceful ideas to DRR (Barlett 2008), and engaged as volunteers in disaster relief and response (Ronan and Johnston 2005). Therefore, this paper observes Indonesian youth as a potential capacity in DRR, instead of as vulnerable group. The population of youth in Indonesia will rapidly increase in 2020–2035, which will be a demographic bonus as the productive population will reach 64% of the total population (Nahrawi 2016). Citing the former Minister of Youth and Sports, the Law 40/2009 on youth has a specific purpose to "hasten the circulation of leadership both in regional as well as national level which has so far been stagnant" (Raviola 2013).

The national statistics (BPS 2017) also recorded 59.86% of youth participation in the labor force. In terms of economic sectors, 25.57% of youth are engaged in agricultural, 20.76% in trade, and 18.18% in industry. No record is available on the percentage of youth engagement in the third sector, e.g., involvement in NGOs as professionals or volunteers. There is no official record on how many youth engage or participate in youth empowerment programs. The national statistics did, however, record some government champion programs on youth empowerment, namely *Pemuda Sarjana Penggerak Pembangunan di Perdesaan* (PSP3),[2] Indonesian Youth Jamboree, Pemuda Siaga Bencana, *Paskibraka* (National Official Flag Hoister), Youth Entrepreneurship and "Youth Pioneer" in five sectors, namely education, social–cultural and tourism, environmental and natural resources management, food and health industry as well as communication and information. There is also no statistical record from the BPS on youth engagement in disaster-related programs or initiatives.

[2]By 2014, the PSP3 program has sent 1,000 graduates from various universities sent to 494 villages across the nation to help with the development.

Nationwide, there are 149 Youth Organizations registered under the Ministry for Youth and Sports Affairs (Raviola 2013). However, this does not represent the whole youth organization's population in Indonesia. Recently, the emerging youth activities, volunteering, and community services in several fields, such as environment, health, climate change, and education, have been established. Raviola (2013) also mentioned in the paper that the current trend of the involvement of youth is not only at student unions of universities, faculties, and departments, or youth organizations related to political or religious purpose, but also in out-of-campus activities, i.e., NGO or independent sector. Nevertheless, the registration of NGOs in either Ministry of Home Affairs or Ministry of Justice and Human Rights does not specifically classify whether or not a particular NGO is a youth-led organization.

In terms of scope of activities, there are some local youth activities established by university students or neighborhood-based youth organization (e.g., Karang Taruna) to bring positive change to the local neighborhood. There are also other types of youth activities which target not only Indonesian youth, but also bridging Indonesian youth with youth from other countries; e.g., *Bina Antar Budaya* (The Indonesian Foundation for Intercultural Learning), AIESEC, and Purnacaraka Muda Indonesia (PCMI) are still gaining momentum to manage youth exchange, promote cultural exchange and enable youth to make a social impact.

14.5 Indonesian Youth Engagement in DRR: Evidence from the Field

14.5.1 Youth-Related Organizations

The study identified 38 recognizable Indonesian youth organizations on DRR or organizations operating in Indonesia that involves youth through internet search done from October 2016 to January 2017. Eighteen of them are NGOs and government-established organizations that recruit youth as member or volunteers. To mention a few, Muhammadiyah Disaster Management Center (MDMC),[3] a faith-based NGO which has assisted more than 12,000 communities nationwide to be developed as disaster-prepared communities including the youth group. The MDMC also recruits youth as volunteers in the organization. Another example is Taruna Siaga Bencana (TAGANA), a community-based organization formed under the regulation of the Ministry of Social Affairs to support the Indonesian Government on social protection in disaster management (Ministry of Social Affairs Regulation No. 29/2012). Although the members are not necessarily youth, their age ranges from 18 to 45 years old. The organization is managed by the government at the national, provincial, and municipal levels. There are also local NGOs that are mainly managed by young professionals or volunteers but do not define themselves as youth organizations and

[3]https://www.mdmc.or.id/index.php/profil-mdmc (last accessed November 11, 2016).

usually have adults as their board of advisers or board of directors, namely IBU Foundation[4] (based in West Java), IDEP Foundation[5] (based in Bali), and KOGAMI[6] (based in West Sumatera).

Out of the 38 organizations, the remaining 20 are youth led and managed by youth on purpose, independently or under the umbrella of a parent organization (university, research center, or nongovernment organizations). They have various objectives, such as increasing disaster awareness for society including children, school, and local communities through training and simulation. Organizations such as Aliansi Pemuda Peduli and Palang Merah Remaja (the Indonesian Youth Red Cross) focus on the emergency response phase, e.g. collecting donations for disaster victims. Exceptional activities are done by a few youth organizations, such as Holi'ana' Foundation, which organized an Integrated Livelihood Recovery, a post-earthquake and tsunami recovery project targeting women and youth organizations in eight villages in Nias and Aceh.[7] Bandung Disaster Study Group (BDSG) initiated a small-scale DRR Youth Exchange Program and DRR-Themed Homestay Program collaborating with its Japanese counterparts from Waseda University (WASEND) and Kyoto University (KIDS) (BDSG 2012). Holi'ana' Foundation, Relawan Nusantara, Youth for Climate Change (YFCC), BDSG, and Gamarsama have youth empowerment as one of their programs. Another exceptional organization was Indonesian Youth on DRR (IYDRR), a youth forum in Jakarta that opens to individuals and representatives of youth organizations working in the field of DRR to establish network and collaboration. IYDRR organizes sharing sessions and workshops on DRR.

14.5.2 *Youth-Related Events and Initiatives*

There were 38 disaster risk reduction-themed events held in Indonesia that involved Indonesian youth from 2010 to 2015. The results indicate the popularity of disaster-related issues among youth. Scope of activities varies from university, municipality, province, national, and international levels, with the number of participants ranging from the small-limited scale of 5 persons up to massive ones which could involve as many as one thousand youths.

Nevertheless, the events were mostly 'organized for youth', rather than 'organized by youth'. Events for youth were held by the government and/or nongovernment agencies. In 2013, the Japan Foundation held the Youth Competition for Disaster Education that attracted 1,276 Indonesian applicants (Prasasti 2014). The event was further developed into a study visit program to Japan called "HANDs! Project" in the following years, i.e., a human resource development program with disaster education

---

[4]https://ibufoundation.or.id/ (last accessed December 27, 2016).

[5]https://www.idepfoundation.org/en/ (last accessed December 27, 2016).

[6]https://gerashiaga.files.wordpress.com/2014/01/profil-kogami_ind.pdf (last accessed December 27, 2016).

[7]https://www.holianaa.or.id/en/about-us (last accessed 11 November 2016).

as the focus, involving youth from Asian countries who will afterward make their own DRR projects in their respective country. The Ministry of Youth and Sports has also conducted massive events or programs in different cities, such as National Jamboree/Camp (about 600–3000 participants) and annual disaster training for youth volunteers, namely, Pemuda Siaga Bencana (Youth Ready for Disaster).[8] The trained volunteers even set up a camp when there was a flood occurred in Sampang, East Java.[9] Since 2012, National Disaster Management Agency (BNPB Indonesia) has started a Disaster Risk Reduction Month (held every October), followed by National Disaster Preparedness Day (held every 26 April) where some of the participants are youth or at least it could attract large number of youth. Many events held by youth are one-time or spontaneous events, such as a seminar on DRR topics by university students-or taking part in emergency response, such as cleaning up post-flooding or trauma healing for children. Other groups of youth conducted activities that aim to enhance capacity and knowledge exchange as well as emergency response with structured training prior to the events, such as Jakarta Rescue, Bandung Disaster Study Group, and Tagana. They are among the aforementioned recognizable youth organizations.

14.5.3 Indonesian Youth During the 3rd UN World Conference on Disaster Risk Reduction

The UNMGCY's record shows that among 446 Indonesian youth applied for the Children and Youth Forum, 244 (54%) are female and 202 (46%) are male. Their average age by March 2015 was 21.14, with the youngest was 16 and the oldest was 30, just within the definition of youth adopted by the UNMGCY. Almost half of them are 21 (106) and 22 years of age (104). In terms of the geographical place of residence, more than half of them live in Java Island (334) followed by Sumatra Island (86), Kalimantan Island (7), Sulawesi Islands (13), and 6 others live in Bali and NTB islands (Fig. 14.1). Interestingly, 84.98% of the youth came from provinces where major disasters occurred within the 2000s, i.e., Central Java, East Java, Jakarta, Riau, West Java, West Sumatra, and Yogyakarta. This was also expressed by the participants, when they mentioned their disaster experience. Some of the famous large-scale disasters were identified, which include the 2006 Yogyakarta Earthquake, the 2009 West Java Earthquake, the 2009 West Sumatra Earthquake, the 2010 Mount Merapi Eruption in Yogyakarta and Central Java, the 2010 Mentawai Tsunami in West Sumatra, the 2014 Kelud Eruption in East Java, the 2014 Riau Forest Fire, and Greater Jakarta Floods in 2012, 2013 and 2014.

Besides the interesting finding, which links the higher number of applicants to post major disaster experience in the applicant's province, the youth's access to information might also be the factor that pushed the number of applicants in some of

[8] https://www.kemenpora.go.id/index/preview/berita/7674 (last accessed December 24, 2019).

[9] https://www.kemenpora.go.id/index/preview/berita/10429 (last accessed December 24, 2019).

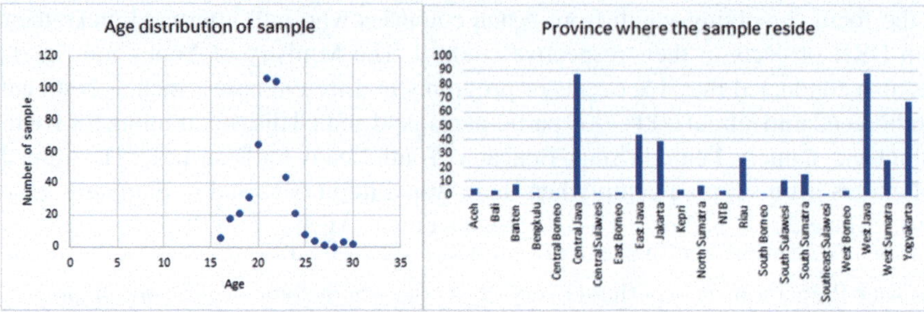

Fig. 14.1 Distribution of participants based on age and province of residence (upper side, obtained from UNMGCY 2015)

the provinces. The provinces mentioned above have reputable national and private universities. Universities are potential institutions to provide youth/students access to information, and reputable universities generally encourage students to have international exposure. Out of the 3,089 higher education institutions in Indonesia, only 37 involved in the University Forum for Disaster Risk Reduction and have education or research program on disaster management. There are 30 out of 37 of those universities located in Java and Sumatra Islands (Sakurai and Bisri 2013), where most of the youth applicants reside (Fig. 14.2).

Before doing an in-depth analysis of the portfolio of youth engagement in DRR, we look at the basic information of their affiliation to one or more organizations. From the record, we learned that 116 applicants (26%) did not have any DRR experience but they were engaged in youth-related organizations; 118 applicants (26.46%) have DRR experiences although limited to only as participants of a program or event, e.g., DRR conferences, trainings, or seminars; and 212 (47.54%) have involved in the past as initiators or organizers in disaster-related programs or initiatives. We further observed in-depth only on these 212 participants' involvement in disaster-related programs or initiatives.

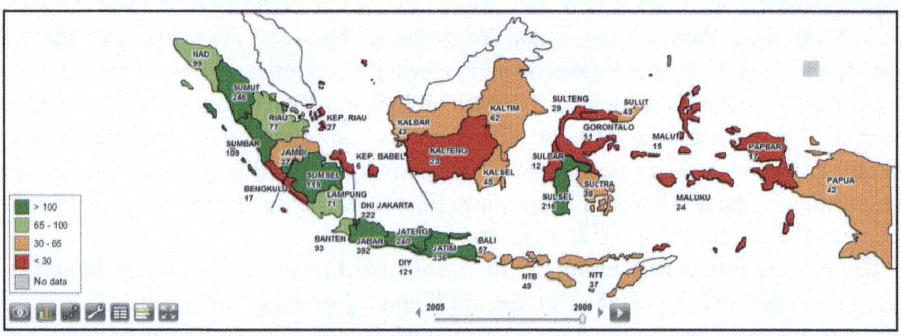

Fig. 14.2 Distribution of higher education institutions in Indonesia

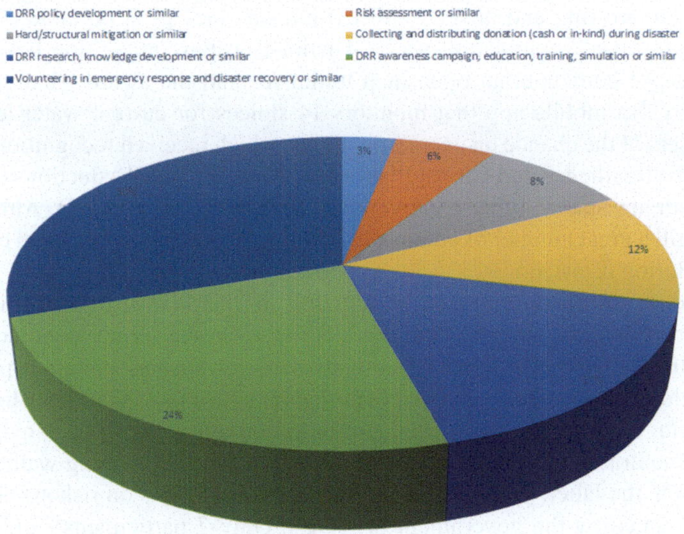

Fig. 14.3 The percentage of youth who (co-)implemented or organized disaster-related initiatives (modified from the record of UNMGCY 2015)

As shown in Fig. 14.3, most of the youth-led programs where they engage as implementers or organizers are in the form of "volunteering on emergency response and disaster recovery" (30.22%). This type of engagement ranges from initiating or participating in search and rescue efforts, emergency food distribution, trauma healing, emergency medical assistance (particularly students in medical field), and debris removal. Most of the participants declared engagement in this category were those who had direct experience of disasters as mentioned earlier. The next category is the youth-led implementation of "DRR awareness campaign, education, training, disaster simulation or similar" (24.00%). This type of initiatives range from the emergency-bag campaign on social media,[10] community education, such as the Gerakan Teknik Mengajar (that channels engineering students to teach the community on DRR) as well as youth involvement in full-scale disaster simulation, such as the 2014 Mentawai Megathrust Disaster Relief Exercise. This category also confirms that youth has great potential as risk communicators (Ronan et al. 2008).

Around 16.89% of participants have been involved in "DRR research, knowledge development or similar". This is mainly attributed to students' work in their thesis or projects in undergraduate or postgraduate programs, involvement in university research, and in notable science competitions in Indonesia and abroad. For instance, the annual PIMNAS (National Students' Scientific Week) organized by the Ministry of Science Technology and Higher Education. Interestingly, no participant has been involved in the annual disaster-competition organized byIndonesia National Disaster Management Agency (BNPB) or other line ministries. Nevertheless, there are some

[10]BDSG Emergency Bag Campaign on Facebook (Last accessed December 27, 2016).

records on interesting and novel youth-led innovation, e.g., a portable evacuation shelter that received a nomination in a 2014 national design competition, development of disaster and humanitarian open map initiatives and the award-winning Jakarta Flood Alert,[11] a mobile app that monitors 14 sluices for current water levels and informs users of the chance on upcoming floods, which received recognition through the Code for Resilience and Global Platform on Disaster Risk Reduction (GPDRR). Truly, youth are able to bring creative ideas on DRR (Barlett 2008). Although, in contrast, still a great number of youths (11.56%) are involved only on cash or in-kind donation during disasters.

Less youth, however, engaged or initiated in "hard or structural mitigation" (8.44%) and "risk assessment or similar" (5.78%). On the former one, most of the youth contributed or initiated natural mitigation efforts, such as mangrove plantation to reduce the risk of tsunamis or tree plantation to reduce the risk of landslides. Very few of them, in fact only one participant based on our records, declared involvement in structural mitigation, such as the construction of a retaining wall to reduce landslide. On the latter, youth mostly participated as analysts on risk assessment in projects initiated by the government or NGOs. Only 7 participants (3.11%) were involved in "DRR policy or framework development or similar". About half of them refer to their participation as youth delegates in the 5th and 6th Asian Ministerial Conference on Disaster Risk Reduction in Yogyakarta-Indonesia and Bangkok-Thailand, respectively. One refers to engagement with Local Disaster Management Agency (BPBD) on formulating DRR policy and another one refers to advocacy to the national government on safe schools. Clearly, the area of strategic partnership for national and local DRR policy and framework formulation is where Indonesian youth lacks engagement.

Furthermore, when asked, "How will you use the knowledge and skills you gained from the Children & Youth Forum?", almost all answered "will re-share the lessons from this forum". However, some of the youth do not belong to any organization, so they would not have the channel to implement potential lessons learned from the forum. On the other hand, some already have potential channels as described:

> I will use my knowledge and skill that I gained to share to the people especially in my area through my English Zone community at my beginners' discussion group by making Disaster Risk Reduction Discussion to discuss, inform and share to them about this Children and Youth Forum in Sendai Japan. I will share my knowledge as well to my campus through Physics English Club in Riau University (female, 22 years of age).

Furthermore, some of the youth are actually able to think strategically and realize the need for a dedicated youth platform on DRR in Indonesia:

> I will ask to Indonesia Student & Youth Forum to make the same program like the Children & Youth Forum do, and I will share my experience, knowledge, and skills to my friends in here. (Male, 19 years of age).

[11] https://www.worldbank.org/en/news/feature/2014/06/30/ (Last accessed December 27, 2016).

14.5.4 U-INSPIRE Roles in Leveraging SETI for DRR in Indonesia

U-INSPIRE Indonesia was first initiated at the end of 2017 by a small group of YYP coming from different backgrounds but shared the same interests on DRR. They realize that there are few numbers of disaster-related organizations that have been established that involve young people. However, most of the existing organizations are either based on single expertise or mostly dominated by mature scientists, resulting in a lack of cross-generational collaboration, discussions, and engagements. Another consideration is that there is no existing formal youth platform focusing on DRR in Indonesia. In short, the absence of a youth platform that can provide a collaborative space for YYP using SETI to reduce disaster risk is the reason behind the establishment of U-INSPIRE Indonesia. The idea was later conveyed to the other stakeholders.

The idea started by young scientists who represented Indonesia at The Global Scientific and Advisory Group (G-STAG UNDRR) who believed there is a need to have a platform for YYP in DRR in Indonesia. They conveyed their ideas to some young scientists, UNESCO Jakarta, The Indonesian Institute of Sciences (ICIAR LIPI), and UNMGCY that responded to the idea positively and later agreed to help a small group of YYP to conduct a consolidation meeting on YYP in SETI for DRR in March 2018. The objective of this meeting was to find and initiate a core of YYP interested in initiating activity to strengthen Indonesia's SETI for DRR. Attended by 31 YYP from 6 cities across Indonesia with various backgrounds and interests, they agreed to establish a platform to accelerate YYP's roles in DRR through SETI, inline with the Young Scientist Roadmap of UNMGCY and G-STAG UNDRR for the implementation of Sendai Framework for Disaster Risk Reduction. The core team further organized several follow-up meetings facilitated by BNPB, the Agency for Meteorology, Climatology, and Geophysics (BMKG) and The Agency for the Assessment and Application of Technology (BPPT) on April 2018, as well as several informal meetings through social media platforms. On May 3, 2018, ICIAR LIPI and UNESCO followed up this initiative through a side event workshop at the Fifth Annual Disaster Scientific Research Meeting (PIT IABI) organized by the Indonesian Disaster Experts Association (IABI) at Andalas University, Padang, West Sumatera. The workshop was titled Inter-generational and Interdisciplinary Dialogue to Strengthen the Role of Youth in Science and Technology in DRR. More than 60 YYP attended the workshop and discussed the gap and opportunities for them to play an active role in contributing to DRR. Some notable findings from the workshop were limited opportunities and platform for YYP to develop and contribute in SETI for DRR; (2) Limited or no platform for YYP to voice their ideas and have transdisciplinary interaction and communication.

As a follow-up action, on May 20, 2018, under UNESCO, ICIAR LIPI and BPPT's advocacy and guidance, the YYP in Indonesia agreed to formalize its initiative by establishing a platform named U-INSPIRE (Youth and Young Professionals on Innovation, Science, and Technology Platform for Resiliency). U-INSPIRE is defined as a platform of YYP in SETI to accelerate the implementation of DRR in line with the Sendai framework to support DRR policy and action at the local, national, and international level, with the vision, is "Indonesian YYP as the generator of SETI for disaster resilience at national and global level". After the formal initiation, U-INSPIRE later conducted the first regional workshop on November 1–4, 2018 by the support from UNESCO Jakarta, ICIAR LIPI, UNDRR, UNMGCY, IRDR, ITB, and other partners with the theme Strengthening, Empowering, and Mobilizing YYP in SETI for DRR in Asia and the Pacific. The workshop was attended by 20 speakers and 53 participants from 32 countries from Asia and the Pacific, Africa, Europe, America and was conducted to address some of the YYP' challenges and potentials in contributing to DRR, including inter alia, limited opportunities and platform contribute in SETI for DRR. Objectives of this workshop were to build understanding among YYP on their role in SETI for DRR as well as various important topics to leverage their roles, to develop joint programs or research to support the implementation of SFDRR and to share lessons learned from the establishment of U-INSPIRE Indonesia. The official launching of U-INSPIRE was held on November 3, 2019, marking the birth of U-INSPIRE as the YYP platform for SETI in DRR in Indonesia. Some notables' outputs from the workshops were the common understanding among participants on limited jobs and opportunities for YYP in Asia and the Pacific to contribute to DRR activites and there is a need to have a platform that can bridge the missing link and answer the demand. Interestingly, participants from other countries saw the model and the establishment of U-INSPIRE in Indonesia is essential in bridging and filling the gap between youth, young professionals and DRR. This later led to the expression of interests from participants in replicating the establishment of U-INSPIRE in their respective countries.

In terms of its division, currently, U-INSPIRE has four working groups (WGs) with its specific goals and activities. The first WG is Knowledge Management WG that aims to create space for data and resources management and utilization for inter-, multi-, and transdisciplinary research, enabling the platform to respond to issues related to SETI. The second one is Innovation Room WG that is focusing on empowering and strengthening capacities and substance of SETI in DRR and acting as a catalyst for advanced inter-, multi-, and transdisciplinary research. The third WG is Science Communication WG with the purpose to translate and utilize scientific knowledge to improve resilience through SETI in DRR, and the last WG is Advocacy, Communication, and Networking WG with the goals to engage in science-based DRR policy development and action at local, national, regional, and international level and to strengthen network and collaboration on research, innovation, and policy. In addition to U-INSPIRE WGs, activities and collaboration are mainly categorized into the following activities: (1) research and dissemination, (2) science communication and education, (3) design and technological innovation, (4) evidence-based policy, and (5) operations (Hanifa et al. 2020).

From the first launching on November 3, 2018 up to October 27, 2019, U-INSPIRE has 188 members, 57% male and 43% female coming from 22 provinces across Indonesia. However, most of the members are still centralized from Java

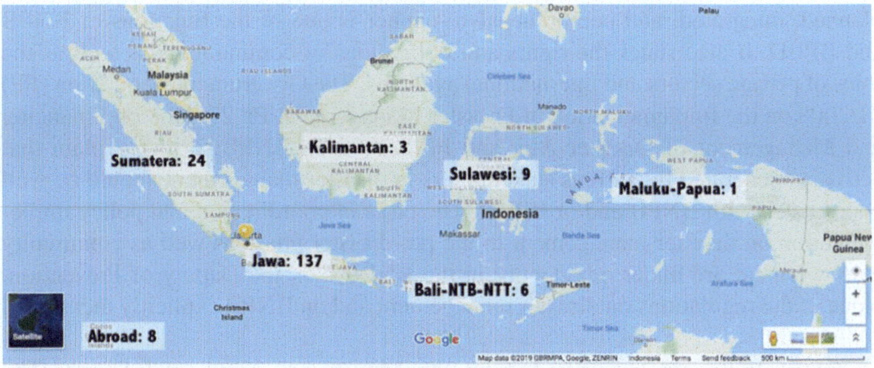

Fig. 14.4 Distribution of U-INSPIRE members as of October 2019

Fig. 14.5 U-INSPIRE members field of expertise

Rank	Field of Expertises	No. of members
1	Geoscience and Natural Sciences	37
2	Disaster Risk and Emergency Management	32
3	Engineering	22
4	Health and Psychology	18
5	Architecture, Urban and Rural Management	14
6	Climate, Biology and Environmental Risk	14
7	Public policy, Social and Political Science	14
8	Mapping and Spatial Analysis	13
9	Business, Finance and Economy	12
10	Arts, Design, Media and Journalism	10

and Sumatera Island. The background of U-INSPIRE members varies from geoscientists, DRR specialists, engineers, environmentalists, planners, to media journalists, and artists. In terms of occupation, the majority of U-INSPIRE members are employees (government officials, professionals at NGOs, international organizations, private institutions), students, lecturers/researchers, entrepreneurs, artists, and freelancers. Regarding the collaboration activity, so far, U-INSPIRE has partnered with 11 government institutions, 7 international agencies, 18 NGOs/CBOs, 4 private sectors, and 3 media channels (Figs. 14.4 and 14.5).

14.5.5 National Policies on DRR and Youth

The role of the Government of Indonesia in disaster management has been mandated in the country's law and regulations. The Law No. 24/2007 on Disaster Management states that coordinating the implementation of disaster management through a well planned, integrated, and comprehensive manner is one of the functions of BNPB and BPBD. It also states the rights and obligations of communities as well as the roles of private sectors and international agencies. The Government Regulation (PP) 21/2008 on the Implementation of Disaster Management, PP 22/2008 on Financing and Management of Disaster Aid, and PP 8/2008 on BNPB further explain that activities generated by various stakeholders in disaster management should be well coordinated with BNPB and/or BPBD. The tasks of coordination and policy implementation on disaster management in the pre-disaster phase as well as community empowerment are under one deputy in the BNPB, i.e., the Deputy of Prevention. None of the regulations on disaster management and on BNPB explicitly mentioned about youth.

Youth affairs are basically led by the Ministry of Youth and Sports. The ministry coordinates policy and program on youth with other ministries and agencies (including nongovernment agencies), local government and other relevant elements of society. Law No. 40/2009 on Youth states that government is responsible for raising awareness, empowering, and developing youth potentials; whereas the youth are expected to actively play roles as those with moral strength, social control, and agent of change in all aspects of national development. Government Regulation No. 4/2011 on Entrepreneurship Development, Youth Pioneering, and Provision of Facilities for Youth remarks that facilitating youth pioneering is the task and responsibility of government at the central, province, and local level, which includes providing infrastructure, budgeting, and planning. To support planning, the government needs to create inventory and identification of youth interest and potentials, required infrastructure and facilities, analyses, and formulation of standards, guidelines, and technical assistance. None of these regulations under this ministry explicitly mention about youth in the context of disaster management, despite the urgent need of disaster management and youth potentials. However, there are programs or activities under Ministry of Youth and Sports focused on disaster, including disaster management training, and relief distributions that can serve as an entry point as an opportunity to promote disaster management related contents in the youth leadership development curriculum that is regulated under the Regulation of Ministry of Youth and Sports No. 59/2013 as one of the basic requirements.

Regulations related to the role of youth in disaster management are available under the Ministry of Social Affairs and Ministry of Health, i.e., Regulation of Minister of Social Affairs No. 29/2014 on Taruna Siaga Bencana and Ministry of Health Decree 406/Menkes/SK/IV/2008 on the Forming of Disaster Ready Youth or *Dasipena* (stands for *Pemuda Siaga Peduli Bencana*). Under the aforementioned regulations, both ministries established voluntary like space nationwide to support disaster management by recruiting youth as volunteers. *Dasipena*, set up by the Ministry of Health aims to support the health aspects of disaster management. It organizes students from Health Polytechniques and representatives of the youth organization at the provincial level, trained students from a health education institution, Saka Bhakti Husada (SBH), trained scouts, trained youth organization representatives. Operations are financed by government budget and other additional funding. Whereas *Tagana* (abbreviation for *Taruna Siaga Bencana*, or Disaster Management

Cadre), set up by the Ministry of Social Affairs, aims to support the government for social protection in disaster management. It mobilizes local communities from a wider range of age, not only youth, i.e. between 18 and 45 years old. The nature of both organizations is to implement health and disaster response, and they occasionally get invited for disaster preparedness programs. There are no modalities to encourage proactive innovation or the provision of potential funding support for their proposal. In addition, the BNPB's policy and programs do not specifically aim for youth as the main driver of the DRR. The recent BNPB's strategic plan of 2014–2019 on prevention and preparedness, however, emphasizes on "Disaster Risk Reduction Movement" (*Gerakan Pengurangan Risiko Bencana*), which has a focus on cooperation with thematic university-student community service program (*KKN Tematik*).

14.6 Discussion

14.6.1 Youth as Agents of Risk Communication and Disaster Risk Reduction

As explained in previous sections, there are quite a number of DRR-related movements and organizations involving youth. This suggests that various stakeholders have recognized youth potentials as risk communication agents. Figure 14.6 summarizes various youth-related organizations and movements, the nature of their activities, and gaps that need to be filled to better engage youth.

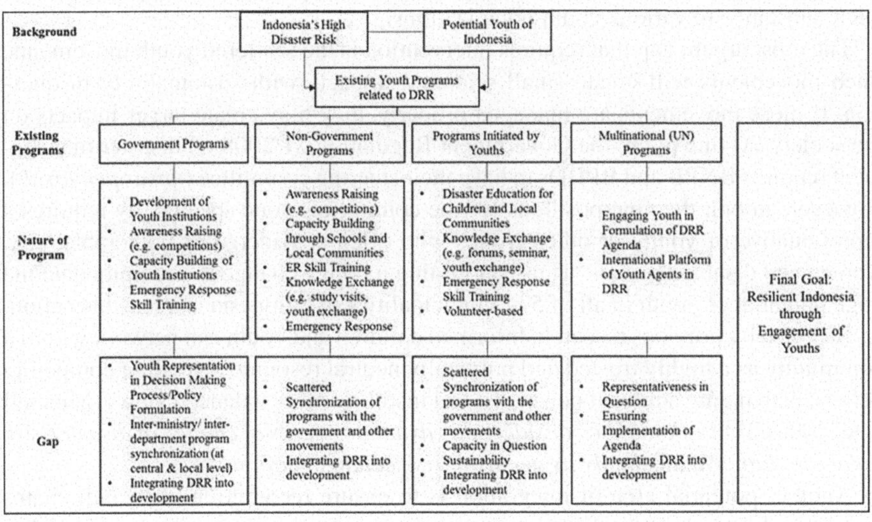

Fig. 14.6 Various youth-related programs and the gaps in youth engagement

Many Indonesian youths nowadays are aware of disaster risks and are willing to contribute to reduce the risks. Due to their huge potential and passion, their capacity should be built according to the needs and should be recognized by the government for disaster risk reduction. For instance, youth organizations that provide disaster education to communities should be ensured that the materials and methods used are effective and are not conflicting with any DRR principles. Another example is that during emergency response, their actions should be well coordinated with BNPB or BPBD, as regulated in the Government Regulations No. 21/2008.

From the samples in this study, most youth organizations and movements are limited to disaster education, training, seminars, and emergency response. The government may want to explore other fields of expertise where youth can take part. Youth who have passion for doing research can maximize their efforts in research related to DRR and resilience that could be useful for society. Another option is to engage youth from various fields of expertise to integrate DRR principles in their activities, thus whichever fields of expertise they are pursuing, DRR, resilience and sustainability as core values will be implemented.

Indonesia has developed several institutionalized youth CSOs (Civil Society Organization) that have a structure from the central level up to the grassroots level, e.g., *Karang Taruna* (Youth Union) and KNPI (Indonesia National Youth Council). These are strategic organizations where DRR could be integrated into their activities as they have access to vast human resources and have great influences to bring about change within the community (Mulyasari and Shaw 2012). For example, Karang Taruna is established in both urban and rural areas, and the networks are nationwide, as they were formed formally under the Ministry of Social Affairs. DRR could be introduced through their capacity building events. Enacting the Ministry of Youth and Sports Affairs Regulation No. 59/2013 that regulates minimum substance to be delivered in the Youth Leadership Education Curriculum, may be a great opportunity to introduce DRR substance to various youth organizations.

The most urgent gap that requires intervention is the scattered youth movements. Such movements will create small scattered impacts with sustainability in question. If those movements are managed properly, they may create larger impacts on the society. At one point, the Government Regulation 21/2008 on disaster management requires BNPB and BPBDs to have their steering committee (*unsur pengarah*). However, among the nine positions for the committee, none specifically requires a representative of youth, or other groups who often considered as vulnerable, e.g., women and disabled groups. Thus, suggesting a policy change to accommodate the huge potential of youth in all of 534 municipalities in Indonesia is a feasible action.

As noted by previous research, informed children and youth can network within a community as a highly trusted and politically neutral resource dispelling competing beliefs, convincing adults of new risks, and instilling more balanced views (Mitchell et al. 2008). *Youth should be considered as dynamic agents of change of disaster risk reduction rather than simply as agents of risk communication.*

Another potential area of integration is to ensure recognition of "youth champion or student ambassador" on disaster management through existing national DRR policy and program. In other fields, such as environmental management or public health, there were examples where a student being recognized as a symbol of awareness. A similar approach can be done as part of national safe school and disaster education strategies. One potential reason is because the safe school initiative has better leverage and visibility in the eyes of policymakers. In both 2015 and 2016 of DRR month in Indonesia (October), there was no specific policy and practical discussions on youth engagement in DRR, but more on safe school or safe hospital. The DRR month recognized the scout movement as DRR education agents that actively conduct self-assessment at schools located in disaster-prone areas. However, as our 446 participants suggest, only 2 scout members actually participated in the C&Y Forum during WCDRR, and upon re-interview with them, it is known that the mainstream DRR programs were unknown to scout movement. Although such a notion has been echoed for quite sometimes, relying on scout could not ensure all-schools implementation of DRR. Instead, it is the potential to include scout as prerequisite elements strategies for ensuring disaster education and school safety in all schools (Bisri and Sakurai 2017).

In summary, even though recently there is greater attention from Ministry of Youth and BNPB/BPBDs, youth are still simply considered as (1) targets of training programs, particularly due to the need of extending arm of the government in emergency response, e.g., *Tagana*, and (2) agents of risk communications instead of the actors of disaster risk reduction. Youth has not been considered as strategic partners in guiding DRR-related policy and formulation of national/local strategies. There is no national or local mapping to genuinely understand the potential of youth for DRR and strategically build partnerships with youth on DRR implementation. Accordingly, one may want to adopt the past advocacy strategies, which put CSOs having a greater role in providing social services. Furthermore, in recent years CSOs no longer just provide services to people that the government has failed to reach, but they are now far more in the mainstream of development activities (Mulyasari and Shaw 2014).

14.6.2 Youth Engagement in DRR: Does Indonesian Institutional Setup Provides Room for Engagement?

In this section, we explore and identify whether there is room for youth to make a significant contribution under the institutional setup related to disaster management and youth development in Indonesia. In general, opportunities to leverage engagement of youth in DRR exist under several national-level policies. Some opportunities are yet unexplored, and there is a disconnection between the two most relevant policies that could act as an enabler for such matter—disaster management policy, which mainly coordinated through the National Disaster Management Agency (BNPB) and youth development policy, which was mainly developed by the Ministry of Youth and Sports (*Kemenpora*). This is somewhat similar with the challenge of partial-connection of disaster management and education policy domains where despite various pilot programs and projects on disaster education and safe schools, until present Indonesia has not yet able to ensure nationwide implementation at all schools due to low institutionalization at education sector (Bisri and Sakurai 2017). Accordingly, on leveraging youth engagement on DRR, it should work both ways where disaster management affairs provide space for youth and youth development policy has to specifically provide programs related to DRR.

As elaborated before, the government has seen the youth affairs as important matters. The abovementioned law and regulations allow various groups of youth to get the opportunity to be facilitated by the government to have their skills, knowledge, and various capacities increased. Despite the Law No. 40/2009 gives importance to have youth as active development actors, they are still not represented in the advisory board of BNPB. Head of BNPB Regulation No. 2/2008 explains the selection mechanism and criteria of disaster management advisory board member in the BNPB, which consists of 19 personnel: 10 from government and 9 from professionals/society or local leaders with a minimum age of 30 years old. This seems to be a strategic position in determining the policy and implementation of disaster management at the national level, and thus youth representation is worth to be considered. Upon looking at policy statements of several line ministries in Indonesia, one can also conclude that basically youth engagement is rather as a potential alternative and auxiliary implementer to achieve government's policy targets. Case in point, the Ministry of Youth and Sport, BAPPENAS, and UNFPA just recently completed the enactment of National Youth Action Plan (*RAN Kepemudaan*) 2015–2019, which serves as a technical document for materializing the 2009 Youth Law. However, BNPB is not within the list of 25 ministries identified in the *RAN Kepemudaan.*

Figure 14.7 illustrates the window of opportunities for better youth engagement in disaster risk reduction in the current institutional setup, summarizing the above discussions.

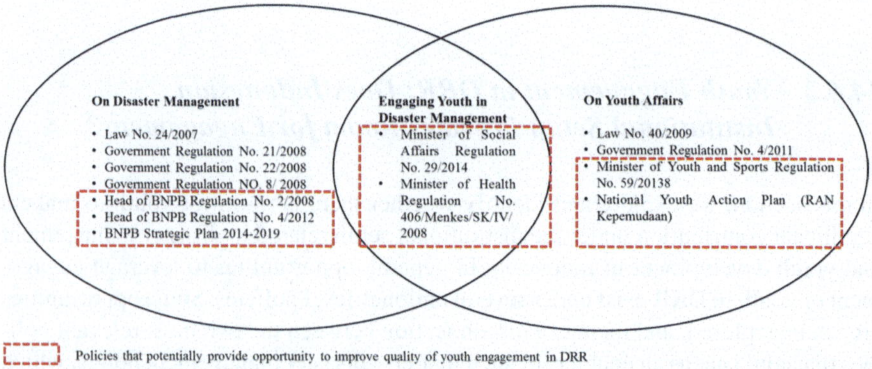

Fig. 14.7 Illustration of windows of opportunities for youth engagement in DRR within the current national legal framework

14.6.3 Science, Engineering, Technology, and Innovation (SETI) as the Medium for Country Resilience Building

Reflecting on the current condition, both in policy development and on the ground situation where there is no formal youth platform on DRR available and the engagement of youth in DRR in Indonesia seems unclear and only serves as complementary in the overall policy target. U-INSPIRE comes as an effort from YYP to contribute to DRR by utilizing SETI as a catalyst. U-INSPIRE also believes that DRR is not a stand-alone subject rather than a multi-, trans-, and interdisciplinary subject where various disciplines and approaches can contribute to reducing risks. Many scholars and the SFDRR itself emphasize the importance of SETI to achieve all of the priority actions (Weichselgartner and Pigeon 2015). Aitsi-Selmi et al. (2015) also mention that a collective consensus after the adoption of SFDRR reveals that DRR requires a more integrated approach and utilization of science-based policy, meanwhile Southgate et al. (2013) in UNISDR report on Using Science for DRR reconfirms that science can impact on DRR and that science can be used to implement effective evidence-based policy and practice Shaw et al. (2016) further highlight that utilization of SETI for DRR has benefitted society in mitigating the risks, and a combination of hard and soft sciences is practically answered the critical disaster problems within the community.

Based on the abovementioned reasons, it is critical to have a platform that will not only able to bridge the multi-, trans-, and interdisciplinary research and collaboration among YYP from various backgrounds and expertise, but also able to serve as an arena to cultivate and utilize SETI for reducing risks as what U-INSPIRE envisions to foster Indonesian YYP as the generator of SETI for disaster resilience at local, national, and global level. This is made actions taken as illustrated in the picture below (1) bringing the voice of YYP in SETI for DRR; (2) creating and endorsing space for YYP to empower their role in SETI for DRR; (2) bridging the valley of death between researchers and government, practitioners, and communities (Fig. 14.8).

More broadly, U-INSPIRE also seeks to take more strategic actions by doing advocacy works on various levels to voice the YYP aspirations and to pursue changes in the DRR policy landscape. U-INSPIRE tries to emphasize the message that YYP is not only the object of policy but also able to serve as a strategic partner for stakeholders in building the resilient of the community. These actions are done through various lobbying and advocacy works since the first establishment of U-INSPIRE at many levels. The highest achievement was the acknowledgment of BNPB Indonesia to U-INSPIRE as an official YYP platfrom on SETI for DRR that was delivered on the Indonesia Country Report on the Sixth session of Global Platform for Disaster Risk Reduction (GPDRR) on October 2019 in Geneva.

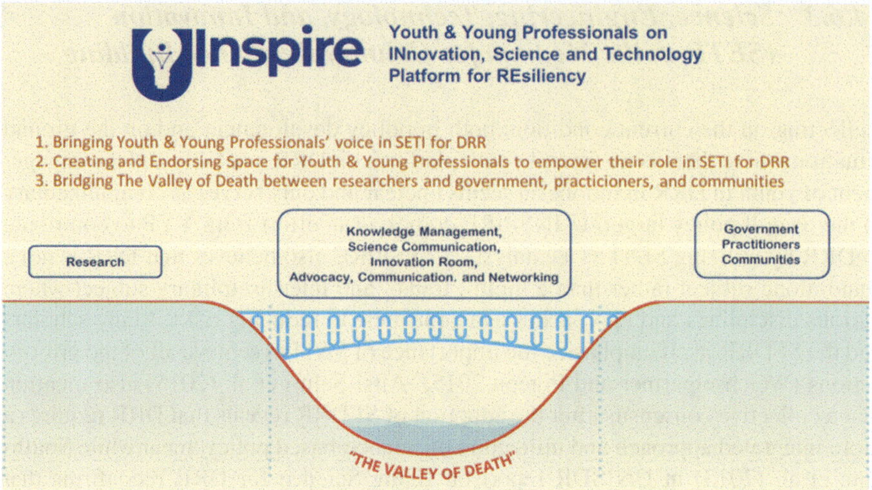

Fig. 14.8 The missions of U-INSPIRE Indonesia

14.7 A Conceptual Framework of Effective Partnership with Indonesian Youth in DRR

Various policies and institutions as a window of opportunities to leverage the engagement of youth in DRR through SETI are already available at the national level. The national government has recognized that youth and disaster management are two important matters, as evident by the forming of ministry and body to deal with such matters. Also, National Research Master Plan 2017–2045 issued by the the Ministry of Research, Technology, and Higher Education, incorporating disaster research as one of the priority research focus, is a promising sign of the government's recognition to the importance of disaster research. However, the combined efforts on youth, DRR, and research, has not been regulated and institutionalized from national down to local level. The Ministry of Health and the Ministry of Social Affairs have tried to combine both youth and DRR matters into one institution, i.e., the forming of youth organizations Dasipena for health issues, and Tagana for social issues. Both are good initiatives in empowering youth in order to support the government in managing disaster risks.

Nevertheless youth engagement in policy formulation has not been accommodated in the Indonesian context. This is contrary to the process in the WCDRR where youth have their own space for capacity building on DRR and youth was given space for contributing to the negotiation to decide the Sendai Framework for Disaster Risk Reduction, i.e., as the key output of the conference. Reflecting from experience from the Philippines for example (Fernandez and Shaw, 2013), youth council in the Barangay (village) level could access resources, including the 10 per cent of the annual Barangay budget. That enables youths to support their political role including that in the context of disaster risk reduction, and to put science and technology interventions in particular. Therefore, it is important to review the context of institutional arrangements to then further suggest how to institutionalize role of youth and young professionals in more systemic and sustainable way, reaching every region and communities at risks throughout Indonesia.

It is worth to observe DRR-themed events held during the DRR month (October in each year). At least in 2015 and 2016, there were no special slots for youth to contribute to formulating DRR Agenda (renewal commitment to the national DRR goal). It was also proven by the Surakarta Declaration, the national commitment made during the DRR month in 2015, which does not mention about disaster risk reduction and youth empowerment at all. For establishing an effective partnership in DRR at the Indonesian level, there must be a space for youth engagement in policy formulation as well. This can be achieved following the example during the Third WCDRR (2015) and during the 2019 Global Platform for DRR, where they engaged youth to provide inputs for developing the DRR framework.

On building an effective youth partnership in DRR, there is also a critical requirement to revise the current policy setup. The Head of BNPB Regulation No. 2/2008 states that the slot for disaster management advisory board member, whether from government or professional/society or local leader representatives within BNPB, must be at least 30 years of age. This is itself does not match with either the definition of youth within Law No. 40/2009 on Youth, or other international definition that requires youth representativeness, e.g., youth definition of UNMGCY, or of the UNICEF. If Indonesia is willing to ensure seamless inter-generational discourse on disaster management, revising the regulation and ensuring youth representativeness is important.

Of the previous discussion, the authors would like to propose institutional setup as an alternative conceptual framework to better leverage youth potentials in Indonesia, as shown in the following figure. It would be an advantage to have the Youth Advisory Board within the body of BNPB or BPBD. The board functions as representing the government in Youth-related Forums, policy formulation and decision-maker, representing youth in the government forum, coordinating with another ministries/department. The Youth Advisory Board will contribute to the Youth Advisory Forum, which consists of many youth representatives from NGOs and youth groups. The forum will function as a coordination space of knowledge and innovation bank, recommending policies, etc. The proposed institutional setup is shown in Fig. 14.9.

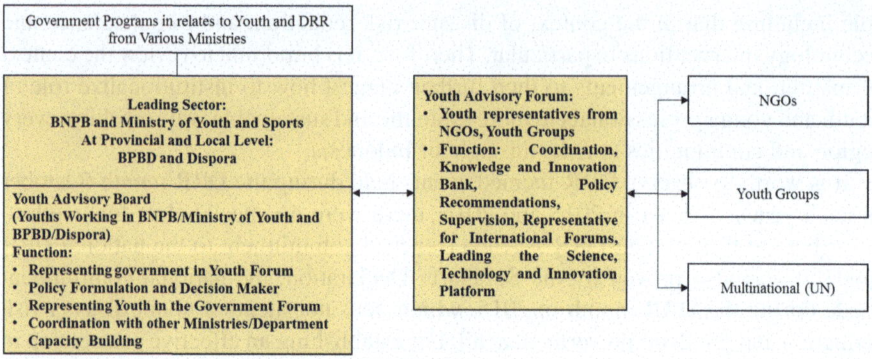

Fig. 14.9 Bridging the gaps: proposed institutional setup

14.8 Effective Youth Engagement in Fulfilling the Targets of SFDRR 2015–2030

It is important to outline the requirement and essential checklists for effective youth engagement in fulfilling the targets of the Sendai Framework for Disaster Risk Reduction 2015–2030 (SFDRR), both for the Indonesian context and with transferability to other developing countries. The guiding framework within the SFDRR has been clear on this: "*a gender, age, disability and cultural perspective in all policies and practices; and the promotion of women and youth leadership; in this context, special attention should be paid to the improvement of organized voluntary work of citizens.* (SFDRR, III,19.d, UNISDR 2015). Furthermore, another essential requirement is to ensure the following notion: "*children and youth are agents of change and should be given the space and modalities to contribute to disaster risk reduction, in accordance with legislation, national practice and educational curricula*" (SFDRR, V.36.a.iii, UNISDR 2015).

While recognizing their leading, regulatory, and coordination role, Governments should engage relevant stakeholders including women, children, and youth, persons with disabilities, poor people, migrants, indigenous peoples, volunteers, the community of practitioners and older persons in the design and implementation of policies, plans, and standards.

There is substantial evidence that youth benefit when given the opportunity to make and act on decisions for the common good in youth organizations and programs. Youth not only receive services, but also provide them. In this way, they change themselves from participants to partners, from being cared for to key resources for their communities.

Agenda setting is clearly needed given the little attention toward youth within the DRR policy setup. This may also put commitment of the Government of Indonesia in question on this area given the Bangkok Declaration, SFDRR, and the Asia Regional Plan for the Implementation of the SFDRR 2016–2030. Immediate action points to show commitment on this can be taken through policy change by including youth advisor as part of the steering committee (*Unsur Pengarah*) of the BNPB and local BPBDs. Consequently, there will be more accountability in sending and promoting Indonesian youth engagement at the global negotiation processes.

This case also shows that youth engagement has been mostly done through soft DRR activities, particularly on communicating risk through various creative channels. Only a few youths involved in DRR through the development of independent hard DRR intervention. There should be more encouragement and facilities for youth from the science and technology side in DRR, such as a small-scale community infrastructure development, small yet essential community technology application, software development for DRR, and other tangible DRR means. At the national level, BNPB can show commitment to this by diversifying their annual DRR competition to include such categories above and not limit themselves on disaster-related photography, video, and essay contests.

Lastly, it is important to discuss how youth incorporate DRR principles in any aspect of their lives or in any field they are pursuing. It is strategic on the intergenerational aspect of development, so that regardless of the youth involvement in any development sector, they will have the risk reduction mentality, protection to the natural environment and ensuring sustainable development.

14.9 Conclusions

This case study showcases the range of Indonesian youth engagement on DRR with an overview of 37 youth organizations and 446 Indonesian participants in the C&Y Forum during the Third UN WCDRR and U-INSPIRE Indonesia platform member database. In general, the records show the range of youth engagement on DRR from the traditional view as volunteers for disaster response and relief to risk communicators and young professionals working on science, engineering, technology, and innovation for disaster risk reduction. Certainly, the genuine potential of Indonesian youth in DRR and nation's resilience building is actually far greater in relation to the existence 61.8 million of youth or around 24.5% of the total population (BPS 2014). The so-called demography bonus for Indonesia must be utilized further on DRR implementation from one end to another, from policy formulation to DRR implementation. Accordingly, given the gap between numbers of youth observed in this paper and the actual number of youth population in Indonesia, there is a need in the near future to systematically record and assess youth contribution to DRR and nation's resilience building.

It also explores how the current national framework deliver an enabling environment for the youth to best contribute to DRR through SETI. From the exploration and as also confirmed with records of youth engagement on DRR, the chapter clearly suggests that very limited opportunities are given to the youth group to engage in national and local DRR framework and policy development, as well as its monitoring and evaluation processes. This is the area where priority must be given for ensuring Indonesia's commitment to all of the SFDRR content, particularly on engagement with children and youth. On this, the paper has identified current policies where the window of opportunity existed for building meaningful partnerships with youth on DRR. It has also suggested the institutional setup on bridging the gaps of youth engagement on DRR in Indonesia. Moreover, the chapter strongly suggests policy change to include youth group as one of the members within the advisory board of BNPB and local BPBDs, for ensuring youth voices within the coordination of DRR in the pre-disaster phase, emergency response, and deliberation for deciding course of recovery in the post-disaster stage.

Generally, there seems to be a disconnection between the agency/ministry who coordinates disaster management policy with that who coordinates youth development policy. Limited opportunities are given to the youth group to engage in national and local DRR framework and policy development. As one of the conclusions, a clear and legal standing for youth engagement on DRR must be established. This study suggests policy change to include youth group as one of the members within the advisory board of the National Disaster Management Agency (BNPB) and the Local Disaster Management Agency (BPBD), to ensure youth voices to be considered throughout the whole disaster management cycle.

Finally, in terms of transferability aspects to the context of other developing countries, each government or youth in a particular country might want to start by a systematic review on youth initiatives on DRR, for understanding the potential and challenges faced by youth. The stock-taking must be able to identify the pockets of youth that potential for mobilization within DRR, and prioritize the empowerment of existing pockets with DRR ideas and innovations instead of creating new ones. This also includes the potential of synergy with investments in safe schools and the role of science, technology, and higher education for DRR. In addition, each country must also conduct a policy analysis of the institutional setup on disaster management and how the SFDRR Article 36-ii-a is interpreted. Thus, based on the system of each country, a clear and legal standing for youth engagement on DRR must be established. Lastly, it is recommended for all states to encourage the peer learning process between youth from various countries to learn from each other on planning, implementing, monitoring, and evaluating their DRR initiatives.

Acknowledgments The authors wish to acknowledge the United Nations Major Group for Children and Youth (UNMGCY) for data sharing of the Children of Youth Forum of the Third UN WCDRR in Sendai, Japan, March 2015. This paper is also integral to the first author mandate as Regional Focal Point for Northeast and Southeast Asia in the DRR Working Group of UNMGCY. Hence, the statement in this paper does not represent the official view of the UNMGCY. We are also indebted to Ms. Meta Indriyani, for her assistance in data compilation. The authors receive no funding in the development of this manuscript.

References

Aitsi-Selmi A, Blanchard K, Al-Khudhairy D, Ammann W, Basabe P, Johnston D, Ogallo L, Onishi T, Renn O, Revi A, Roth C, Peijun S, Schneider J, Wenger D, Murray V (2015) UNISDR STAG 2015 report: science is used for disaster risk reduction. https://preventionweb.net/go/42848

Barlett S (2008) After the tsunami in Cooks Nagar: the challenges of participatory rebuilding. Child Youth Environ 18(1):470–484

BDSG (2012) BDSG activity report: youth exchange program in DRR. Bandung: Bandung Disaster Study Group (BDSG). https://www.preventionweb.net/publications/view/32432

Bisri MB, Sakurai A (2017) Disaster education and school safety governance after the 2004 Indian ocean tsunami in Indonesia: from national policy to local implementation. In: Djalante R, Garschagen M, Thomalla F, Shaw R (eds) Disaster risk reduction in Indonesia: progress and challenges in managing risks, reducing vulnerability and building resilience. Springer, Geneva.

BPS (2018) Statistical yearbook of Indonesia 2018. BPS (Badan Pusat Statistik/National Statistical Agency), Jakarta

BPS (2017) Statistik Pemuda Indonesia (Indonesian Youth Statistics). BPS (Badan Pusat Statistik/National Statistical Agency), Jakarta. https://deputi1.kemenpora.go.id/assets/uploads/dokumen/74908-spi-2017.pdf

BPS (2014) Statistik Pemuda Indonesia (Indonesian Youth Statistics). BPS (Badan Pusat Statistik/National Statistical Agency), Jakarta. https://bps.go.id/website/pdf_publikasi/Statistik-Pemuda-Indonesia-2014.pdf

Brown CV (2015) Global Volcanic Hazards and Risk. Cambridge University Press, Cambridge

Chatterjee R, Shiwaku K, Gupta RD, Nakano G, Shaw R (2015) Bangkok to Sendai and Beyond: implications for disaster risk reduction in Asia. Int J Disaster Risk Sci 6:177–188

Cumsikey L, Hoang T, Suzuki S, Pettigrew C, Herrgard M (2015) Youth participation at the third UN world conference on disaster risk reduction. Int J Disaster Risk Sci, 150–163. https://doi.org/10.1007/s13753-015-0054-5

Djalante R (2016) Recent trends on hazards, disasters, risk reduction and climate change in Indonesia: a systematic literature review. Nat Hazards Earth Syst Sci, 1–48. https://doi.org/10.5194/nhess-2016-112

Fernandez G, Shaw R (2014) Youth council participation in disaster risk reduction in Infanta and Makati, Philippines: a policy review. Int J Disaster Risk Sci 4(3):126–136

Hanifa NR, Dwiyani R, Hamidah N, Bisri MF, Shidiq F, Mulyadi TM (2020) The role of youth in science and technology for disaster risk reduction in Indonesia: capturing through the Lens of U-INSPIRE

Haynes K, Tanner TM (2013) Empowering young people and strengthening resilience: youth-centered participatory video as a tool for climate change adaptation and disaster risk reduction. Child Geogr 13(3):357–371

Mitchell T, Haynes K, Hall N, Choong W, Oven K (2008) The role of children and youth in communicating disaster risk. Child Youth Environ 18(1):254–279

Mulyasari F, Shaw R (2012) Civil society organization and disaster risk reduction in Indonesia: role of women, youth, and faith-based groups. In: Shaw R (ed) Community-based disaster risk reduction. Bingley, Emerald Insight, pp 131–150

Mulyasari F, Shaw R (2014) Risk communication through community-based society organizations as local response to disaster in Bandung, Indonesia. In: Neef A, Shaw R (eds) Risks and conflicts: local responses to natural disasters (community, environment and disaster risk management). Emerald Group Publishing, Bingley, pp 231–250

Nahrawi I (2016, October 28) Pemuda Indonesia (Siap) Menatap Dunia—Indonesian youth ready to face the world. https://www.tempo.co/read/kolom/2016/10/28/2415/pemuda-indonesia-siap-menatap-dunia

Prasasti KS (2014, August) Meet and learn from active Indonesian youth working for disaster education! Report from the youth competition for disaster education 2013. Retrieved December

2016, from Wochi Kochi Magazine, Japan Foundation.https://www.wochikochi.jp/english/report/2014/08/how-1276-indonesian-youth-competed-for-disaster-education-trip-to-japan.php
Raviola M (2013) Capacity building for the empowerment and involvement of youth in Indonesia. Retrieved from Youthpolicy.org:https://www.youthpolicy.org/national/Indonesia_2013_Mapping_Youth_Organisations.pdf
Ronan KR, Johnston DM (2005) Promoting community resilience in disasters: the role for schools, youth and families. Springer, New York
Ronan KR, Crellin K, Johnston DM, Finnis K, Paton D, Becker J (2008) Promoting child and family resilience to disasters: effects, interventions and prevention effectiveness. Child Youth Environ 18(1):332–353
Simkin T, Tilling RI, Vogt PR, Kirby SH, Kimberly P, Stewart DB (2006) This dynamic planet: world map of volcanoes, earthquakes, impact craters, and plate tectonics. United States Geological Survey, Denver. Retrieved May 29, 2017, fromhttps://pubs.usgs.gov/imap/2800/
Shaw R, Izumi T, Shi P (2016) Perspectives of science and technology in disaster risk reduction of Asia. Int J Disaster Risk Sci 7(4):329–342. https://doi.org/10.1007/s13753-016-0104-7
Southgate RJ, Roth C, Schneider J, Shi P, Onishi T, Wenger D, Amman W, Ogallo L, Beddington J, Murray V (2013) Using science for disaster risk reduction. www.preventionweb.net/go/scitech
UNDESA (2015) The world youth report on youth civic engagement. United Nations Department of Economic and Social Affairs (UNDESA), New York. https://www.unworldyouthreport.org/
UNISDR (2000, October 11) United Nations 2000 World Disaster Reduction Campaign to focus on disaster prevention, education and youth. Retrieved September 19, 2016, from United Nations International strategy for Disaster Reduction. https://www.unisdr.org/2000/campaign/pa-camp00-kit-eng.htm
UNISDR (2005) Hyogo framework for action 2005–2015: building the resilience of nations and communities to disasters. United Nations International Strategy for Disaster Reduction, New York. Retrieved May 28, 2016, from https://www.unisdr.org/2005/wcdr/intergover/official-doc/L-docs/Hyogo-framework-for-action-english.pdf
UNISDR (2015, March 18) Sendai framework for disaster risk reduction 2015–2030. Retrieved December 5, 2016, from Preventionweb: https://www.preventionweb.net/files/43291_sendaiframeworkfordrren.pdf
United Nations (2013, January 23) Definition of youth. Retrieved November 15, 2016, fromhttps://www.un.org/esa/socdev/documents/youth/fact-sheets/youth-definition.pdf
Weichselgartner J, Pigeon P (2015) The role of knowledge in disaster risk reduction. Int J Disaster Risk Sci 6:107–116. https://doi.org/10.1007/s13753-015-0052-7

Chapter 15
Understanding Gender Dimensions of Disaster Impacts on Agriculture in the Global South

Zubaria Andlib, Givemore Munashe Makonya, and Kumbirai Ivyne Mateva

Abstract Climate change is recognized as inevitable and one of the most complex challenges that humankind faces now and, in the future. Simulation models have predicted increased atmospheric temperatures, an increase in intense rain events as well as a reduced number of rain days, leading to increased drought risk and hampered crop production. Indeed, the effects of drought and heat stress associated with climate change are expected to be more severe in the tropics and sub-tropics where temperatures are already quite high. Poor and marginalized segments of society are especially vulnerable to the adverse effects of climate change. They tend to have limited resources, the capacity to adapt, and hence their livelihoods tend to be highly dependent on natural resources which are sensitive to climatic variability. Just as different countries and groups in society have varying degrees of susceptibility to the adverse impacts of climate change, men and women also have different coping and adaptive capacities that lend themselves to different vulnerabilities and responses. Therefore, gender equality is a fundamental part of increased resilience to disasters for improved livelihoods. This review chapter is centred on cases of marginalized third world countries in Southern Africa and the Asia Pacific with the specific objective of understanding the nexus between gender, climate change adaptation and disaster risk reduction with a specific focus on agriculture. With this review, we intend to provide insight into gender dimensions for recovery in developing countries that is often overlooked by national and international policymakers, as well as bridging the gap between the policymakers, researchers and farmers.

Z. Andlib (✉)
Science and Technology, Department of Economics, Federal Urdu University of Arts, Islamabad 44000, Pakistan
e-mail: economist243@gmail.com

G. M. Makonya
Department of Biological Sciences, University of Cape Town, Private Bag X3, Rondebosch 7701, South Africa

K. I. Mateva
School of Biosciences, University of Nottingham Malaysia, Selangor Darul Ehsan, Jalan Broga, 43500 Semenyih, Malaysia

Crops for the Future, Selangor Darul Ehsan, Jalan Broga, 43500 Semenyih, Malaysia

R. Djalante et al. (eds.), *Integrated Research on Disaster Risks*, Disaster Risk Reduction,
https://doi.org/10.1007/978-3-030-55563-4_15

Keywords Agriculture · Climate change adaptation · Disaster risk reduction · Gender equality

15.1 Introduction

In the Disaster Risk Reduction (DRR) community, an understanding that gender equality is fundamental in increased resilience and adaptation to climate change is broad (FAO 2015a). This chapter explores the environmental and climatic vulnerabilities faced by less affluent communities, especially women in Southern Africa and the Asia Pacific. Particular focus is on the climate-related shocks to their agricultural productivity, adding pressure to their already precarious livelihoods. It means that there is a need for a strong focus on these agriculturally based livelihoods directly being affected by climate change, with 82% of crop and livestock production losses said to be due to drought, floods and extreme climatic events (FAO 2015b). With strengthened global interest to improve DRR in the agricultural sector, absorbing more than 22% of economic losses due to disasters, the need to ensure that gender-responsive approaches are part of the adaptation and resilience cannot be underestimated (FAO 2015a, b). The chapter is centred on cases of marginalized regions through understanding the nexus between gender, climate change adaptation and disaster risk reduction in agriculture. This chapter will feed into the goals of the Sendai Framework for Action through strengthening resilience, awareness and adaptive capacity to climate-related hazards and disasters globally, as well as decreased direct economic losses relative to Gross Domestic Product (GDP) caused by disasters. Its objectives are also in line with the Integrated Research on Disaster Risk (IRDR) Science Plan, with one of its important objectives being an understanding of decision-making in complex and changing risk contexts (ICSU 2008). Therefore, an understanding of the decision-making choices that contribute to agricultural disasters and how they may be influenced, as well as how this knowledge can better lead to disaster risk reduction, would significantly aid in providing an insight into gender dimensions in recovery in the regions concerned.

It is generally observed that women in developing countries are considered to be vulnerable as compared to their men counterparts. The main reason behind this vulnerability is the high dependence of women on natural resources, such as agriculture, fisheries, biofuels and forest resources. Climate change impacts agriculture directly by decreasing the crop yields, heat or cold stress that may cause low productivity among women-headed farm owners, may affect their health, whereas the indirect impacts include stressed water sources, changes in crops pattern and undermine the supply of essential agricultural crops that put stress on the price of agricultural commodities. The chapter is centred on cases of marginalized third world countries in Southern Africa and the Asia Pacific with specific objectives; to understand the nexus between gender, climate change adaptation and disaster risk reduction in agriculture.

The rest of the review study is organized as follows: Sect. 15.2 discusses the theoretical framework. Section 15.3 will present the case studies, whereas Sect. 15.4 highlights the importance of gender mainstreaming in the climatic analysis. Section 15.5 will conclude the study.

15.2 Review of Literature: Disasters, Climate Change and Gender

Disasters are characterized by sudden and extreme events, caused by different environmental and climatic factors that usually hurt people and damage property. The most important disasters are earthquakes, floods, landslide, hurricanes, droughts and diseases. *Source* EMDAT (2017).

Global climate change is one of the greatest environmental challenges facing the world today. In the twentieth century, the increase in the global average temperature reached 0.74 °C, the average sea level increased by 17 cm and the Northern Hemisphere experienced a considerable decrease in snow cover). Eleven of the warmest years since records have been kept, have occurred during the last 12 years, representing an accelerating warming trend. The Intergovernmental Panel on Climate Change (IPCC) projects additional global warming over the twenty-first century from 1.8 to 4.0 °C. *Source* IPCC (2007).

Gender refers not to male and female, but to masculine and feminine—that is, to qualities or characteristics that society ascribes to each sex. People are born female or male, but learn to be women and men. Perceptions of gender are deeply rooted, vary widely both within and between cultures, and change over time. But in all cultures, gender determines power and resources for females and males. Gender is a central organizing factor in societies, and it can significantly affect the processes of production, consumption and distribution. In fact, the influence of gender on rural people's lives and livelihoods is so substantial that by any indicator of human development, female power and resources are lowest in rural areas of the developing world. *Source* FAO (2011).

Gender inequalities constrain the survival of households when faced with climate change. Due to existing socio-economic conditions, cultural beliefs and traditional practices, women are more likely to be disproportionately affected by disasters, including increased loss of livelihoods, gender-based violence and even loss of life during, and in the aftermath of, disasters. For example, there is a strong correlation between gender inequalities and women's survival rate in disasters such as typhoons, droughts and floods (Murillo and Tan 2017). The World Health Organization (WHO) has estimated that women are up to 14 times more likely than men to die as a result of disasters (WHO 2003; Neumayer and Plumper 2007). Hence, the empowerment of women is a critical ingredient in building disaster resilience. International communities over the years have increased acknowledgement on the different experiences and skills women have. At the Seventh Conference of Parties under the United Nations Framework Convention on Climate Change (UNFCCC) held in Marrakech, Morocco, participants from Samoa argued for increased representation of women within the organizational and decision-making structure of the UNFCCC (UNFCCC

2001). This serves as an avenue for women's voices on climate change adaptation to be incorporated into the international policy framework on climate change mitigation and adaptation. More so, international frameworks are beginning to incorporate a gender dimension into action on climate change. For instance, advocating for stronger engagement and empowerment of women and gender mainstreaming in policy planning and implementation of DRR programmes is central to United Nation for Disaster Risk reduction (UNDRR), formally known as (UNISDR) work. UNDRR facilitates and coordinates global and regional actions on disaster risk reduction and works closely with governments at national and local levels to strengthen resilience. In 2014, four Regional platforms on DRR organized by UNDRR with member states participation and in collaboration with regional partners in Africa, the Asia Pacific, Arab states and the Americas, included dedicated sessions on gender-sensitive DRR. They explained the role of women in a post-2015 Framework for DRR and adopted commitments on the importance of a leading role for women in DRR. In addition, the UNFCCC (2001) emphasizes gender balance and increased participation of women in its processes and in national delegations, as well as the development of gender-responsive climate policies at all levels.

The Sendai Framework for Disaster Risk Reduction (2015–2030) the successor agreement to the Hyogo Framework for Action (2005–2015), was adopted at the 3rd united nation (UN) World Conference on Disasters Risk Reduction in March 2015. The Framework is a 15-year international blueprint for saving lives and reducing the economic impact of natural and man-made hazards. One of the key distinctions of the Sendai Framework from past frameworks is its emphasis on the need for a paradigm shift from managing disasters to managing risk and to addressing the underlying causes and drivers of disasters. In this spirit, the framework recognizes the significance of gender dimensions in DRR and calls for the inclusion of all members of the society. It calls for "a gender, age, disability and cultural perspective in all policies and practices; and the promotion of women and youth leadership; in this context". Furthermore, the framework underscores that "women participation are critical to effectively managing disaster risk and designing, resourcing and implementing gender-sensitive disaster risk reduction policies, plans and programmes; and the need for taking adequate capacity-building measures to empower women for preparedness as well as build their capacity for alternate livelihood means in post-disaster situations". The Integrated Climate Risk Management Programme (ICRMP), 2013, supported six countries (Honduras, Armenia, Kenya, Uganda, Nepal and Indonesia) in: (i) formulating gender-responsive integrated climate risk management frameworks, which promote linkages between disaster risk reduction and climate change adaptation and mainstreaming into national and sector-specific development plans; (ii) strengthening and expanding the evidence base for climate risk management using gender- and sex-disaggregated information; (iii) building risk reduction and adaptive capacities of institutions, administrative/sectoral agencies and communities, which include women organizations and associations; and (iv) implementing gender-responsive climate risk management interventions to build resilience. For example: in Indonesia, United Nation Development Programme (UNDP) supported the government to incorporate gender concerns in the formulation of its Disaster

Risk Reduction-Climate Change Adaptation (DRR-CCA) framework and policies through gender reviews and participation of the Ministry of Women's Affairs in the DRR National Platform. UNDP supported the Government of Indonesia to improve disaster and climate risk assessments by developing guidelines for gender analysis along with collecting gender-disaggregated data based on UNDP piloting experience in conducting climate change vulnerability assessment and developing urban climate risk management plan in the cities of Kupang and Makassar. In Uganda, a participatory study, 'Better understanding men and women's vulnerability to disaster and climate risks in Uganda', was conducted. The study analysed the differentiated vulnerabilities of women and men to climate and disaster risks, their coping mechanisms in the face of disasters, and their roles and participation in disaster risk management measures. The findings and recommendations from the study have been used as Uganda proceeds with the completion of the Resilience and Disaster Risk Management (DRM) Strategic Framework and Investment Programme (to operationalize the National Development Plan 2015–2020) and in the development (and eventual passing) of a Resilience and DRM Act. In Nepal, UNDP assisted in improving awareness of and understanding of the importance of addressing gender concerns in (DRR-CCA) policies and programmes through training and conducting gender analyses within climate risk assessments. The purpose of this was to better understand the gender-related impacts of climate change to develop appropriate DRR and CCA solutions to gender-related vulnerabilities. In October 2015, the Food Security Sector in Myanmar co-led by Food and Agriculture Organization (FAO) and the World Food Programme (WFP) released an Agriculture and Livelihood Flood Impact Assessment Report. The assessment was jointly led by the Ministry of Agriculture and Irrigation; the Ministry of Livestock, Fisheries and Rural Development; FAO and WFP with the partnership of UN Women and several non-governmental organizations (NGOs). By following a gender-responsive process, the assessment was able to reveal how female-headed households suffered most from the impact of floods on their livelihoods with large losses of small livestock and fishing equipment, deprivation of income by lack of casual labour demands, and limited access to land and productive assets for recovery. Twice as many women as men reported reduced food intake. The assessment allowed planners to address these gender-specific needs in its response and recovery interventions, and to incorporate gender analysis for promoting resilient livelihood recovery.

15.3 Theoretical Framework: Gender Risk and Vulnerability

The existing literature on gender impacts of agriculture in the context of climate change found multiple linkages, such as men and women of different ages have different user characteristics, access to natural resources—land and water—access to the necessary information to cope with climate change events and relationship with institutions (Carvajal-Escobar et al. 2008). It is commonly observed that women are holding fewer assets, less access to information thus they have less bargaining power as compared to men, whereas access to these characteristics is generally identified essential for adaptation. Moreover, given the gender difference in resources, assets holding and decision-making authority at the household level, it is more likely to be expected that men and women have different priorities, needs and preferences towards climate change adaptation (UNWomen 2015).

Climate change includes long-term and average change in rainfall and temperature such as the intensity and frequency of extreme weather events such as heatwaves and rainfall spells. Due to biological differences, men and women experience climate change effects differently even if they live in the same household or broadly speaking in the same region (Murillo and Tan 2017). Therefore, these gender-differentiated factors as well as adaptation decisions made in the specific context determine the ways in which individual and communities experience and are affected by climate change events.

However, focusing on the gender perceptions of climate change events from agriculture households is the first and foremost step to understand the potential differential impacts of climate change events in the long run on men and women. Different studies on catering to the impacts of extreme climatic events on gender-differentiated asset dynamics indicate that the impacts of climatic shocks are unexpected and also its impact actually depends on the type of climatic shock and its local context (Rakib and Matz 2014).

The vulnerability context takes into account three different dimensions; exposure, sensitivity and adaptive capacity of households and communities to different climatic shocks. Generally, sensitivity and adaptive capacity is influenced by biophysical characteristics, besides this user characteristics and institutional context also impact sensitivity and adaptive capacity of individuals. Moreover, access to information at the household and community level and access to Information and Communication Technologies (ICT) tools also enhance the adaptive capacity of households. However, different studies indicate that in a household biophysical and institutional context may be the same for men and women but their impact on women is usually different. Access to information and communication technologies affects climate change perceptions and adaptation strategies (Bernier et al. 2015). Together these factors influence the different adaptive options at the household level available to household members. There are certain individuals or groups which have limited coping capacity towards climate change events, most of the available literature on household level focused on a specific context of vulnerability at social and cultural domains (Nielsen and Reenberg 2010), risk preferences (Grothmann and Patt 2005), or the most salient factors influencing adaptation, such as access to extension services, information, and credit (Bryan et al. 2013; Nhemachena and Hassan 2008). Instead of focusing on gender-disaggregated impacts of extreme climatic events, these studies explore household characteristics, i.e. head of the household age, education, their access to information and their risk preferences. The most important inference of interest is that adaptation decisions are strongly associated with the sex of the head of the household, and female headed households are less likely to adapt climate change because they have to face so many constraints including less access to information, fewer assets and limited access to credit (Nabikolo et al. 2012). It is also noted that gender-differentiated perceptions to climatic shocks, role within the household and at the community level and perceptions towards the same climatic event may differ between men and women (Perez et al. 2015).

Mostly in underdeveloped countries of the global south, farmers use traditional and out of dated technologies to predict weather patterns, however, it is important to make them aware of the new and advanced technologies to predict the weather for better agriculture production. Moreover, in the recent era, access to new and latest technologies has become one of the essential tools of adaptation among agricultural communities in the developed world (Tambo and Abdoulaye 2012). However, there is a need to understand the difference between men and women's needs and access to information for better adaptation among female farm owners (Peterman et al. 2014; Ragasa et al. 2013).

We cannot ignore the role of social and cultural norms with respect to adaptation context. Social norms restrict women to adopt certain practices for instance agroforestry. Some certain cultures and traditions prohibit women to take the ownership of land in different African and Asian countries. Simultaneously in traditional societies, women are not preferred to extend long-term loans for advanced agricultural equipments or they may not encourage adapting new technologies for agricultural production (Peterman et al. 2011).

In nutshell, people whose livelihood depends on agriculture, climate change events alter their ability to manage natural resources, which consequently affecting their income levels (Bernier et al. 2015). Women farm owners are largely neglected by service providers, and at the same time, they have less access to information. Thus, it is important to address the needs of both genders in the policy arena, it is even more important to equip women with the necessary tools and knowledge that helps them to adapt climate variability (FAO 2011; Peterman et al. 2014; UNWomen 2015). In essence agricultural impacts of gender is an important area of concern for policymakers as a large number of women are engaged in the agriculture sector in the global south. Therefore, present research adds on the existing literature on gender and climate change that it is particularly focusing on the selected region from the global south and will be an attempt to motivate the policymakers to focus on the selected regions in the future.

15.4 Agricultural Impacts from Climatic Disasters: Understanding Gender Dimensions in Southern African and the Asia Pacific Countries

The degree to which individuals are influenced by climate change and extreme weather conditions is distributed differently among different genders (IPCC 2001). In many national and worldwide disaster reduction approaches, information on how climate change identifies with social imbalance is frequently downplayed

(Dankelman 2010). Women and men are vulnerable against the impact of extreme weather events, and this is determined not only by biology but by differences in their social roles and duties (Easterling 2000; Wisner et al. 2004; Sultana 2014). Women are flanked by gender-differentiated circumstances, social-based hindrances, low monetary status and a muted voice in shaping decisions in developing countries, frequently having less assets to fall back on than men when faced with extreme weather events. Ribeiro and Chaúque (2010) observed that Mozambique women have no influence over natural resources. Moreover, they are excluded from decision-making in view of unequal participation in positions of authority—further aggravating their vulnerability. Along these lines, when adapting to climate change, women usually have less adaptive strategies than men and therefore, both sexes experience the impacts of climate change differently. However, seeing how both genders settle on decisions makes it simpler to apply adaptation programmes that will work over the long haul (Nabikolo et al. 2012). Studies indicate that women and men's decisions differ along taking risks, choice of coping techniques, versatility, as well as getting information. A study by Lunduka et al. (2019), indicated that educated male-headed households were found to be risk-takers, more innovative when deciding on using better adapted drought-tolerant (DT) maize varieties in Zimbabwe. In other reports, gender-differentiated circumstances, for example, poverty, were identified as compounding the increasing effects of the impacts of climate change after some time (Chersich et al. 2018). For instance, 70% of the 1.3 billion individuals living in poverty are ladies and gender-related inequalities are unavoidable in the developing world. In spite of the fact that women represent approximately 80% of the agricultural sector in Africa and the Asia Pacific, they remain vulnerable and poor (Denton 2002; ADB 2017). Gender inequalities and women's intermittent socio-economic disadvantages continue to be disregarded and remain a basic test to adaptation efforts. In particular, supporting women empowerment and drawing from their abilities and experiences could make climate change responses increasingly effective at the household and national levels. As the world keeps on managing what a post-Kyoto climate framework should look like, it is crucial that mitigation and adaptation efforts incorporate gender issues. This is an ideal opportunity to ensure that both women and men's roles are understood in order to reinforce the resilience of communities, and put them at the centre of disaster risk reduction approaches. This will reduce risks to women yet guarantee more pronounced efforts in addressing climate change. In the rest of this section, we focus on highlighting the example from Southern Africa and the Asia Pacific regions.

Agriculture is an important sector, contributing billions into the economies of the Southern African and Pacific regions (IUCN 2010; Baudoin et al. 2017b, a). However, the sector's nature hinges on specific climatic conditions, which, with increases in climate change impacts, could make it more difficult for the marginalized communities in these regions to participate in their traditional agricultural activities (FAO 2008). Climate change is recognized as inevitable and one of the most complex challenges that humankind faces now and, in the future, with simulation models predicting an increase in atmospheric temperatures, intense rain events as well as a reduced number of rain days, leading to increased drought risk and hampered crop production (IPCC 2007). Literature on the economics of climate change impacts have suggested that although crop production may be boosted in the short run, it will ultimately turn negative in the long term as the impact of climate change grows (Bruinsma 2003; IPCC 2007). Globally, 82% of crop and livestock production losses are said to be associated in one way or another to drought, floods and extreme climatic events (FAO 2015), with those most at risk being low-lying and atoll countries, as well as coastal areas (ABD 2013). The IPCC has also stressed that people who are socially, culturally, politically, institutionally, economically or otherwise marginalized are most at risk and vulnerable to climate change (IPCC 2014). There is more evidence showing that these marginalized groups of people suffer disproportionately from both direct and indirect effects of climate hazards, whereas earlier alluded to, loss of crops as an example is a direct loss, but subsequent rise in food prices and degraded diets are possible indirect impacts (Ishiwatari et al. 2019). Detailed reports of estimates by country and region have indicated that the Asian region and Africa would be the two regions most harmed by changing climate (UNFCCC 2007; AMCEN 2011).

Southern Africa is one of the world's poorest and most rainfall dependent regions, with agricultural producers operating with limited resources in fragile environments sensitive to minor shifts in temperature and rainfall patterns on seasonal and decadal time scales (UNFCCC 2007; AMCEN 2011; Awojobi and Tetteh 2017). The high levels of dependence on precipitation for the viability of the region's agriculture, combined with observed crop sensitivities to maximum temperatures during cropping seasons further highlight risks to the sector from changing climate (Asseng et al. 2011; Lobell et al. 2011; Schlenker and Roberts 2009). The IPCC has reported that the high levels of confidence that the overall effect of climate change on yields of major cereal crops in the region is very likely to be negative, with strong regional variation (Niang et al. 2014; Awojobi and Tetteh 2017).

Drought in South Africa has been reported to have resulted in maize imports being required in the year 1994 leading up to the year 1995, while maize export earnings fell, resulting in a 27 and 2.4% decline in agricultural and national gross domestic products (GDPs), respectively (FAO 2015), with a similarly severe drought characterized by a strong El Niño event experience in 2014–2015 (Baudoin et al. 2017b, a). In Malawi, climate change models have painted a bleak image due to the predicted 2–3 °C increases in atmospheric temperatures, with a decline in rainfall and water availability (IPCC 2007; FAO 2015). The combination of elevated temperatures and reduced rainfall has been noted to affect more than 90% of smallholder farmers dependent on rain-fed production. This has resulted in the projections that the production potential of maize, the main smallholder food crop (source of three-quarters of calorie consumption) would fall by over 10%. With almost one million people in the country living with HIV/AIDS, the superimposing effect that climate change and malnutrition has been having on the country cannot be overstated (FAO 2015; Bickton 2016). The 2001/2002 drought forced Malawians to cut back on meals, sell household goods and also eat seeds that would have been planted in the next cropping season. In 2005, drought caused food shortages to 4.7 million of the total 13 million people in the country (Phiri et al. 2005; FAO 2015). The 1991/1992 drought in Zimbabwe had a significant effect on maize, cotton and sugar cane productivity, negatively affecting agro-processing and textiles, causing manufacturing output to fall by 9% by the end of 1992. Agriculture sector growth in Zimbabwe fell by 23% in real terms in 1992 resulting in more than 5.3 million people in the country registering for drought relief (FAO 2015). A study on the effects of climate change on hydrology and irrigation in portions of the four riparian countries of Botswana, Mozambique, South Africa and Zimbabwe 9 years ago showed that water resource of the Limpopo River Basin is already stressed under the present-day climate conditions (Zhu and Ringler 2010; Awojobi and Tetteh 2017). Livestock production in Southern Africa is also vulnerable to climate change through drought, despite its importance as a source of food, income, or insurance against crop failure.

(Seo and Mendelsohn 2007). This industry is very significant to the economy of Botswana and it depends on local biomass production (Masike and Ulrich 2008), with a strong correlation having been found between drought due to changing climate and animal death (Thornton et al. 2009; Awojobi and Tetteh 2017).

Agricultural production significantly contributes to the GDP of some of the countries of the Asia and Pacific regions and is a source of employment for a large part of the population in many of the developing member countries (ADB 2017). Declining soil productivity, groundwater depletion, as well as increased pest incidence and salinity associated with changing climates threaten the sustainability and food security of the region (Wassmann et al. 2009; ADB 2017). Assessments of climate change impacts on agricultural production in Asia and the Pacific region have shown that despite regional variations, both biophysical climate impacts, as well as ensuing impacts on development, are likely to be greatly substantial (Hijioka et al. 2014; Schellnhuber et al. 2014; Vinke et al. 2016). Winter monsoons in Arid Asia, India and Southeast Asia have been predicted to decline significantly affecting agricultural production (Preston et al. 2006). Any increase in rainfall amounts in the region due to climate change will be offset by rising temperatures, posing a challenge to agriculture, stressing crops and reducing yields (Mendelsohn 2005; Preston et al. 2006). Climate change estimates suggest relatively large increases in temperatures, with the potential for warming in excess of 5 °C for continental Asia (Preston et al. 2006).

The low lying and very flat country of Bangladesh is one of the countries most severely affected by climate change in the region (Vinke et al. 2016). High geographic vulnerability coincides with higher populations consisting of mostly poor income groups heavily reliant on subsistence agriculture. Severe floods/flash flooding (almost 80% of the country prone to flooding), droughts and cyclones/storms on the South and south-eastern parts of the country have been prevalent, causing crop losses (Preston et al. 2006; Wassman et al. 2009b; Chen et al. 2012; IPCC 2014; Sikder and Xiaoying 2014). These extreme climate events occur in Bangladesh almost annually and sometimes more than once a year, affecting crop agriculture, particularly rice production (MoEF 2005; Yamin et al. 2005). Rice plays a leading role by contributing 95% of the total food production in Bangladesh (GoB 2010), with any changes in climate thus having a profound effect on its productivity. Other climatic change impacts like extreme temperatures and salinity intrusion have also contributed to a significant reduction in crop yields in Bangladesh, which have also come with a reduction in arable land. The salinity intrusion in the coastal area has created a serious implication for coastal land which traditionally was used for rice production (Sikder and Xiaoying 2014). A study also found that an increase in maximum temperatures by 1 °C at vegetative, reproductive and ripening stages decreased Aman rice production by 2.94, 53.06 and 17.28 tons, respectively (Islam et al. 2008). With changes in temperatures by (2 to 4 °C), prospects of growing wheat and potato would severely be impaired. The Philippines ranks highest globally in terms of vulnerability to tropical cyclones, third in terms of people's exposure to seasonal events and is the fourth most affected by extreme weather events (Sönke Kreft et al. 2016). Agriculture is an integral part of the economic sector of the Philippines, with a 12% contribution to the country's GDP and a source of employment for around 32% of the economically active population (The World Bank 2017). Rice, being one of the integral sources of revenue as well as nutrition in the Philippines, has had its yields and revenue declining, with a link established with extreme weather events (Preston et al. 2006). Between 1990 and 2006, damages to agricultural production in the country were attributed to typhoons (70%), droughts (18%), and floods (5%), with average typhoon related damages to the sector estimated at USD 136 million (CCC 2011). According to the Pakistan economic survey (2011–12), the country's agricultural sector absorbs 45% of the labour force, has an export share of 18% and a 21% contribution to the country's GDP. Therefore, the impact that changing climates have on the agricultural sector of the country cannot be underestimated. Wheat, being a staple food crop in Pakistan is deemed important due to its extensive use as food and a relatively cheaper source of animal feed. A decrease of 1.9% in wheat production from 25. 979 million tons to 25.478 million tons during the period from 2013 to 2015 was partly attributed to disruptions in the rainfall patterns and prolonged winter periods due to changing climate (GOP 2015; Ali et al. 2017).

The principal climatic hazards affecting Bangladesh—floods and cyclones—are likely to increase in frequency, intensity, duration, and extent. The summer monsoon rainfall is projected to increase, swelling the main river systems in the wider catchment, and boosting the rainfall impact within the country. How these increased hazard impacts will affect women, in particular, is extremely difficult to predict.

As a primary factor of social organization, gender shapes the social worlds within which natural events occur. Not surprisingly, gender differences are found in studies of emergency preparedness, voluntary action, emergency communication, the division of labour, post-traumatic stress, and coping strategies, among other areas. Rural women make up the majority of the world's poor population. In all developing regions, female-headed rural households are among the poorest of the poor.

Until recently, international climate change policymakers have neglected the gender dimension of climate change. A focus on technical solutions has ignored social and political factors. The successful implementation of climate change policies and projects requires an understanding of the gender-based roles and relationships vis-à-vis natural resources, as well as the gender-differentiated impacts of climate change and the different risks and vulnerabilities of women and men. With this established negative link between agricultural damage and climate change (Zilberman et al. 2004;

Cline 2007), reversing this trend would be an important component for Asian and African strategies for coping with the potential risks from the expected changes.

15.5 Mainstreaming Gender into Adaptation Programmes and Measures

Climate change impacts women's lives differently than men. Consequently, adaptation policies and measures need to be gender-sensitive: restricted access to economic resources may lead to fewer possibilities to adapt to climate change (e.g. to changing weather conditions and their impacts on agricultural activities), less possibilities of protection against disasters and recovering in the aftermath. Lacking control over land/property affects the ability to crop changes or switching to other fields of income generation; lack of transport systems leads to more time consumption and physical burdens (like transportation of heavy loads over long distances). The lacking access to political resources, like influence, networks, information, skills or control over decision-making leads to poor participation in the project and programme development and poor recognition of the practical and strategic needs of women and girls.

Gender aspects currently are poorly addressed in climate change debates. This is due to lacking gender sensibility of those shaping the debates, leading to a strong technical and economical bias in the contents—best to be seen in the Kyoto Protocol. Social issues and gender issues do not have space in these discussions and documents. In 1997, at UNFCCC COP3 in Kyoto, a female environmentalist asked via a mailing list why women's organizations did not take part in the process leading to the Kyoto Protocol. The answer she gave was that '"the arguments used here are almost entirely economic. Decisions are made mostly with little consideration being given to survival. Perhaps women felt they could not penetrate this masculine perspective—and stayed at home" (Sargent 1997)'. This assumption is backed by experiences in other UN processes: If women's organizations are not actively involved, gender and women's aspects will not be addressed. Thus, it is like a vicious circle: because women's organization or gender experts are not involved, women's/gender aspects are hardly addressed. And because they are not addressed, women's organizations don't take part. Recently some efforts have been undertaken to sensitize the parties of the UNFCCC and Kyoto Protocol as well as the constituencies taking part in the process on gender issues. Because negotiations were coming into a new phase, because the post-Kyoto regime may provide an entry point for gender perspective, the situation is improving slowly. There are more women's organizations interested in the UNFCCC process now, and thus hopefully the vicious circle will be opened.

15.6 Conclusions

There is substantial interest and merit in pursuing gender-sensitive integrated approaches to the three post-2015 agendas: the Paris Agreement, the Sustainable Development Goals and the Sendai Framework. To effectively achieve the goals of all three agendas, understanding the nexus between gender, climate change adaptation and disaster risk reduction in agriculture will help to enhance coherence between the frameworks and more efficiently utilize limited resources in the face of economic impacts of disasters.

Gender issues intersect with many aspects of DRR and ultimately contribute in shaping the lives of both men and women, hence should, therefore, be a basis of DRR efforts by making sure that views and needs of both men and women must are identified and addressed while building their resilience to disasters so that hidden gender biases do not limit the impact of any intervention.

DRR and disaster risk management, which are critical to resilience and adaptation should be gender-sensitive and systematically embedded into the agricultural sector and sub sectoral development plants and investment, especially in countries of the Southern Africa and the Asia Pacific, where disasters are recurrent and agriculture is the backbone of the economy and source of livelihood and nutrition.

This chapter, highlighting the importance of gender-oriented policies in DRR frameworks at national and international levels will set the stage for further research, policy development and action on gender-sensitive integrated approaches to pursuing adaptation, sustainable development and disaster risk reduction in Southern Africa and the Asia Pacific regions.

References

Ali S, Liu Y, Ishaq M, Shah T, Ilyas A, Din I (2017) Climate change and its impact on the yield of major food crops: evidence from Pakistan. Foods 6(6):39

ADB (2017) Key indicators for Asia and Pacific. Asian Development Bank, Manila, Philippines

Asseng S, Foster I, Turner NC (2011) The impact of temperature variability on wheat yields. Glob Change Biol 17:997–1012. https://doi.org/10.1111/j.1365-2486.2010.02262.x

Awojobi ON, Tetteh J (2017) The impacts of climate change in Africa: a review of the scientific literature. J Int Acad Res Multidiscipl 5(11):39–52

Bernier Q, Meinzen-Dick RS, Kristjanson P, Haglund E, Kovarik C, Bryan E,… Ringler C (2015) Gender and institutional aspects of climate smart agricultural practices: evidence from Kenya (CCAFS Working Paper No. 79). CGIAR Research Program on Climate Change, Agriculture and Food Security (CCAFS), Copenhagen. Retrieved from https://cgspace.cgiar.org/rest/bitstreams/53043/retrieve

Baudoin MA, Vogel C, Nortje K, Naik M (2017a) Living with drought in South Africa: lessons learnt from the recent El Niño drought period. Int J Dis Risk Reduc 23:128–137
Baudoin M-A, Vogel C, Nortje K, Naik M (2017b) Living with drought in south Africa: lessons learnt from the recent El Niño drought period. Int J Dis Risk Reduct 23:128–137
Bickton F (2016) Climate change as the biggest threat to public health in southern Africa and measures to reduce its impacts. Malawi Med J 28(2):70–72
Bruinsma M, Kowalchuk GA, Van Veen JA (2003) Effects of genetically modified plants on microbial communities and processes in soil. Biol Fertil Soils 37(6):329–337
Bryan E, Behrman J (2013) Community-based adaptation to climate change: a theoretical framework, overview of key issues and discussion of gender-differentiated priorities and participation (CAPRi Working Paper 109). International Food Policy Research Institute (IFPRI), Washington, DC
Carvajal-Escobar Y, Quintero-Angel M, Garcia-Vargas M (2008) Women's role in adapting to climate change and variability. Adv Geosci 14:277–280
Climate Change Commission (CCC) (2011) National climate change action plan. Manila. Retrieved from. https://extwprlegs1.fao.org/docs/pdf/phi152934.pdf
Chersich M, Wright C, Venter F, Rees H, Scorgie F, Erasmus B (2018) Impacts of climate change on health and wellbeing in South Africa. Int J Env Res Public Health 15(9):1884
Chen J, Li M, Wang W (2012) Statistical uncertainty estimation using random forests and its application to drought forecast. Mathematical. Problems in engineering, 915053, doi:https://doi.org/10.1155/2012/915053
Cline WR (2007) Global warming and agriculture: impact estimates by country. Center for global development and Peterson institute for international economics, Washington
Dankelman I (ed.) (2010) Gender and climate change: an introduction. Routledge
Denton F (2002) Climate change vulnerability, impacts, and adaptation: Why does gender matter? Gen Devel 10(2):10–20
Easterling DR, Evans JL, Groisman PY, Karl TR, Kunkel KE, Ambenje P (2000) Observed variability and trends in extreme climate events: a brief review. Bull Am Meteor Soc 81(3):417–426
EMDAT (2017) Natural disasters: annual disaster statistical review, available from: https://www.emdat.be/annual-disaster-statistical-review-2017
FAO (2008) An assessment of the impact of climate change on agriculture and food security in the pacific: a case study in the republic of the Marshall islands. Prepared for FAO SAPA Apia, Samoa by Muliagatele Iosefatu Reti, Pacific Environment Consultants ltd (PECL)
FAO (2011) The state of food and agriculture 2010–11. Women in agriculture: closing the gender gap for development. Author, Rome
FAO (2015a) Climate change vulnerability in fisheries and aquaculture: a synthesis of six regional studies. FAO fisheries and aquaculture circular no. 1104
FAO (2015b) Input paper for the global assessment report 2015 by UNISDR: Mainstreaming Disaster Risk Reduction in Agriculture: An Assessment of Progress Made Against the Hyogo Framework for Action
Grothmann T, Patt A (2005) Adaptive capacity and human cognition: the process of individual adaptation to climate change. Global Env Chang 15(3):199–213
Hijioka Y, Lin E, Pereira JJ, Corlett RT, Cui X, Insarov GE, Lasco RD, Lindgren E, Surjan A (2014) Asia. In: Climate change 2014: impacts, adaptation, and vulnerability. Part B: regional aspects. contribution of working group II to the fifth assessment report of the intergovernmental panel on climate change, Barros VR, Field CB, Dokken DJ, Mastrandrea MD, Mach KJ, Bilir TE, Chatterjee M, Ebi KL, Estrada YO, Genova RC, Girma B, Kissel ES, Levy AN, MacCracken S, Mastrandrea PR, White LL (eds.). Cambridge University Press, Cambridge, United Kingdom and New York, NY, USA, pp 1327–1370

International Council for Science (2008) A science plan for integrated research on disaster risk: addressing the challenge of natural and human-induced environmental hazards. Retrieved from https://council.science/cms/2017/05/irdr-science-plan.pdf

IPCC (2001) intergovernmental panel on climate change. Impacts, adaptation and vulnerability. Contributions of the working group III to the third assessment report of the intergovernmental panel on climate change. Cambridge University Press

IPCC (2007) The synthesis report of the intergovernmental panel on climate change. Cambridge University Press, Cambridge

Centritto M, Loreto F (2005) Photosynthesis in a changing world: photosynthesis and abiotic stresses. Agric Ecosys Env 106:115–117

IPCC (2014) Summary for policymakers. In: Field CB, Barros VR, Dokken DJ et al. (eds.) Climate change

Ishiwatari M, Djalante R, Mavrodieva A, Gómez OA, Prabhakar SVRK, Wataya E, Shaw R (2019) Climate Fragility Risks (CFR) in development sectors: six principles for managing synergies and trade-offs, the university of Tokyo, The United Nations University—Institute for the Advances Study of Sustainability (UNU-IAS), Keio University, Ritsumeikan Asia Pacific University, Institute for Global Environmental Strategies (IGES), Integrated Research on Disaster Risk (IRDR), Japan, 28 pages

Lobell DB, Schlenker W, Costa-Roberts J (2011) Climate trends and global crop production since 1980. Science 333:616–620. https://doi.org/10.1126/science.1204531

Lunduka RW, Mateva KI, Magorokosho C, Manjeru P (2019) Impact of adoption of drought-tolerant maize varieties on total maize production in south Eastern Zimbabwe. Climate Dev 11(1):35–46

Masike S, Ulrich P (2008) Vulnerability of traditional beef sector to drought and the challenges of climate change: the case of Kgatleng district, Botswana. J Geo Reg Plan 1(1):12–18. https://www.academicjournals.org/JGRP

Murillo MQ, Tan S (2017) Discovering the differential and gendered consequences of natural disasters on the gender gap in life expectancy in Southeast Asia. Nat Hazards Earth Sys Sci. https://doi.org/10.5194/nhess-2017-370

Nabikolo D, Bashaasha B, Mangheni MN, Majaliwa JGM (2012) Determinants of climate change adaptation among male and female headed farm households in eastern Uganda. Afr Crop Sci J 20(s2):203–212

Nhemachena C, Hassan R (2008) Determinants of African farmers' strategies for adapting to climate change: Multinomial choice analysis. African J Agr Res Econo 2(1):83–104

Neumayer E, Plümper T (2007) The gendered nature of natural disasters: the impact of catastrophic events on the gender gap in life expectancy, 1981–2002. Ann Assoc Am Geogr 97(3):551–566

Niang I, Ruppel O, Abdrabo M, Essel A, Lennard C, Padgham J, Urquhart P (2014) Africa. In: Climate change 2014: impacts, adaptation and vulnerability. Contribution of Working Group II to the Fifth Assessment Report of the Intergovernmental Panel on Climate Change. Cambridge University Press, Cambridge. Retrieved from https://www.ipcc.ch/pdf/assessmentreport/ar5/wg2/WGIIAR5-Chap22_FINAL.pdf

Nielsen JØ, Reenberg A (2010) Cultural barriers to climate change adaptation: a case study from northern Burkina Faso. Global Env Change 20(1):142–152

Perez C, Kristjanson P, Förch W, Thornton PK, Cramer L (2015) How resilient are farming households, communities, men and women to a changing climate in Africa? Global Environ Change 34:95–107

Peterman A, Quisumbing A, Behrman J, Nkonya E (2011) Understanding the complexities surrounding gender differences in agricultural productivity in Nigeria and Uganda. J Devel Stud 47(10):1482–1509

Phiri MG, Ibrahim, Saka RA (2005) The impact of changing environmental conditions on vulnerable communities of the Shire valley, southern Malawi. Lilongwe. AMCEN, 2011: addressing climate change challenges in Africa; a practical guide towards sustainable development

Preston BL, Suppiah R, Macadam I, Bathols J (2006) Climate change in the Asia/Pacific region: a consultancy report prepared for the climate change and development roundtable. CSIRO, Aspendale, Victoria

Quisumbing AR, Meinzen-Dick R, Raney TL, Croppenstedt A, Behrman JA, Peterman A (2014) Gender in agriculture: closing the knowledge gap. Springer, Netherlands. https://doi.org/10.1007/978-94-017-8616-4

Ragasa C, Berhane G, Tadesse F, Taffesse AS (2013) Gender differences in access to extension services and agricultural productivity. J Agri Educ Exten 19(5):437–468

Rakib M, Matz JA (2014) The impact of shocks on gender-differentiated asset dynamics in Bangladesh (IFPRI Discussion Paper 01356). International Food Policy Institute, Washington, DC. Available at SSRN. doi:https://doi.org/10.2139/ssrn.2483974

Ribeiro N, Chaúque A (2010) Gender and climate change: mozambique case study. Heinrich Böll Stiftung Southern Africa, Cape Town

Schellnhuber HJ, Frieler K, Kabat P (2014) The elephant, the blind, and the intersectoral intercomparison of climate impacts. Proc Natl Acad Sci 111(9):3225–3227

Schlenker W, Roberts MJ (2009) Nonlinear temperature effects indicate severe damages to US crop yields under climate change. Proc Natl Acad Sci 106(37):15594–15598

Seo N, Mendelsohn R (2007) An analysis of livestock choice: adapting to climate change in Latin American farms world bank policy research working paper 4164, World Bank, Washington DC

Sikder R, Xiaoying J (2014) Climate change impact and agriculture of Bangladesh. J Env Earth Sci 4(1):35–40

Sargent M (1997) How the women stayed at home. United Nations Third Conference on Climate Change

Sönke Kreft S, Eckstein D, Dorsch L, Fischer L (2016) Global climate risk index. Who suffers most from extreme weather events? Weather-related loss events in 2014 and 1995 to 2014. Briefing Paper. Bonn, Germanwatch

Sultana F (2014) Gendering climate change: geographical insights. Profession Geograp 66(3):372–381

Tambo JA, Abdoulaye T (2012) Climate change and agricultural technology adoption: the case of drought tolerant maize in rural Nigeria. Mitig Adapt Strat Glob Change 17(3):277–292

The World Bank (2017) World development indicators. World Bank, Washington, DC

Thornton PK, van de Steeg J, Notenbaert A, Herrero M (2009) The impacts of climate change on livestock and livestock systems in developing countries: a review of what we know and what we need to know. Agric Syst 101(3):113–127

UN Women (2015) The cost of the gender gap in agricultural productivity in Malawi, Tanzania and Uganda. New York: UN Women, United Nations Development Programme–United Nations Environment Programme Poverty-Environment Initiative (UNDP-UNEP PEI) Africa, and the World Bank

Wassmann R, Jagadish SVK, Sumfleth K, Pathak H, Howell G, Ismail A, …, Heuer S (2009) Regional vulnerability of climate change impacts on Asian rice production and scope for adaptation. Adv Agron 102:91–133

Wisner B, Blaikie PM, Blaikie P, Cannon T, Davis I (2004) At risk: natural hazards, people's vulnerability and disasters. Psychology Press

World Health Organization (2003) Diet, nutrition and the prevention of chronic diseasess. (WHO Techical Report series 916). Author, Geneva, Switzerland

Yamin F, Rahman A, Huq S (2005) Vulnerability, adaptation and climate disasters: a conceptual overview. IDS Bull 36(4):1–14
Zilberman D, Liu X, Roland-Holst D, Sunding D (2004) The economics of climate change in agriculture. Mitig Adapt Strat Glob Change 9(4):365–382
Zhu T, Ringler C (2010) Climate change implications for water resources in the Limpopo River Basin. International Food Policy Research Institute (IFPRI)

Chapter 16
Inclusive Resilience: Incorporating the Indigenous into the Picture of Resilient Reconstruction

Diocel Harold M. Aquino, Suzanne Wilkinson, Gary M. Raftery, and Sandeeka Mannakkara

Abstract Traditional structures are among the cultural treasures that stand as testament to the ingenuity of people who have come before us. In this chapter, we look at the case of Navala, the sole remaining village in Fiji that has managed to preserve their indigenous housing structures, called bure. In 2016, Fiji was struck by TC Winston, a Category 5 cyclone that severely damaged over 30,000 houses in the country, among these, a number of bure houses in Navala. Post-Winston recovery saw the construction of contemporary houses in lieu of the traditional ones. Herein, we unpack the policy-related barriers towards reconstructing the Fijian traditional housing structure and propose measures on how these structures can be incorporated into the post-disaster reconstruction framework. Three key things were found to cause the predominance of contemporary housing structures over the bure in the reconstruction phase: perspectives of key stakeholders on resilient housing, recognition of indigenous materials and methods in the code, and the inflexibility of government programmes and policies as regard to informal practices. Ultimately, empowering indigenous institutions to construct their traditional housing enhances the ability to achieve a timely and efficient recovery through the complementation of modern and traditional knowledge and contributions.

Keywords Reconstruction · Traditional housing · Indigenous knowledge · Fiji · TC winston

D. H. M. Aquino (✉) · G. M. Raftery · S. Mannakkara
The University of Auckland, Auckland, New Zealand
e-mail: aquino.harold@gmail.com

D. H. M. Aquino
University of the Philippines Diliman, Quezon City, Philippines

S. Wilkinson
Massey University, Auckland, New Zealand

R. Djalante et al. (eds.), *Integrated Research on Disaster Risks*, Disaster Risk Reduction,
https://doi.org/10.1007/978-3-030-55563-4_16

16.1 The Traditional Housing Structures of Navala Village, Fiji

Navala is an indigenous Fijian village situated at the bank of Ba River, located in the province of Ba, in the north-central part of Viti Levu, Fiji's biggest and most populous island. It is known to be the sole remaining traditional village in Fiji, having been able to preserve their traditional housing structures. Until recently, all the houses at the core of the village were of traditional construction.

The village has a population of 630 and has an average household size of 4.8. There were 132 houses in the village prior to Cyclone Winston: 99 of which were traditional houses, while 33 were of modern construction.

Fijian indigenous housing structures are called bure. They are characterised by low, breathable walls, high-pitched, hipped, thatch roof with short eaves, built over an earthen plinth (Fig. 16.1).

Materials used for the different components of the bure vary depending on where the village is located. For an inland village such as Navala, the posts are made of hardwood, walls and flooring are lined with matted pandanus leaves, and the roof thatch is made from local reeds or ferns, lined underneath by bamboo slats. Coastal communities, on the other hand, use mostly mangrove species for the components of their bure houses.

Traditionally, the construction of a bure is typically directed by a traditional builder called *mataisau*. Everyone in the village, however, contributes to building the bure. This practice of doing work communally in an iTaukei (indigenous Fijian) village is called *solesolevaki*.

Fig. 16.1 A bure house in Navala Village (*Source* Authors)

Navala is one of the top tourism destinations in Fiji. Tourists wanting a glimpse of authentic Fijian culture and traditions can live in a bure house accommodation and experience the Fijian way of life. Because of this, tourism has become one of the major income sources of the people in the villages. The influx of tourists also boosts their income from small-scale agriculture. Bure homeowners also get income by offering homestays for tourists. As a result, most of the village residents choose to stay in the village rather than move to urban centres or agricultural plantations because they already have a stable income source.

16.2 Cyclone Winston and Its Impacts

Tropical Cyclone Winston is the strongest storm on record to ever make landfall in Fiji and in the entire Southern Hemisphere. The cyclone was first detected as a tropical disturbance off Vanuatu on 7 February 2016. It developed into a Category 5 cyclone on 20 February and reached its maximum intensity just before making landfall on the north-eastern part of Viti Levu. The cyclone brought strong winds reaching a peak 10-min sustained intensity of 280 kph (APDRC 2016).

The storm caused a massive catastrophe in Fiji (National Disaster Management Office 2016). Over 60% of the Fijian population was affected by cyclone. Fifteen percent, or around 131,000 people, were left homeless in Winston's aftermath. Over 9,000 houses were destroyed, 17,000 were severely damaged and 29,000 sustained minor damage. Power and communication systems infrastructures were also extensively damaged.

A post-Cyclone Winston reconnaissance study reveals the inadequacies of Fijian housing construction. Predominant failure modes are the lack of a proper lateral-load-resisting system, roof tear-offs and window breakages leading to a build-up of internal pressure within the house, which then leads to collapse (Aquino et al. 2019). Such is underpinned by a shortage in the number of skilled builders in the country, limited knowledge and compliance to the building code, as well as the inadequacy in the code provisions for designing against a tropical cyclone as intense as Winston (Aquino et al. 2018). These studies, however, focused on villages that predominantly comprise contemporary housing structures and had limited insights on traditional housing construction.

16.3 Introduction: The Need for Preserving Heritage Structures Amid Disaster Reconstruction

Traditional societies respond differently in the face of natural hazards (Gaillard 2006b). Some societies have managed to resist changes, while others have recovered from the effects of disasters through changes in their way of life. The transformations also vary in degree and longevity. Some changes are extensive and drastic while others are minimal. Some adapt changes on the short-term in order to cope with temporary post-disaster scarcity and then revert when they have already restored normalcy. Other changes are observed in the longer haul and appear to be permanent.

Gaillard (2006b) argues that, if resilience is to be viewed as the ability to overcome the impact of disasters or prevent the potential adverse impacts of a hazard, then it is resilience whether a traditional society achieves that by preserving their pre-disaster social fabric or by embracing substantive changes to their lifestyle, as long as they survive. For instance, the Aetas of the Philippines who were relocated in the aftermath of the 1991 Mt. Pinatubo eruption have shifted to contemporary housing when they built their new homes (Gaillard 2006a).

On the other hand, housing is one of the defining features of a traditional society. According to Lozar and Rapoport (1970), "house-form is not simply the result of the physical forces or any single causal factor, but it is also the consequence of a whole range of socio-cultural factors—the specific characteristics of culture—the accepted way of doing things."

Traditional societies have also demonstrated that it is possible to preserve traditional settlements and house-forms in the aftermath of a disaster. For instance, the Quechua Indians resisted the government-initiated resettlement following an earthquake that struck Peru in 1970 (1977). Instead, they rebuilt their town by themselves in the vicinity of the ruins of their former houses. People of the Maheshpur village in Jhapa region of Nepal have managed to preserve the Rajbanshi houses while residents of the Sarankot village in the Kaski region have been able to retain their Gurung rounded houses (Gautam et al. 2016).

Likewise, the community of Kotagede in Yogyakarta, Indonesia was able to restore its heritage structures severely affected by a M5.9 earthquake in 2006 (Rahmi 2018). This was achieved through a collaboration of the community, the civil society, the government and the academia. Funding came from a plethora of sources: the government, multilateral agencies, the private sector and foreign governments.

In Fiji, the number of bure structures has steadily declined over the years. In the late 1970s, bures comprised 12% of the Fijian housing structures. In 1989, the number was down to 9%. In 2013, the number of bure structures was further reduced to 2.2% (NDMO 2013). As of 2016, bures only comprised a little less than 2% of the Fijian housing stock (NDMO 2016).

If traditional housing structures, as housing and cultural treasures, are to be preserved then there is a need for a systematic approach to further curb the decline in number.

Fujieda and Kobayashi (2013) identified three challenges that undermine the construction of new bures in Fijian rural villages: knowledge and skills, workforce, and natural resources.

There is little to no documentation of indigenous construction practices. Furthermore, there is no formal training on indigenous construction methods; it is a trade that is informally passed on through generations for builder clans. Additionally, while the village can rely on the practice of *solesolevaki* (communal work) in building their bures, such work would be too massive in the aftermath of a major disaster that had caused a significant number of the village population to lose their homes. In the same situation, a scarcity in the available resources for housing construction is also expected.

Beyond these challenges, which are mostly internal to the villages, Fiji's Cyclone Winston experience in 2016 had surfaced more overarching factors that affecting the construction of bure houses, such as government policy, codes of practice and people's desire for modern houses.

The aim of this study is to identify the policy-related factors that influence the reconstruction of indigenous housing structures after a disaster. This prompts us to ask the question: What systemic factors result in the construction of contemporary houses in place of damaged bure houses in the post-disaster setting? How do we enable communities to reconstruct their bure houses if they prefer to do so?

This study contributes towards Priority 3 of the Sendai Framework for Disaster Risk Reduction (UNISDR 2015), which espouses the protection and preservation of cultural heritage sites and structures. Additionally, it is very much in line with the third objective of the IRDR agenda, which is to reduce risk and curb losses through knowledge-based actions. The case study contributes towards filling the relatively narrowly studied field of post-disaster heritage preservation.

16.4 Methodology

A case study approach was used in the study. Case studies are beneficial when trying to answer questions of "how" or "why" and when dealing within real-life contexts wherein the investigator has little control (Yin 2006). Herein, we investigate the post-Cyclone Winston housing reconstruction process in the village of Navala. Navala is the sole remaining village in Fiji that has managed to predominantly consist of traditional housing structures.

Grounded theory methodology was used to unravel the factors affecting the rebuilding of bure houses. Grounded theory is a research approach that involves a systematic collection of empirical data and then coding and generating hypotheses (Glaser and Strauss 1967). Orlikowski (1993) suggests that grounded theory method is advantageous when investigating processes.

For this study, the data collection methods included a direct observation of the houses in the village, focus group discussion in the community, interviews with the pertinent stakeholders and a review of relevant documents, which included post-disaster assessment reports, policies, plans, the National Building Code of Fiji and related scientific literature.

A focus group discussion was conducted with the Navala residents (G1). The focus group had a total of 12 people, which included the *ratu* (chief), the *turaga ni koro* (village headman), clan leaders and village elders. Of the twelve, two were female and ten were male. The focus group discussion ran for approximately 90 min and was centred around their post-disaster housing reconstruction experience.

Table 16.1 Interview and focus group discussion respondents

ID	Affiliation
G1	Navala Village
P1	Rural Housing Unit—MHCD
P2	Ministry of iTaukei Affairs
P3	Ba Provincial Government
P4	National Disaster Management Office
P5	Habitat for Humanity Fiji
P6	Fiji Red Cross Society

Interview respondents come from the government sector (P1–P4), from institutions that are largely influential in the post-disaster reconstruction policies and programmes, and the development sector (P5–P6), institutions that have worked in the communities extensively in the aftermath of Cyclone Winston, and hence are highly knowledgeable of the key issues. A summary of interview respondents is shown in Table 16.1.

The respondents were interviewed about their involvement in post-Cyclone Winston housing reconstruction, the challenges they encountered, the gaps they observed and the best practices in relation to rebuilding or repairing a traditional housing structure.

Data collected on the ground were systematically coded, categorised and integrated into propositions on the challenges to rebuilding bure houses after a major disaster.

16.5 Navala after Cyclone Winston

In the village of Navala, 32 out of 132 or 24% of the houses were rendered uninhabitable (P3). An additional 30 houses suffered moderate to severe damage. Miyaji et al. (2017) quantified the structural performances of the two different types of housing structures, i.e. traditional and contemporary, through a post-disaster survey questionnaire ($n = 72$) as follows (Table 16.2).

Table 16.2 Effect of Cyclone Winston on housing structures in Navala

Damage state	Traditional ($n = 58$) (%)	Contemporary ($n = 14$) (%)
Collapsed	29.3	14.3
Severe	6.9	0.0
Moderate	5.2	7.1
Minimal	51.7	35.7
Undamaged	6.9	42.9

Statistically, contemporary houses have performed better in the wake of Cyclone Winston with 21.4% sustaining considerable damage as compared to 41.4% of the traditional bures. The collapse rate of the bure is also twice as much as that of the contemporary housing structure. It would be inappropriate, however, to make generalised comparisons regarding their structural resistance against tropical cyclones for two reasons.

Firstly, the bure structures in the village are far older than the modern-style houses, which have only emerged and grown in number over the last two decades (G1). Figure 16.2 shows an aerial photo of Navala in 2003. It can be observed that aside from some miscellaneous structures (light coloured roofs) such as the drying shed in the middle of the village, all the structures in the village core are bure houses. There were a few contemporary structures situated on the fringes of the village. They were mostly community facilities and sheds. Bures are only rethatched when the roof had been weathered (G1); the structural framing is not typically replaced or maintained.

Secondly, the location of the structures in the village plays an important part. Navala had a standing rule that modern houses cannot be built on the core, along the cross-street, of the village (G1). They can only be constructed off the core or behind the bure houses. This can be observed in Figs. 16.2 and 16.3, where the

Fig. 16.2 Navala in 2003 (*Source* Google Earth)

Fig. 16.3 Navala in 2015 (*Source* Google Earth)

main cross-upstreet, which is a T-shaped clear area at the heart of the village is only lined with bure houses. As a result, the modern houses had the leverage of shielding effect by the bigger bure houses. Additionally, Fig. 16.3 also shows that the contemporary structures in the main part of the village had the benefit of the terrain effect of the trees in the peripheries of the village. Both the shielding and terrain effects cause a reduction in the wind load on a structure because of upwind obstructions. Figure 16.4 shows that the community retained adherence to this rule. Amid the newly constructed contemporary houses after Cyclone Winston, the main cross-street still had all bures.

16.6 Challenges to Rebuilding Bure Houses

Three major challenges in rebuilding bure houses in the aftermath of a disaster were identified: the divergent constructs on what a resilient house is, the absence of

Fig. 16.4 Navala in 2018 (*Source* Google Earth)

provision for traditional housing in the building code and the lack of consideration for informal construction in post-disaster reconstruction support programmes.

16.6.1 Reconciling Views on Resilient Housing

Proposition 1: There are divergent constructs on what a resilient housing structure is among key stakeholders
Government institutions are generally more inclined towards building contemporary housing structures. The provincial administration of Ba, the local government unit that has jurisdiction over the village of Navala, encouraged the building of modern houses in lieu of the damaged bure houses because "*a modern house could resist cyclones better than a bure*" (P3). Additionally, the house design circulated by the Rural Housing Unit after Cyclone Winston is that of a timber-framed house with wood-panel walls and corrugated iron roof (P1).

Development sector respondents (P5, P6) recognize that both modern and traditional structures have their corresponding pros and cons and that the decision should be deferred to the person who will reside in the house. During the reconstruction phase, however, the information developed and disseminated by these civil society organisations as part of the Build Back Safer programme only contained guidance for timber-framed construction. No guidance on bure housing reconstruction was released.

The Ministry of Internal Affairs (P2), which is tasked to oversee that indigenous Fijian villages and their affairs, corroborates the stand of the civil society groups. This is because the Ministry is tasked with promoting the indigenous culture and traditions. They, however, had very little involvement in the reconstruction process.

Navala village residents, however, are unanimous in the belief that the bure is more resilient than a contemporary house. Drawing this belief from their experience in living in bures for decades, they claim that it was the first time that bures in their village sustained severe, irreparable damage (G1). Normally, strong winds would just cause the bure to lean towards one side and just had to be pushed back. They further add that in the event of a cyclone, "*the bure will not kill you, he will give you time to find shelter*". This pertains to the bure's gradual, as opposed to sudden and instantaneous, failure in the face of severe loads. The high-pitched roof design also enables the occupants to take shelter in the roof by moving towards the centre underneath the roof ridge.

16.6.2 Recognition in the Building Code

Proposition 2: There is a lack of recognition of indigenous construction design, methods and materials in building codes and standards

The National Building Code of Fiji (1990) governs the design and construction of structures in Fiji. It is largely based on Australian and New Zealand building codes and referral standards. There is also the Home Building Manual of Fiji (Ayyar 1990), which is a simplified version of housing construction. None of these standards and publications are compatible with indigenous construction.

Firstly, materials used for indigenous construction such as thatch, bamboo, mangrove timber, smoke-treated local timber, are not recognised in the code.

Likewise, indigenous housing structural designs are beyond the scope of the Home Building Manual. The maximum roof pitch covered by the manual is 25 degrees, whereas the roof slopes of a typical bure are at least 45 degrees, or even steeper. The roofing structure is also prescribed to be of a beam-and-rafter construction and having lightweight, presumed to be corrugated iron, cladding. This is as opposed to the bamboo matted underlay of the bure's thatched roofs.

While these standards are not mandatory for general rural housing construction at present, code compliance is required for all aid projects and this includes post-disaster housing provided by aid agencies. This means that, in the absence of sufficient proof of concept on the structural integrity of a bure house, they cannot build such housing structure for aid recipients.

Additionally, whether geared towards compliance or not, there is a need to provide guidance for resilient housing design and construction through codes and standards.

Cyclone Winston had elicited efforts to upgrade the National Building Code of Fiji (Lacanivalu 2018). Primarily, the aim is to consider upgrading the design wind speeds to resist Category 5 cyclones. However, this also opens possibilities to look into the other aspects of the Fijian Building Code, which has never been updated since its inception in 1991 (Fiji Building Standards Committee 1990).

16.6.3 Regard for Informal Practices in Government Policies and Programmes

Proposition 3: Post-disaster reconstruction policies and programmes are inflexible towards informal and indigenous practices

On 14 April 2016, the Fijian Government launched the Help for Homes Initiative (Ministry of Communications 2016). The programme was intended to support low-income Fijians, i.e. those earning FJD 50,000 (approx. USD 23,000) or less, who either lost their homes or had significant housing damage due to Cyclone Winston in the rebuild process. Homeowners were issued electronic cards that can be used at government-approved hardware stores to purchase materials for repair or reconstruction.

People whose houses sustained minor damage were eligible for a grant amounting to FJD 1,500 (USD 700). Those with roofs completely damaged were granted FJD 3,000 (USD 1,400). Homeowners whose homes were destroyed by the cyclone were eligible to receive FJD 7,000 (USD 3,250). Meanwhile, houses situated in informal settlements were only entitled to a maximum of FJD 1,500 (USD 700) financial assistance regardless of the level of damage sustained. Programme recipients were able to use the electronic cards to purchase materials from the hardware starting 9 May 2016, which marks the eightieth day since Cyclone Winston's onset.

The programme is coupled with the Build Back Safer programme of the government and comes with guide on how to reconstruct a house developed by the Shelter Cluster, which is handed to the beneficiary upon claiming the electronic card. The guide, however, shows a house with a contemporary design. This concern is closely tied to the concern on the recognition of indigenous structures in the code because the non-government organisations involved in the shelter cluster are compelled to comply with the code.

The financial support released through e-cards could only be used in accredited hardware stores. While this particular measure is good in regard to ensuring that the grant is used for house rebuilding, such is restrictive towards the reconstruction of bure structures. The materials necessary for rebuilding or repairing a bure cannot be bought from hardware stores because they are typically sourced from forestland or mangrove areas near the village. And while these materials can be harvested for free, monetary assistance is needed in terms of logistics and transportation. Significant costs are incurred in harvesting the materials and transporting them to the village.

Even with the communal practice of rebuilding houses in the village, it still entails substantive costs to feed the village men and women partaking in the *solesolevaki*.

Those who wanted to rebuild or repair their bures needed to do so out of their own funds (G1). Villagers who were eligible for the Help for Homes grant claimed their e-cards and constructed contemporary houses to live in while saving money to repair or rebuild their bures (G1). Those who did so can be seen spending the day in their damaged bures and then moving to the contemporary houses to sleep in at night.

16.7 Prospects

The Sendai Framework for Disaster Risk Reduction (2015) recognises the worth of indigenous knowledge in addressing the challenge of risk reduction and developing resilience. In the absence of formal means to develop resilience, indigenous societies have thrived and developed resilience by learning through experience, mostly in a trial and error method. The bure design, conceived by indigenous Fijians, is a living design that had evolved with the various environmental and socio-cultural factors that the community had been exposed to. This Cyclone Winston experience could be used to enrich the bure-building practice towards resilience, by deriving the lessons learnt, and should not be seen as a reason to abandon tradition.

16.7.1 Recommendations

In view of the policy-related challenges in rebuilding the bure in the aftermath of a disaster, there is a need to reconcile the various notions of resilience. There need not be only one absolute interpretation of what resilience means. UNISDR (2015) defines resilience as "the ability of a system, community or society exposed to hazards to resist, absorb, accommodate, adapt to, transform and recover from the effects of a hazard in a timely and efficient manner, including through the preservation and restoration of its essential basic structures and functions through risk management." Taking off from this, a resilient house could be something that is sturdy and could resist the most intense of winds, if we are to focus on the first part of the definition, or it could also be a house that could easily adapt to changes in the environmental condition, or one that might be feeble but is quick and easy to reconstruct in the aftermath of a disaster.

Ultimately, there is a need to dispel the notion that modern is necessarily better. In a similar sense that contemporary housing had advanced through technological breakthroughs, so can traditional ones. After all, the Sendai Framework espouses the integration of scientific and traditional practices in formulating solutions towards resilience (UNISDR 2015).

In the context of a bure, we can analyse its vulnerabilities and devise solutions. The shape of the bure, with its rounded corners, and the hipped roof style enables it to have better aerodynamic performance, in theory, as compared to the sharp corners, and flat gable roofs that typify contemporary housing structures (Holmes 2001). This could be validated by means of wind tunnel testing and/or numerical simulations. Furthermore, there is a need to evaluate the quality of the materials and the construction details used therein. This could be done by performing tests on indigenous materials. These combined would form a proof of concept which could then enable the integration of the indigenous housing design and practice into the code.

In the aftermath of the 2005 Kashmir earthquake, more than 100,000 dhajji dewari homes were built (Hicyilmaz et al. 2012). Dhajji dewari is an indigenous construction method prevalent in the Western Himalayas, particularly in Kashmir. It uses timber framing with intricate connection details packed with masonry infill materials. To date, this type of structure and construction method has been extensively studied and its design further developed towards seismic resilience (for instance, Ahmad et al. 2012, 2017; Ali et al. 2017; Dar and Ahmad 2015) and is now recognised as a seismic-resistant housing structure.

While globally there is little to no recognition of indigenous materials and methods in building codes, some countries have already started doing so. For instance, Peru was able to incorporate provisions for earthen housing into its building code (Vargas et al. 2006). In 2015, Nepal has also been able to do so for masonry (NBC 203) and earthen constructions (NBC 204) (Government of Nepal 2015a, b).

Finally, there is a need to be more conscientious in formulating post-disaster reconstruction policies and programmes that includes everyone in the picture. There needs to be more focus on how we can leverage traditional knowledge and skills in developing resilience.

16.7.2 Self-build and Post-disaster Recovery

Fiji is a country of a little over 100 inhabited islands scattered over a 1.3 million square kilometre area in the South Pacific region. In the aftermath of Cyclone Winston, one of the biggest challenges in humanitarian aid was the logistics of transporting the materials and skilled people to the far-flung and isolated islands (P5, P6). Thus, there arises the need to determine those which can be sourced from within the locality.

Miyaji et al (2017) had demonstrated the contribution of the practice of self-build in post-disaster recovery. The bure, which is typically constructed communally from locally available materials, promotes resilience by decreasing the community's reliance on external assistance. Not only does it address the issue of materials needing to be transported from the major islands, it also addresses the manpower gap post-disaster, which may then lead to a faster recovery.

In the event of a major disaster, the same may be the case for other nations in the South Pacific, on in the similarly climate-vulnerable nations in the Northwest Pacific, the Caribbean and the Indian Ocean, countries consisting of multiple islands spread over vast geographical areas. Creating space for indigenous knowledge, skills and materials into post-disaster reconstruction allows for empowerment of the indigenous peoples and the integration of indigenous practices into the general framework of disaster reconstruction.

References

Ahmad B, Alam A, Bhat MS, Ahmad S, Shafi M, Rasool R (2017) Seismic risk reduction through indigenous architecture in Kashmir Valley. Int J Disaster Risk Reduct. https://doi.org/10.1016/j.ijdrr.2016.11.005

Ahmad N, Ali Q, Umar M (2012) Simplified engineering tools for seismic analysis and design of traditional Dhajji-Dewari structures. Bull Earthq Eng. https://doi.org/10.1007/s10518-012-9364-9

Ali Q, Ahmad N, Ashraf M, Rashid M, Schacher T (2017) Shake table tests on single-story Dhajji Dewari traditional buildings. Int J Architect Herit. https://doi.org/10.1080/15583058.2017.1338789

Aquino DHM, Wilkinson S, Raftery GM, Potangaroa R (2019) Building back towards storm-resilient housing: lessons from Fiji's Cyclone Winston experience. International Journal of Disaster Risk Reduction 33:355–364. https://doi.org/10.1016/j.ijdrr.2018.10.020

Aquino DHM, Wilkinson S, Raftery GM, Potangaroa R, Chang-Richards A (2018) Challenges to building housing resilience: The case of Fiji post-cyclone Winston. Procedia Eng 212. https://doi.org/10.1016/j.proeng.2018.01.061

Asia-Pacific Data Research Center (2016) South Pacific Enhanced Archive of Tropical Cyclones (SPEArTC)

Ayyar KFBSC (1990). Home building manual of Fiji

Dar MA, Ahmad S (2015) Traditional earthquake resistant systems of Kashmir. Int J Civil Struct Eng Res

Fiji Building Standards Committee (1990) National building code of Fiji. Government of Fiji

Fujieda A, Kobayashi H (2013) The potential of Fijian traditional housing to cope with natural disasters in rural Fiji. J Disaster Res 8(1):18–27. https://repository.kulib.kyoto-u.ac.jp/dspace/bitstream/2433/173637/1/JDR_8(1)_18.pdf

Gaillard J-C (2006a) Traditional societies in the face of natural hazards: the 1991 Mt. Pinatubo eruption and the Aetas of the Philippines. Int J Mass Emerg and Diast

Gaillard J-C (2006b) Traditional societies in the face of natural hazards. Int J Mass Emerg Disast

Gautam D, Prajapati J, Paterno KV, Bhetwal KK, Neupane P (2016) Disaster resilient vernacular housing technology in Nepal. Geoenviron Disast. https://doi.org/10.1186/s40677-016-0036-y

Glaser B, Strauss A (1967) Grounded theory: the discovery of grounded theory. Sociology the Journal of the British Sociological Association

Government of Nepal (2015) Guidelines for earthquake resistant building construction: earthen building (NB204)

Government of Nepal (2015) Guidelines for earthquake resistant building construction: low-strength masonry (NB203)

Hicyilmaz KMO, Wilcock T, Izatt C, da Silva J, Langenbach R (2012) Seismic performance of Dhajji Dewari. In: Proceedings of the 15th world conference on earthquake engineering. Lisbon, Portugal

Holmes J (2001) Wind loading of structures.https://doi.org/10.4324/9780203301647

Lacanivalu L (2018, April 24) Our government's re-looking at national building code. Fiji Sun. Retrieved fromhttps://fijisun.com.fj/2018/04/24/our-governments-re-looking-at-national-building-code/

Lozar C, Rapoport A (1970) House form and culture. J Aesthetic Educ. https://doi.org/10.2307/3331293

Ministry of Communications—Government of Fiji (2016) Help for homes initiative—information background. Retrieved May 13, 2019, from https://www.fiji.gov.fj/Media-Center/Press-Releases/HELP-FOR-HOMES-INITIATIVE.aspx

Miyaji M, Fujieda A, Waqalevu SV, Kobayashi H (2017) Challenges for self-recovery from cyclone disasters in a traditional Fijian village: the case of Navala village after tropical cyclone Winston. In WIT transactions on the built environment, vol 173. pp 161–172).https://doi.org/10.2495/DMAN170161

National Disaster Management Office (2013) Tropical cyclone Evan, 17th December 2012 : post-disaster needs assessment, Fiji

National Disaster Management Office (2016) Tropical cyclone Winston, February 20, 2016: post-disaster needs assessment, Fiji. Retrieved fromhttps://www.gfdrr.org/sites/default/files/publication/Post%20Disaster%20Needs%20Assessments%20CYCLONE%20WINSTON%20Fiji%202016 (Online Version).pdf

Oliver-Smith A (1977) Traditional agriculture, central places, and postdisaster urban relocation in Peru. Am Ethnolog 4(1):102–116. https://doi.org/10.1525/ae.1977.4.1.02a00060

Orlikowski WJ (1993) CASE tools as organizational change: investigating incremental and radical changes in systems development. MIS Q 17(3):309. https://doi.org/10.2307/249774

Rahmi DH (2018) Building resilience in heritage district: lesson learned from Kotagede Yogyakarta Indonesia. In: IOP conference series: earth and environmental science. https://doi.org/10.1088/1755-1315/99/1/012006

UNISDR. (2015). *Sendai Framework for Disaster Risk Reduction.*

Vargas J, Blondet M, Tarque N (2006) The Peruvian building code for earthen buildings. In: Proceedings of Getty Seismic adobe project 2006 colloquium

Yin RK (2006) Case study reserach—design and methods. Clin Res. https://doi.org/10.1016/j.jada.2010.09.005

Chapter 17
Adapting to Climate Change by *Building Back Better* in Disaster Recovery: Case Study of Rarotonga, Cook Islands

Sandeeka Mannakkara

Abstract The world has witnessed increasing magnitudes and frequencies of disaster events, of which the exacerbation of weather-related events can certainly be attributed to the impacts of climate change. A large component of responding and adapting to climate change lies in attempts to successfully prepare for and recover from severe weather events such as hurricanes, storms and cyclones, and acclimate to gradual changes such as sea-level rise. The solutions to address these issues can be found in the intersection between climate adaptation practices as recommended by the UNFCCC Secretariat, and disaster risk reduction mandates set by the UNISDR Sendai Framework. This paper presents the concept of *Building Back Better (BBB)* as a fitting solution to adapt to climate change. BBB is a holistic concept that aims to build resilience in a community's physical, social and economic environments, enhancing the ability to adapt to climate change impacts. A case study looking at the climate change challenges of coastal businesses on the island of Rarotonga, Cook Islands was conducted using the "BBB Framework" developed by Mannakkara and Wilkinson (2014) to identify best practices for climate adaptation. Findings were obtained through qualitative semi-structured interviews with coastal businesses and relevant Government authorities. The results showcased that framed under BBB, a combination of approaches is necessary to prepare for and respond to climate change impacts such as innovating and re-designing the built environment, re-thinking land-use, appropriate education, effective early warning and evacuation practices, community cohesion, social and business networks, adaptive business practices, collaboration and alignment between all stakeholders involved, appropriate legislation and long-term monitoring and evaluation. The findings advocate for policymakers to adopt BBB as an effective concept to guide climate adaptation practices.

S. Mannakkara (✉)
Department of Civil and Environmental Engineering, University of Auckland, Auckland, New Zealand
e-mail: s.mannakkara@auckland.ac.nz

R. Djalante et al. (eds.), *Integrated Research on Disaster Risks*, Disaster Risk Reduction,
https://doi.org/10.1007/978-3-030-55563-4_17

17.1 Introduction

Our climate is changing as a result of increasing concentrations of greenhouse gases found in the Earth's atmosphere. Greenhouse gases like carbon dioxide, methane, nitrous oxide and chlorofluorocarbons trap heat resulting in a global warming effect, which is causing Earth's temperature to rise rapidly (Watts 2013; IPCC 2018). Some of the impacts of this warming include the destabilisation and melting of ice sheets resulting in sea-level rise, the increase in frequency and magnitude of extreme events such as storms, floods, droughts and heatwaves, increasing ocean acidification and changes to rainfall (IPCC 2014). Climate change thus poses a threat to communities through increased high-impact disaster events as well as slow-creep impacts such as sea-level rise that cause gradual changes over time.

Climate change adaptation (CCA) is referred to as the actions are taken to manage the imminent impacts of climate change to reduce the vulnerability of communities (IPCC 2014; McCarthy et al. 2001). To respond to extreme disaster impacts and slow-creep impacts resulting from climate change, both climate adaptation practices and disaster risk management are necessary (O'Brien et al. 2012). Adapting to climate change requires adaptation and risk reduction initiatives to minimise and mitigate anticipated risks, and reduce exposure and vulnerability to existing risk levels resulting from previous urban planning and building decisions (Lavell 2011). The recently published Global Assessment Report on Disaster Risk Reduction iterates that in fact, CCA can be characterised as a subset of disaster risk reduction (DRR) and that CCA needs to be integrated within DRR (UNDRR 2019). CCA and DRR are also necessary to ensure the long-term sustainability of communities (Wilbanks and Kates 2013).

The United Nations Framework Convention on Climate Change (UNFCCC), as part of the 2011 Cancun Adaptation Framework (Kato 2011) and the 2015 Paris Agreement (UNFCCC 2015), has set adaptation as a global challenge to change processes, practices and structures to moderate potential damages and benefit from opportunities associated with climate change. Adaptation activities are to be adopted in all sectors of a community. DRR priorities set by the 2010 United Nations Sendai Framework (UNISDR 2015) include understanding disaster risk, strengthening disaster risk governance to manage disaster risk, investing in disaster risk reduction for resilience and enhancing disaster preparedness for effective response and to "Build Back Better" in recovery, rehabilitation and reconstruction. There is a need to look at these two issues synergistically (ICSU 2008; UNDRR 2019).

Folke (2006), Walker and Salt (2006) have proposed that a systems approach based on the concept of resilience is most fitting to address the challenges posed by climate change, where resilience is defined as the "ability to absorb the effects of a disruptive event, minimise adverse impacts, respond effectively post-event, maintain or recover functionality, and adapt in a way that allows for learning and thriving, while mitigating the adverse impacts of future events" (Stevenson et al. 2015). Creating resilience through acceptance of the change (whether sudden or slow, and whether known or unknown) as an inherent characteristic is key. The concept of resilience already largely underpins the field of disaster risk reduction and recovery and is the core product of using a "Build Back Better" (BBB) approach. BBB, which is the fourth priority of action in the Sendai Framework (UNISDR 2015), is a strategy used in pre-disaster planning and post-disaster recovery to improve a community's physical, social, environmental and economic conditions to create a more resilient community in an effective and efficient way (Clinton 2006; Mannakkara et al. 2018; Monday 2002).

This chapter explores the concept of Building Back Better as a model for looking at climate change adaptation using the case study of Rarotonga, Cook Islands. The next section examines what BBB is, and uses Mannakkara and Wilkinson's BBB Framework to understand how this concept can be useful to look at climate change adaptation and resilience-building. Then, the methodology used to obtain findings from the case study is explained, followed by the results, showing how the BBB Framework was used to identify climate adaptation best practices for coastal businesses in Rarotonga who regularly face climate change impacts. The chapter concludes iterating that BBB can be used as an effective strategy to reduce risks and adapt to climate change.

17.2 What is "Building Back Better"?

Building Back Better (BBB) is a concept that became popularised following the catastrophic 2004 Indian Ocean Tsunami disaster, which impacted 14 countries and caused devastation in vulnerable coastal communities (Clinton 2006). The premise of BBB was the recognition that post-disaster reconstruction and recovery can be used as an opportunity to address pre-existing vulnerabilities and embed resilience into communities (Khasalamwa 2009; Twigg 2007). Although BBB was sometimes understood to only relate to improving the structural resilience of the built environment, there has been a strong sentiment that BBB can be advocated as a holistic approach with the aim of improving a community's physical, social, environmental and economic conditions to successfully address all vulnerabilities and improve overall resilience (James Lee Witt Associates 2005; Kennedy et al. 2008; Mannakkara and Wilkinson 2014). Authors such as Clinton (2006), Iglesias et al. (2009), DN and PA (2008), and Rotimi et al. (2009) added that understanding recovery processes and stakeholders and factoring in features such as clear role allocation, effective partnerships, transparent communication, equity, tailored solutions, participatory approaches and supportive legislation are also necessary for building back better.

Mannakkara and Wilkinson's "Build Back Better Framework" (Fig. 17.1) is a useful tool that defines the scope of the BBB concept (Mannakkara 2014; Mannakkara and Wilkinson 2014). The BBB Framework has been practically tested and aligned with the Sendai Framework and can be used to pre-plan and implement post-disaster recovery programmes in order to build back better (Mannakkara et al. 2018).

Fig. 17.1 The build back better framework. *Source* © Mannakkara and Wilkinson, University of Auckland, 2010, 2016

The BBB Framework uses a three-pronged approach for BBB comprising of (1) Disaster Risk Reduction, (2) Community Recovery and (3) Effective Implementation reflecting international findings and best practices (Mannakkara and Wilkinson 2014). The Framework illustrates the importance of considering all these three elements together in recovery programmes as they each interrelate and contribute significantly to successful recovery and resilience.

17.3 Disaster Risk Reduction

Disaster Risk Reduction in the BBB Framework refers to ways in which the level of risk posed to the built environment and the community can be reduced through building resilience. The Framework shows that the first DRR initiative for BBB is the improvement of the built environment's structural resilience. Practical implementation of structural resilience is not limited to updating building codes for higher capacity structural designs but needs to consider accurate and up-to-date multi-hazard risk information, consideration of how design changes can be implemented in the absence of building codes and regulations in developing countries, funding, resources, skills and quality assurance for BBB (Mannakkara and Wilkinson 2013a; Lyons 2009; Palliyaguru et al. 2010). Structural resilience efforts alone are insufficient in reducing risk.

The second DRR initiative is multi-hazard based land use planning, which is also an effective risk reduction method where changes in land-use are considered to match risk levels along with appropriate building solutions (Mannakkara and Wilkinson 2012b; Baradan 2006). The relocation of community settlements and infrastructure from high- to low-risk areas is the last resort option in terms of BBB due to the disruptions relocation causes to livelihoods and social structures (Oliver-Smith 1991). BBB theory proposes that if relocation is opted for, it needs to be implemented under a comprehensive resettlement strategy developed with community consent and input which considers multi-hazard risks of new lands, infrastructure, community preferences, and livelihood and lifestyle opportunities (Mannakkara et al. 2018).

The third DRR initiative, early warning and DRR education is often the most practical form of reducing risk in disadvantaged communities that are unable to effectively upgrade the built environment due to funding and resource constraints or change land-use due to land shortages and social and cultural pressures. Early warning and DRR education bridge the gap between DRR knowledge and implementation (UNISDR 2015). BBB advocates for early warning to provide timely, effective information to take action to avoid or reduce risks (UNISDR 2004; de León et al. 2006) and DRR education to "use knowledge, innovation and education to build a culture of resilience" in stakeholders as well as the local community (UNISDR 2005; Shaw et al. 2009).

17.4 Community Recovery

BBB stresses the importance of addressing and improving the psychosocial and economic environments of communities along with the risk reduction process to create resilience (Colten et al. 2008; Chang and Rose 2012). The Community Recovery component of the BBB Framework first advocates for psychological and social well-being (Mannakkara and Wilkinson 2015b). Communities need to be supported through the recovery process by providing personalised advisory services, special assistance for vulnerable populations, psychological support and counselling services, keeping the community informed, ensuring transparency, and having activities to retain and build community cohesion and social capital (Matanle 2011; Florian 2007). Another important aspect contributing to psychosocial recovery is the involvement of communities in the recovery process. People who actively take part in recovery through involvement in community consultation, community recovery and rebuilding projects and owner-building are often more satisfied with recovery outcomes, and develop personal resilience and adaptive capacity to cope with future events (Gibbs et al. 2015; Reich 2006; Gordon 2009).

Economic prosperity is as vital to a community's recovery and resilience, as a safe built environment and good social and cultural welfare. A community cannot thrive and retain and attract residents without economic benefits. The second Community Recovery component in the BBB Framework is economic recovery and recommends the development of a tailored economic recovery strategy that meets the needs and constraints of the local community based on comprehensive data (Mannakkara and Wilkinson 2012a; Mannakkara et al. 2018). BBB findings state that traditional livelihoods need to be empowered and supported, and new livelihood options introduced if appropriate, that make use of local resources and skills. The rebuilding process should be coordinated with business recovery to fast-track infrastructure and service access to businesses, and funding mechanisms need to be provided to support business recovery. Business and livelihood recovery needs to also be encouraged through business advisory services, up-skilling people and putting in place initiatives to boost the economy, create new jobs and attract residents and tourists.

17.5 Effective Implementation

Effective implementation in the BBB Framework includes selecting an appropriate institutional mechanism, the use of legislation and regulation, and monitoring and evaluation to ensure that recovery processes are productive and efficient. Firstly, it is necessary to adopt an appropriate institutional mechanism to plan and implement the recovery process. The institutional mechanism chosen needs to be locally suitable; whether recovery requires the creation of a dedicated authority or utilises existing government structures and the level of centralisation/decentralisation most fitting for the community needs to be carefully considered (Mannakkara and Wilkinson 2015a). The recovery process has to facilitate coordination and collaboration between the various stakeholders present in recovery efforts, include local councils and grassroots-level input, consult experts, put in place an effective funding mechanism, have a good information management system, have a well-formed transition strategy to move out of recovery back to normal operations, and address many factors to ensure BBB.

The second factor under Effective Implementation in the BBB Framework is legislation and regulation. Legislation and regulation need to be used to enforce BBB-based recovery through mechanisms such as building regulations and planning regulations, as well as facilitate the recovery process by fast-tracking and simplifying processes (Mannakkara and Wilkinson 2013b). The final BBB component is monitoring and evaluation, which needs to be put in place to ensure that BBB-based initiatives are properly implemented, address ongoing recovery problems and obtain lessons for the future (Clinton 2006; Wiles et al. 2005; Norio et al. 2011). It is important for recovery experiences to be retained and the knowledge transferred through capacity building, education and training locally and internationally (Olsen et al. 2005; Mikko 2009).

BBB with its holistic, resilience-focused approach provides a good pathway for reducing risks and adapting to climate change impacts through an all-of-community approach. The BBB Framework addresses all the areas where a community needs to adapt and build resilience to face known and unknown risks arising from climate change, whether they are extreme disaster events or slow-creep events such as sea-level rise. The BBB concept is already used in pre-disaster planning as per the Sendai

Framework (UNISDR 2015), and it can also be utilised as an effective tool for climate change adaptation planning and implementation.

17.6 Methodology for Case Study: Climate Adaptation and Business Resilience in Rarotonga

This research was undertaken as a qualitative case study looking at the climate change resilience of coastal businesses located on the island of Rarotonga, in the Pacific nation of Cook Islands. The Cook Islands consists of 15 islands spread over a large area in the South Pacific. Rarotonga is Cook Islands' capital and biggest island, 32 km in circumference, and is the economic hub of the country. Rarotonga is most popular for its scenic beaches, making tourism a large component of the island and country's economy. Along with the rest of the Pacific, the Cook Islands and the island of Rarotonga are impacted by annual tropical cyclone seasons. The country has been witnessing a clear increase in the magnitude of cyclones with the highest magnitude category 5 cyclones becoming more frequent in the past few years (De Scally 2008; Kumar 2018). Being an island nation, the country is also experiencing the effects of climate change, with rises in sea-level recorded at 4 mm per year since 1993, increasing temperature and rainfall, and increased ocean acidification (Australian Bureau of Meteorology and CSIRO 2011; Australian Aid et al. 2011). One of the key industries impacted by these changes is the coastal tourism business industry in Rarotonga that has to face the physical challenges of coastal hazards such as cyclones and sea-level rise as a result of climate change. Therefore, the aim of this research was to explore how BBB can be used to examine the climate adaptation capabilities of coastal businesses.

A qualitative case study approach was chosen for this research to enable in-depth exploration and comprehension of the situation (Yin 2009). Data for this study was collected through semi-structured open-ended interviews conducted with popular and prominent coastal businesses in Rarotonga. The coastal businesses were largely tourism-based and included large, medium and small hotels, small and medium restaurants, and other businesses (see Table 17.1 for business and interviewee profiles). Most of the businesses had previously been affected by cyclone events and coastal hazards. Participants from Emergency Management Cook Islands (EMCI), who is the designated government body responsible for disaster risk management in the Cook Islands, and the Tourism Corporation and Tourism Industry Council were also interviewed to understand the steps taken to ensure the resilience of a large number of tourism businesses vulnerable to climate change impacts.

The interviews were structured to understand the challenges faced by the businesses and identify their business practices under each of the BBB Framework components. This data was transcribed and analysed using an inductive approach based on grounded theory and the constant comparative method (Maxwell 2005;

Table 17.1 Interviewee profiles (*Source* Author)

Business/organisation type	Number of interviewees	Business/organisation name	Size
Accommodation	7	Pacific resort	Large
		Edgewater	Large
		Sunset resort	Medium
		Muri beach club resort	Medium
		Club Raro	Medium
		The islander hotel	Small
		Little polynesian	Small
Cuisine	3	Trader jacks	Medium
		Palace takeaways	Small
		Raro fried chicken	Small
Other	2	Rehab nightclub	Medium
		Captain Tama's Lagoon Cruises	Small
National-level organisations	3	EMCI	N/A
		Tourism corporation	N/A
		Tourism Industry Council	N/A

Dick 2005). The NVivo programme was used to code the data under the BBB Framework. The limitations of this study include the number of businesses that were able to take part in the study and the lack of representation from Government organisations who could contribute due to time constraints. Data obtained from businesses were validated through the information collected from Government organisations such as EMCI and the Tourism Industry Council and written documents such as reports, but further validation opportunities would have been desirable.

17.7 An Analysis of Climate Adaptation and Business Resilience in Rarotonga Using the Build Back Better Framework

The findings from the research study conducted in Rarotonga were categorised under "Disaster Risk Reduction", "Business Resilience" and "Effective Implementation", following the format of the BBB Framework.

17.8 Disaster Risk Reduction in Rarotonga

The businesses from this study had all experienced some form of impacts from cyclones and coastal hazards, whether major or minor. Being situated along the coast and familiar with annual tropical cyclone seasons, the businesses had a cultural attitude of expecting coastal hazards and often relied on knowledge passed down generationally on ways of adaptation. These businesses preferred to adapt to risks over relocation despite the susceptibility to climate change challenges due to the economic advantages offered by their beach-side locations in terms of attracting tourists. Furthermore, the Cook Islands practices customary land-ownership traditions where land is transferred through family and cannot be sold. This factor is also a large contributor to the reason coastal businesses have had to remain where they are and learn to adapt. Therefore, in terms of the BBB Framework and risk reduction, the primary ways of adaptation and resilience possible in Rarotonga are through structural resilience and early warning and DRR education.

The accommodation businesses which were major operators in Rarotonga's tourism industry had adopted sturdy concrete and steel structures that could withstand higher category cyclones. The hotels in particular had a very good understanding of the importance of building to higher standards to ensure the safety of their guests. Restaurants with less capital to invest in structural upgrades had opted for open-plan simple structures with portable designs allowing fast evacuation of assets in the event of cyclone warnings. Smaller businesses had innovatively adopted a flexible approach to risk reduction, where their priority was to safeguard key assets so that the businesses could operate despite building damages. Most businesses were in touch with the meteorological department and relied heavily on early warnings for cyclone events. They adopted traditional preparatory exercises such as placing shutters on windows and doors, tying roofs down and putting in place evacuation plans in the event of cyclones. Having comprehensive and well-practiced evacuation plans was a consistent feature among businesses, and there was very good awareness of evacuation routes and sites.

Many businesses had also made changes such as switching to lighter roofs and changing to stainless steel screws to counteract rusting. Some businesses added seawalls and groins to manage sea impacts. It was evident that these businesses had a very good understanding and awareness of the risks they are facing, and made an effort to adapt in different ways. The main focus of their adaptation activities seemed to be to minimise impacts from cyclone events, but the adaptive capacity displayed by the businesses will no doubt assist with slower climate change impacts such as sea-level rise.

There was consistent feedback from the businesses that they would prefer more information and guidance from government authorities around the most effective measures to combat the risks they are facing, and the enforcement of building codes to assist. Most businesses are adapting based on experience, and the larger businesses have chosen to adopt New Zealand building code standards. The EMCI commented that local building code revision is underway and that they are attempting to distribute this knowledge more widely. The Tourism Industry Council commented that they have an accreditation standard based on the disaster preparedness of businesses. The businesses still felt that better DRR and CCA information and training programmes are necessary for better adaptation.

17.9 Business Resilience in Rarotonga

The coastal tourism businesses understood their importance to the economy of the country and strived to rise above the challenges posed by climate change impacts. Apart from individual adaptive structural measures taken by businesses, the businesses remained aligned with the social and cultural values of the Cook Islands and took advantage of the strong village networks and relationships as a way of being resilient. Businesses and community members have a culture of assisting each other in times of disasters. Rarotonga displayed unique attributes such as neighbouring businesses that normally compete with each other coming together to share resources and manage guests post-disaster. There was an underlying principle that bound the business community in working for the collective good of the tourism industry and economy, which is an impressive indicator of resilience. The businesses found that assisting each other at the grassroots level was largely more effective than awaiting Government level or external help.

The businesses had not considered the slow-creep impacts of climate change in terms of business continuity planning but placed a lot of effort into monitoring their business recovery progress following disaster events based on rebuilding progress, aesthetics, number of guests, turnover and speed of general business recovery. The businesses prioritised making a "good impression" to keep attracting guests and tourists and thus ensured that they contributed to business resilience and growth through being open to change and having good relationships with the community and other businesses.

The resilience and adaptation capabilities of businesses in terms of social and economic well-being were impressive and aligned with BBB as presented in the earlier sections of this chapter; however, there was a lack of input and facilitation evident from government organisations. For instance, bodies such as the Tourism Corporation and Tourism Industry Council could add value by assisting these businesses with further resilience-building. Businesses can improve their adaptive capacity through receiving training on business continuity measures, fiscal and non-fiscal support for climate change adaptation and post-disaster recovery, the creation of local business networks, and more direct links with the Tourism Corporation and Tourism Industry Council.

17.10 Effective Implementation in Rarotonga

The main stakeholders relevant for post-disaster recovery and climate change adaptation activities of the coastal businesses in this study seemed to be EMCI, the Tourism Corporation, and the Tourism Industry Council. However, there seemed to be little direct support provided to businesses due to a lack of communication and coordination between these organisations and businesses. Culturally, Pacific nations such as the Cook Islands have built their resilience primarily through grassroots capabilities, which could be the reason for the inadequacy in government interventions. However, the businesses interviewed in this study all consistently indicated that they would like better relationships with these organisations to receive guidance on preparedness, especially for issues like climate change which aren't well understood.

In terms of legislation, all interviewees stated that there is no legislation that they know of that provides adequate assistance for tourism businesses to recover post-disaster or build resilience for future events. EMCI stated that they are working on better multi-hazard risk models and land-use planning to develop policies for restricting infrastructure on high-risk locations and enforce appropriate building design controls. It is also necessary for EMCI to work with the tourism organisations to develop solutions for coastal businesses that will likely remain in high-risk zones. Regulations and practices are also needed to prioritise climate adaptation infrastructure that can reduce risks imposed on coastal businesses, as well as prioritise the rebuilding of infrastructure servicing coastal tourism businesses following disaster events.

The last component of the BBB Framework, monitoring and evaluation, is implemented at an individual level by businesses to address issues and improve upon them, but there doesn't seem to be any significant higher level monitoring of the recovery and resilience of coastal businesses from the government level. The importance of tourism businesses to the country's economy needs to be acknowledged and more resources need to be directed at supporting these businesses and their ability to be resilient to climate change impacts.

17.11 Conclusions

This chapter explored how the holistic concept of Building Back Better can be an effective tool to look at climate change adaptation. Climate change is resulting in higher magnitude and frequency disaster events and slow-creep changes such as sea-level rise, higher rainfall and higher temperatures. Climate change adaptation and disaster risk reduction are therefore a combined problem, with resilience being the best solution. Resilience is the ability for a community, its built environment, people and businesses to be able to cope with known and unknown changes, recover and adapt. The IRDR Science Action Plan (ICSU 2008) calls for integrated approaches and in particular suggests the need to look at DRR and CCA priorities together.

BBB has been a popular concept developed for post-disaster recovery and pre-disaster planning to embed resilience into communities. BBB has evolved from focusing on developing the resilience of the built environment, to looking at developing the resilience of all sectors and aspects of a community in order to create absolute resilience. The BBB Framework developed by Mannakkara and Wilkinson offers a three-pronged approach for looking at pre-disaster planning, post-disaster recovery and climate adaptation to reduce risks and create resilience in communities.

The research case study documented in this chapter looked at the climate adaptation and resilience capacity of coastal businesses in Rarotonga using the BBB Framework. Rarotonga being an island in the Pacific is very exposed to the impacts of climate change, including destructive tropical cyclones and sea-level rise. The BBB Framework enabled an analysis of the CCA and resilience capabilities of the coastal businesses in this study, looking at their built environment, land-use, education and awareness of people, social and economic factors, and implementation processes. The businesses showed very positive indications of resilience from an individual and community-level perspective. The study showed that improvements were needed particularly in implementation and input from government-level organisations to assist these businesses.

The study illustrated the applications of the BBB Framework to examine climate change adaptation and identify gaps. BBB explicitly covers DRR, and is already a priority for action in the Sendai Framework, and can be used to look at both extreme disaster events as well as slow-creep events. The remaining community recovery and effective implementation components are useful for looking at resilience and adaptation capabilities. Therefore, it is recommended that the BBB concept and framework are used as an effective guide for climate change adaptation practices.

References

Australian Aid, Australian Government, CSIRO, Cook Islands Metereological Service (2011) Pacific climate change science program: current and future climate of the cook islands. Pacific climate change science program partners

Australian Bureau of Meteorology, CSIRO (2011) Climate change in the Pacific: scientific assessment and new research (volume 1: regional overview)

Baradan B (2006) Analysis of the post-disaster reconstruction process following the Turkish Earthquakes, 1999. In: GROUP IR (ed) International conference on post-disaster reconstruction meeting stakeholder interests. University de Montreal, Florence

Chang SE, Rose AZ (2012) Towards a theory of economic recovery from disasters. Int J Mass Emergencies Disasters 32(2):171–181

Clinton WJ (2006) Lessons learned from tsunami recovery: key propositions for building back better. Office of the UN Secretary-General's Special Envoy for Tsunami Recovery, New York

Colten CE, Kates RW, Laska SB (2008) Three years after Katrina: lessons for community resilience. Environment 50(5):36–47

de León JCV, Bogardi J, Dannenmann S, Basher R (2006) Early warning systems in the context of disaster risk management. Entwicklung and Ländlicher Raum 223–225

De Scally FA (2008) Historical tropical cyclone activity and impacts in the cook islands1. Pac Sci 62(4):443–460

Dick B (2005) Grounded theory: a thumbnail sketch. http://www.scu.edu.au/schools/gcm/ar/arp/grounded.html. Accessed 1 May 2019

DN, PA (2008) Disaster and development in South Asia: connects and disconnects. South Asia Disaster Report. Duryog Nivaran & Practical Action, Colombo

Florian S (2007) Housing reconstruction and rehabilitation in Aceh and Nias, Indonesia—rebuilding lives. Habitat Int 31(1):150–166

Folke C (2006) Resilience: the emergency of a perspective for social-ecological systems analyses. Glob Environ Change 16(3):253–267

Gibbs L, Harms L, Howell-Meurs S, Block K, Lusher D, Richardson J, MacDougall C, Waters E (2015) Community wellbeing: applications for a disaster context. Aust J Emerg Manage 30(3):20–24

Gordon R (2009) Community impact of disaster and community recovery. Australian Psychological Society, InPsych

ICSU (2008) A science plan for integrated research on disaster risk: addressing the challenge of natural and human-induced environmental hazards. International Council for Science, Paris

Iglesias G, Arambepola NMSI, Rattakul B (2009) Mainstreaming disaster risk reduction into local governance. In: National symposium on creating disaster free safer environment. National building research organisation and ministry of disaster management and human rights, Cp;p,bp

IPCC (2014) Summary for policymakers. In: Edenhofer O, Pichs-Madruga R, Sokona Y, Farahani E, Kadner S, Seyboth K, Adler A, Baum I, Brunner S, Eickemeier P, Kriemann B, Savolainen J, Schlomer S, Stechow CV, Zwickel T, Minx JC (eds) Climate change 2014: mitigation of climate change. Contribution of working group III to the fifth assessment report of the intergovernmental panel on climate change. Intergovernmental Panel on Climate Change, New York

IPCC (2018) Global warming of 1.5°C. An IPCC Special Report on the impacts of global warming of 1.5°C above pre-industrial levels and related global greenhouse gas emission pathways, in the context of strengthening the global response to the threat of climate change, sustainable development, and efforts to eradicate poverty. In: Masson-Delmotte V, Zhai P, Portner HO, Roberts D, Skea J, Shukla PR, Pirani A, Moufouma-Okia W, Pean C, Pidcock R, Connors S, Matthews JBR, Chen Y, Zhou X, Gomis MI, Lonnoy E, Maycock T, Tignor M, Waterfield T (eds) Intergovernmental Panel on Climate Change, Geneva

James Lee Witt Associates (2005) Building back better and safer: private sector summit on post-tsunami reconstruction. James Lee Witt Associates, LLC, Washington DC

Kato M (2011) The Cancun adaptation framework. United Nations Framework Convention on Climate Change Secretariat, Geneva

Kennedy J, Ashmore J, Babister E, Kelman I (2008) The meaning of 'build back better': evidence from post-tsunami Aceh and Sri Lanka. J Contingencies Crisis Manage 16(1):24–36

Khasalamwa S (2009) Is 'build back better' a response to vulnerability? Analysis of the post-tsunami humanitarian interventions in Sri Lanka. Norwegian J Geogr 63(1):73–88

Kumar R (2018) Cooks' elevated risk of cyclones. Cook Islands News, 25 October 2018

Lavell A (2011) Unpacking climate change adaptation and disaster risk management: searching for the links and the differences: a conceptual and epistemological critique and proposal. IUCN-FLACSO project on climate change adaptation and disaster risk reduction

Lyons M (2009) Building back better: the large-scale impact of small-scale approaches to reconstruction. World Dev 37(2):385–398

Mannakkara S (2014) A framework for building back better during post-disaster reconstruction and recovery. Doctor of philosophy in civil engineering (Disaster Management), University of Auckland

Mannakkara S, Wilkinson S (2012a) Build back better principles for economic recovery: the victorian bushfires case study. J Bus Continuity Emerg Plan 6(2):164–173

Mannakkara S, Wilkinson S (2012b) Build back better principles for land-use planning. Urban Des Plan 166(5):288–295

Mannakkara S, Wilkinson S (2013a) Build back better principles for post-disaster structural improvements. Struct Surv 31(4):314–327

Mannakkara S, Wilkinson S (2013b) Post-disaster legislation for building back better. Constr Law J 29(8):495–506
Mannakkara S, Wilkinson S (2014) Reconceptualising "Building Back Better" to improve post-disaster recovery. Int J Managing Projects Bus 7(3):327–341
Mannakkara S, Wilkinson S (2015a) Governance for building back better (unpublished). The University of Auckland
Mannakkara S, Wilkinson S (2015b) Supporting post-disaster social recovery to build back better. Int J Disaster Resilience Built Environ 6(2):126–139
Mannakkara S, Wilkinson S, Potangaroa R (2018) Resilient post disaster recovery through building back better. Routledge, Taylor and Francis Group, London
Matanle P (2011) The Great East Japan Earthquake, tsunami, and nuclear meltdown: towards the (re)construction of a safe, sustainable, and compassionate society in Japan's shrinking regions. Local Environ 16(9):823–847
Maxwell JA (2005) Qualitative research design—an interactive approach. Sage Publications
McCarthy JJ, Canziani OF, Leary NA, Dokken DJ, White KS (2001) Climate change 2001
Mikko K (2009) Managing for innovation in large and complex recovery programmes: tsunami lessons from Sri Lanka. Int J Project Manage 27(2):123–130
Monday JL (2002) Building back better: creating a sustainable community after disaster. Natural Hazards Informer. http://www.colorado.edu/hazards/publications/informer/infrmr3/informer3b.htm. Accessed 27 Feb 2013
Norio O, Ye T, Kajitani Y, Shi P, Tatano H (2011) The 2011 eastern Japan great earthquake disaster: overview and comments. Int J Disaster Risk Sci 2(1):34–42
O'Brien K, Pelling M, Patwardhan A, Hallegatte S, Maskrey A, Oki T, Oswald-Spring U, Wilbanks T, Yanda PZ (2012) Toward a sustainable and resilient future. In: Field CB, Barros V, Stocker TF, Qin D, Dokken DJ, Ebi KL, Mastrandrea MD, Mach KJ, Plattner GK, Allen SK, Tignor M, Midgley PM (eds) Managing the risks of extreme events and disasters to advance climate change adaptation. A Special Report of Working Groups I and II of the Intergovernmental Panel on Climate Change (IPCC). Cambridge University press, New York
Oliver-Smith A (1991) Successes and failures in post-disaster resettlement. Disasters 15(1):12–23
Olsen SB, Matuszeski W, Padma TV, Wickremeratne HJM (2005) Rebuilding after the tsunami: getting it right. AMBIO: J Hum Environ 34(8):611–614
Palliyaguru R, Amaratunga D, Haigh R (2010) Integration of "disaster risk reduction" into infrastructure reconstruction sector: Policy vs practise gaps. Int J Disaster Resilience Built Environ 1(3):277–296
Reich JW (2006) Three psychological principles of resilience in natural disasters. Disaster Prev Manage: An Int J 15(5):793–798
Rotimi JO, Myburgh D, Wilkinson S, Zuo K (2009) Legislation for effective post-disaster reconstruction. Int J Strateg Property Manage 13(2):143+
Shaw R, Takeuchi Y, Rouhban B (2009) Education, capacity building and public awareness for disaster reduction. In: Sassa K, Canuti P (eds) Landslides—disaster risk reduction. Springer Berlin Heidelberg, Berlin, Heidelberg
Stevenson JR, Vargo J, Ivory V, Bowie C, Wilkinson S (2015) Resilience benchmarking and monitoring review. Natl Sci Challenge, Christchurch
Twigg J (2007) Characteristics of a disaster-resilient community—a guidance note. DFID Disaster Risk Reduction Interagency Coordination Group, London
UNDRR (2019) Global assessment report on disaster risk reduction. United Nations Office for Disaster Risk Reduction
UNFCCC (2015) Paris agreement. United Nations Framework Convention on Climate Change, Geneva
UNISDR (2004) International strategy for disaster reduction. UNISDR, Geneva
UNISDR (2005) Hyogo framework for action 2005–2015: building the resilience of nations and communities to disasters. United Nations Office for Disaster Risk Reduction, Hyogo

UNISDR (2015) Sendai framework for disaster risk reduction 2015–2030. United Nations Office for Disaster Risk Reduction, Sendai
Walker B, Salt D (2006) Resilience thinking: sustaining ecosystems and people in a changing world. Island Press, Washington DC
Watts RG (2013) Engineering response to climate change. Taylor & Francis c2013, Boca Raton
Wilbanks TJ, Kates RW (2013) Beyond adapting to climate change: embedding adaptation in responses to multiple threats and stresses. Routledge, Geography of Climate Change
Wiles P, Selvester K, Fidalgo L (2005) Learning lessons from disaster recovery: the case of Mozambique. The World Bank, Washington, D.C.
Yin RK (2009) Case study research—design and methods. SAGE Inc., California

Chapter 18
Rethinking Infrastructure Network *Criticality* for Climate Resilience: Inputs from Complexity Sciences and Disaster Risk Theory

Sarah Lindbergh and John Radke

Abstract Critical infrastructure (CI) functions and spatial boundaries are often implicit premises which can result in misalignments between CI resilience goals and their societal utility. Furthermore, the historical legacy and institutionalization of CI, powered by national security issues, has resulted in technocentric approaches of infrastructure systems that can be counterproductive to disaster risk reduction. Through a literature review, this paper argues that rethinking concepts of infrastructure network criticality quantitatively and qualitatively is a necessary step to address this misalignment and to apply the current paradigm shift which defines CI, beyond hardware, as sociotechnical and socioecological systems. Rethinking these concepts of criticality requires a reexamination of infrastructure networks functions and spatial boundaries. Complementary approaches to CI resilience are harnessed within two large and sometimes disconnected bodies of literature: complexity sciences and disaster risk theories. Complexity sciences have significantly advanced quantitative definitions of infrastructure criticality by modeling interconnectivity and extracting network-level metrics of vulnerability and resilience. At a conceptual level, complexity sciences further revolutionized our understanding of CI as complex adaptive systems through the idea of ecological resilience and panarchical cycles of stasis and change. Nevertheless, the application of ecological resilience to CI risk management still struggles to recognize and assess sociotechnical functions of complex systems, for which disaster risk theories have a latent potential to address, notably within concepts of social construct of risk and the vulnerability paradigm. This theoretical bridge helps realign infrastructure network criticality with CI services

The original version of the book was inadvertently published with an incorrect version of Chapter 18. The correct version of the chapter has been incorporated and the book has been updated with one page reduced. The correction to this chapter is available at https://doi.org/10.1007/978-3-030-55563-4_20

S. Lindbergh (✉) · J. Radke
Department of Landscape Architecture and Environmental Planning, University of California Berkeley, Berkeley, CA, USA
e-mail: sarahlindber@gmail.com; sarah_lindbergh@berkeley.edu

J. Radke
e-mail: ratt@berkeley.edu

R. Djalante et al. (eds.), *Integrated Research on Disaster Risks*, Disaster Risk Reduction,
https://doi.org/10.1007/978-3-030-55563-4_18

and ecological resilience goals. In a practical level, it converges contributions from multiple fields and provides direction for assessing sociotechnical network exposure to climate change threats and identifying network-level organizational stakeholders to support CI collaborative risk governance.

Keywords Critical infrastructure · Conceptual gaps · Vulnerability paradigm · Sociotechnical systems · Ecological resilience · Bridging disciplines

18.1 Introduction: The Institutional Legacy of Critical Infrastructures (CI)

Infrastructure criticality is intricate to modern cities metabolisms which owe their urbanization processes to innovations in water, energy, food supply, sanitation, transport systems, communication, security, safety, and other services. Debilitating effects from disruption of these utilities make it easy to understand why they are considered the "Achilles heel of urbanized societies" (Graham 2011). The institutionalization of critical infrastructure (CI) and the characterization of their *criticality* has historically been powered by national order and security concerns. Notably, governments around the world growing terrorism threat concern, in the mid 1990s, institutionalized CI risk management and governance amidst a security crisis (DHS 2013; Rinaldi et al. 2001; Moteff and Parfomak 2004; Miller 2009).

In the U.S., the first official definition of CI dates from 1996, Executive Order 13,010, signed 2 years after the Oklahoma City bombing, establishing the President's Commission on Critical Infrastructure Protection. Central governmental guidelines on CI protection came from the National Plan for Critical Infrastructure and originated from Presidential Decision Directive 63. After the 2001, 9/11 terrorist act, Executive Order 13,228 established the Department of Homeland Security (DHS) (DHS 2013) with responsibilities to protect a list of CI sectors. Similarly, in Europe, the development of CI policies followed the Madrid and London bombings (March 2004 and July 2005 respectively), originated the European Program for CI Protection in 2005 (Lazari 2014).

In the U.S., CI planning documents borrow CI definition from the Homeland Security Act of 2002:

> critical infrastructure includes systems and assets whether physical or virtual, so vital to the US that the incapacity or destruction of such systems and assets would have a debilitating impact on security, national economic security, national public health or safety, or any combination of those matters

Social studies of technology and disaster risk have well examined the legacies and issues of top-down infrastructure development and institutional standardization of CI, as from the first half of the twentieth century (Edwards 2002; Carse 2017; Wakefield 2018; Forino et al. 2018). Historical definitions of "vital systems security", developed by military strategists and intelligence economists, rose throughout the twentieth-century interwar offensive airpower and derived tactics of strategic bombing, nuclear risk, and protection of "*critical* targets" (Collier and Lakoff 2008). This modern awareness of infrastructure *criticality*, formerly designed for "emergency states of exception" (Collier and Lakoff 2015), has been criticized to enabled mechanisms of political administration to exert extraordinary executive powers. The advent of CI protection and the use of the resilience concept within U.S. security policies has also been critiqued as an excuse for the intense privatization of public utilities at the end of the 1970s (Lovins and Lovins 1982). One other example is the transfer, in the early 2000s, of regulatory power from the civil sectors of CI to the DHS agency, whose risk management strategies "ignore(s) the boundaries between military threat of terrorist attack and civil contingencies such as natural disasters" (Walker and Cooper 2011).

While the institutionalization of CI in the US triggered nation-wide investigations on CI asset design for robustness and reliability, the science of integrating those assets into a whole interacting network with systemic behavior parameters was considered fragmented at the turn of the twenty-first century (NAS et al. 2005). Network science re-emerged then in multiple disciplines as a promising method that allows the abstraction of global behavior from complex systems (Gorman 2005; Lei et al. 2010; Barabási 2016) and it was concomitantly integrated into national security research to understand CI connectivity, dynamics, and graph-based risk assessment, including quantitative approaches to *criticality*.

The institutional roots and historical legacy of infrastructure criticality concepts lead us to question the influence of the CI institutional framework on criticality metrics, its impact on infrastructure modeling and how it is aligned with CI resilience goals. These questions also point to CI resilience gaps which derive from siloed disciplinary approaches to infrastructure system resilience and technocentric approaches to infrastructure risk management, i.e., driven by top-down securitization interests and protective measures overfocused on technical dimensions of CI and the hardening of *central* assets.

Technocentric approaches have also been critiqued in disaster risk reduction (DRR) theories and associated with hazard-centric risk management. These undermine intrinsic vulnerabilities and the interconnectedness of underlying sociotechnical and socioecological problems of disasters. Developed to counter hazard-centric technological fixes in DRR, the vulnerability paradigm is being integrated into CI protection policies, to expand to and strive for CI resilience policies since the turn of the twenty-first century (Hellström 2007; George Mason University 2007; Hosseini et al. 2016; Shen et al. 2016). Dialectic approaches to risk considering both hazard and vulnerability determinants, as well as the hybrid technical and social dimension of infrastructure systems, are key for addressing CI climate resilience. These approaches, rooted on theories of social construct of risk, are incipient in the current paradigm shift of CI as sociotechnical and socioecological systems, and as necessary to investigate their *criticality* from qualitative and quantitative perspectives.

Considering the historical legacy of CI as a military defense issue and institutional concepts of CI we raise the question on how the concept of *criticality* responds to current expectations of CI climate resilience policies. In other words, what is the influence of the CI institutional framework on criticality metrics, and how does this align or misalign with CI resilience goals? What gaps are uncovered by these misalignments? Based on DRR, CI and complexity sciences literature, what inputs can be harnessed to address this misalignment? In Sect. 18.2 we present common metrics derived from network science used to define *criticality* of network infrastructure from a quantitative perspective. We explain how "local" and "global" parameters allow fundamental assessments of CI network vulnerability and resilience but are highly sensitive to the *function* and *spatial boundaries* of the infrastructure network under study. We argue that the omission of this sensitivity at a conceptual level reinforces misalignments between infrastructure criticality definitions and CI climate resilience goals. In Sect. 18.3 we highlight contributions from complexity sciences and disaster risk theory to address current conceptual gaps of CI climate resilience related to loose definitions of CI function and spatial boundaries. We explore how the definition of CI function is relevant within complexity sciences as it is a variable in the definition of ecological resilience, however, whereas ecological function is more easily translatable to socioecological systems, translating stability and collapse in terms of sociotechnical system is ambiguous. We end Sect. 18.2 by visiting root theories of social construct of risk and the vulnerability paradigm in DRR which have high potential to better explore sociotechnical and socioecological functions of CI. In Sect. 18.4 we illustrate the embryonic ingression of CI resilience goals in DRR and climate change policies in the U.S. These policies highlight future directions and current challenges related to CI risk governance and shared risk. We argue that there is a latent potential for theories of social construct of risk and ecological resilience to address the misalignment between criticality metrics for infrastructure networks and CI resilience goals that contribute to collective and collaborative risk governance.

18.2 Infrastructure Networks *Criticality* Quantification and Lingering Conceptual Gaps

Quantitative and spatial concepts of infrastructure *criticality* in the U.S. dates at least from the first half of the twentieth century. Graph theory and systems engineering were then applied to measure high profile utility system vulnerability and criticality such as in the energy sector (Cagley 1964) and the oil industry (Thayer and Shaner 1960; Osleeb 1979). Understanding the criticality of interconnected networks is directly related to our ability to model their mutually dependent properties (Vespignani 2010). CI interdependence describes a bidirectional relationship between two infrastructures through which the state of each infrastructure influences or is correlated to the state of the other, i.e., when two infrastructures depend on each other for the supply of vital services or commodities (Rinaldi et al. 2001). Within complexity sciences, network science was applied to model connectivity and capture the dependencies and interdependencies of infrastructures from a spatial perspective (Barthélemy 2011; Murray et al. 2008; Grubesic et al. 2008; Hasan and Foliente 2015; Dunn et al. 2016). Coinciding with the latest revolutions in information technology and the emergence of "big data" (Li and Milburn 2016; Barabási 2012), network science was used as a tool for extracting explanatory and predictive power of spatial phenomena both from geospatial and topological perspectives (Miller 2004). Mathematically, this studies the relationship between elements named *nodes*, and their interconnections named *links* which together form discrete structures that are also referred to as graphs (Bollobás 1998).

Complexity sciences have helped identify universal parameters of network criticality, which derive from the relationship between network structure (topology or connectivity) and network dynamics (flow of goods and/or services). These parameters are commonly subdivided in local and global. Local parameters pertain to characterizing the smallest units of the network, i.e., nodes and links; while global parameters describe full network characteristics such as its topological structure, giant connected component, or diameter (developed below).

Node or link criticality (local parameter) can often be attributed to CI failure cascade size through load redistribution, or contagion models; while network topology (global parameter) is used to inform network accessibility, path availability, and CI disruption propagation dynamics. Local and global parameters are complementary and are frequently used in conjunction to properly assess CI network structural and dynamical organization (Murray et al. 2008; Grubesic et al. 2008; Boccaletti et al. 2014; Gao et al. 2016; Ouyang 2014).

18.2.1 Centrality Metrics and Local Parameters to Quantify Criticality

Local parameters, such as centrality metrics, assist in the observation that the removal of certain *critical* nodes or links aids rapid degradation of the network function (Gorman et al. 2004). While centrality metrics were originally applied in social sciences and social physics gravity models in the 1930s and 1940s (Scott 2014; Barnes and Wilson 2014), they have become key in CI protection plans. These metrics calculate the position, influence, and overall importance of a node or link in relation to the full topological structure to which they belong and enable a ranking process of assets. Many different centrality formulas have been developed based on graph theory (Lewis 2006; Zhuge and Zhang 2009). The two most common metrics rely on calculating the number of links attributed to each node such as degree and eigenvector centrality; and shortest paths calculations such as betweenness and closeness centrality (Freeman 1978; Brandes et al. 2016). A critical node failure, for example, would increase the load or contagion on neighboring nodes which might exceed their robustness, cause their failure, and possibly trigger a cascading failure (Zio and Sansavini 2011; Ouyang 2014; Roukny et al. 2013). Centrality metrics are also applied in post-failure scenario to measure network infrastructure recovery process and how to maximize the restoration of network functionality. Centrality-based metrics have been mostly applied for single transportation infrastructure networks and have been recently tested to identify efficient recovery strategies for multimodal transportation networks. Percolation thresholds and phase transitions measure topological effects from disabling specific network elements to measure CI robustness and resilience (Barthélemy 2011).

18.2.2 Topological Structure and Global Parameters to Model Critical Infrastructure Networks

Global parameters such as topological structures are essential to understand the way in which nodes and links are interrelated and arranged and how their sequencing can facilitate or impede the transmission of CI information, goods, services, etc. Global metrics can be as simple as the network diameter (maximum length among all shortest paths) and or as complex as universal parameters that help the identification of probability distributions of nodes and links and emergence proprieties. The giant connected component of a CI network is commonly used to assess the functionality of the network. The most relevant theoretical models used to describe real-world networks are small-world and scale-free networks.

Small-world networks, or the Watts and Strogatz model (Watts and Strogatz 1998), have two important unique characteristics: short average paths and short diameter with a high clustering coefficient, or a high ratio between the number of existing links and maximum number of possible links in a graph. Small-world models illustrate the key role of shortcuts in network; especially in relation to the dynamics of communication or contagion spread. Watts and Strogatz provide empirical evidence of emergence of small-world patterns in a series of real-world networks such as neural networks, film actors collaborative networks, and in CI such as the U.S. power grid. Electrical power grids are typically the only physical infrastructure network that have shown to fit the small-world parameters, mostly because their optimization is tightly related to providing shortest distances between power generation, stations, substations, and the consumer (Lewis 2006).

Most CI and social networks follow scale-free properties (Watts and Strogatz 1998; Barabasi and Albert 1999). Scale-free network models, conceived by Barabasi and Albert (1999), present a highly heterogenous degree distribution forming a power law. This degree distribution indicates that most real-world networks, such as CI, are the result of a simple organizing principle like the law of increasing returns, or what Barabasi and Albert named preferential attachment. This principle states that new links are not randomly connected to exiting nodes, there is a higher likelihood that they are linked to high degree nodes, reinforcing the existence of hubs or clustered critical nodes and topological heterogeneity. Small-world networks have a power law degree distribution, but scale-free networks typically present even smaller diameters and high clustering coefficients. The Internet's hardware and the World Wide Web (software) (Albert and Barabási 2002); the US railway and interstate highway system with Chicago as the major hub (Lewis 2006); the global airline network (Molly and Reed 2001 in (Barthélemy 2011)); social network based on mobile phone data (Lambiotte et al. 2008 in (Barthélemy 2011)), and transnational corporation ownership networks (Vitali et al. 2011) all fit scale-free network models.

These empirical applications of small-world and scale-free models near the turn of the twenty-first century allowed the development of fundamental theories of CI network vulnerability and resilience with relation to random failures and targeted attacks (Barthélemy 2011; Barabási 2016). More generally within DRR and CI climate adaptation, network models and criticality metrics pushed the development of methods that evaluate indirect exposure to climate hazards as opposed to purely direct exposure (topological vs. geographical proximity to hazards) (Nelson et al. forthcoming; Altshuler et al. 2015; Amoaning-Yankson and Amekudzi-Kennedy 2017; Bono and Gutiérrez 2011; Taylor 2003).

18.2.3 Lingering Conceptual Gaps: Omissions of Network Function and Spatial Boundaries

The twenty-first-century re-emergence of network science found empirical strength in the engineered protection and military securitization of CI. Knowledge of topological structure is the baseline for interconnectivity modeling as it allows for the quantification of node criticality of specific physical assets as a guiding principle for protection policies (DHS 2013; Lewis 2006). Protection decision-making is fundamentally based on network interdiction simulations that identify worst-case interdiction scenarios (Murray et al. 2008; Grubesic et al. 2008). U.S. CI protection plans are technocentric, which means they focus prioritizing hardening of central nodes or hubs (physical assets) that provide higher return in terms of decreasing the consequences of failure to the entire system in a cost–benefit analysis (Theoharidou et al. 2010). However, it is important to underline that criticality quantification varies immensely based on how a facility or infrastructure network *function* is characterized and too, based on where the *spatial boundaries* of the system are drawn. Very often CI functionality is modeled based on connectivity properties such as centrality metrics and giant connected component, but challenges to incorporate spatial constraints to these properties remain (Table 18.1). Furthermore, in most CI literature, the relationships between current metrics of criticality and institutional definitions of CI functions are rarely addressed.

Better incorporating geospatial constraints to purely topological structures is a well know gap in network science, especially in studies of real-world networks, where the length of edges is relevant and will have significant effects on the network structure (Barthélemy 2011). Nevertheless, the acknowledgement of border effects on criticality results is rare. Some examples of local and global parameters deployed in network science to quantify criticality of infrastructure networks are presented below in Table 18.1 together with notes, if any, on spatial boundary effects on these metrics.

Table 18.1 Examples of graph-based criticality metrics for CI networks

Reference	CI type	Criticality approach	Global/local parameters	Notes on spatial constraints and boundary effects
Gao et al. (2011)	Interdependent CI or network of networks (theoretical)	System robustness	Global (giant connected component size, critical percolation threshold for first and second-order transitions of cascading failures)	Recognizes gaps from the lack of incorporating spatial constraints of real-world networks No notes on spatial or system boundaries
Gorman et al. (2004)	Information networks	Critical nodes, network survivability	Global (small-world hierarchy) and local (nodal hierarchy based on accessibility index)	Recognizes drawbacks from border effect on nodal hierarchy results
Grubesic et al. (2008)	CI network (theoretical review)	Facility importance and network operability	Global (average length of shortest path, gamma index) and local (node accessibility/degree centrality)	Recognizes issues of evaluating node or arc criticality in an *aspatial* manner and independently of network dynamical organization No notes on spatial or system boundaries
Holmgren (2006)	Electric power delivery networks	Network vulnerability	Global (clustering coefficient, average path length, small-world, and scale-free sensitivity to attacks)	Recognizes the shortcomings of not incorporating spatial aspects of the network
Kasmalkar et al. (2020)	Road network	Traffic resilience	Local (metric reach of a road segment within a service radius)	Includes spatial components such as network density. No notes on spatial or system boundaries
Matisziw et al. (2009)	Internet backbone	Network vulnerability	Local (origin–destination path availability)	No notes on spatial or system boundaries
Nuss et al. (2016)	Supply-chain networks	Resource criticality and important actors	Local (node degree, closeness, and betweenness centrality)	No notes on spatial or system boundaries

(continued)

Table 18.1 (continued)

Reference	CI type	Criticality approach	Global/local parameters	Notes on spatial constraints and boundary effects
Yadav et al. (2020)	Urban rail systems	Infrastructure node criticality, network resilience, and robustness	Global (giant connected component size, average shortest path length, greedy algorithm) and local (node betweenness, degree, and eigenvector centrality)	Verifies the need to incorporate spatial constraints to real-world networks resilience metrics
Zio and Sansavini (2011)	Power transmission network	Component criticality for cascading failures	Global (cascade propagation and size) and local (node degree, closeness, betweenness, and informational centrality)	No notes on spatial or system boundaries

While these metrics revolutionized our ability to quantify *criticality of* infrastructure networks, their application in real-world CI systems illustrates conceptual gaps of CI resilience related to loose definitions of system *functions* and *spatial boundaries*. Some of the gaps in CI functionality assessments can be related to the difficulties of assessing CI network dynamic data (flow and usage, i.e., CI services and/or goods) which is less studied due to data accessibility issues and computational complexity.

Considering that one of the goals of complexity sciences is understanding the connections between system functions and topology (Cumming 2011), it is counterproductive to CI resilience that criticality concepts are ill defined. In other words, explicitly articulating CI functions and boundaries is a necessary step to reexamine criticality based on CI services, which is urgent in the context of changing climate landscapes. We argue that failure to properly assess CI network functions and spatial boundaries prior to assessing *criticality* through multiple local or global parameters can cause misalignments between CI resilience goals and societal utility of CI, i.e., CI s*ervices.*

In Sect. 18.3 we highlight conceptual contributions from ecological resilience which frame CI as complex adaptive systems with panarchical cycles of stability and change; then we underline latent contributions from disaster risk theory with special attention to the vulnerability paradigm to better understand shared risk and qualify CI societal functions.

18.3 Covering Conceptual Gaps of CI Climate Resilience Bridging Complexity Sciences and Disaster Risk Theories

18.3.1 Inputs from Ecological Resilience: CI as Complex Adaptive Systems

Complex system theories are the backbone to conceptual and quantitative definitions of resilience. Although resilience as a development goal gained momentum at the turn of the twenty-first century (Zhang and Li 2018), the foundations of the resilience concept stem from engineering (seventeenth Century), biology (1950s), and ecology fields (1970s). The term resilience has been traced to the seventeenth-century Hooke's law in physics, due to the observation that bodies described as being elastic regain their shape after being deformed (Etkin 2016). This origin explains why traditional "engineering resilience" emphasizes robustness, rapidity, and speedy return to a *previous stable state* (Vatn et al. 2012) within CI literature. Studies that adopt sociotechnical and socioecological framework of CI, aspire to ecological resilience, especially in the realm of climate risk management where there are tight couplings between socioecological systems and sociotechnical transitions (Hellström 2007; Smith and Stirling 2008; Monstadt 2009; Haimes 2012; Pescaroli and Alexander 2016; Markolf et al. 2018; Alderson 2019; Amir and Kant 2018). "Ecological resilience", originating in the middle of the twentieth-century fields of biology and ecology, challenges the adequacy of equilibria, or "stable state" to describe the wellbeing of complex adaptive systems (Folke 2006) which are inherently adaptive and evolutionary (Holling 2001).

The advent of resilience in complex systems theory has been traced to Ludwig von Bertalanffy, originator of the open system theory in biology which mathematically describes an organism's growth (Etkin 2016). Different from closed systems, which are stability- and conservation-centered such as the laws of thermodynamics, open systems are holistic and evolutionary. Bertalanffy characterizes an organism as a complex organization and integration of physiological functions in which the whole determines the character and function of its parts (Bertalanffy 1950). Officially coined by ecologist C. S. Holling in the 1970s, "ecological resilience" challenges the stable-centered and equilibria theories of mature organisms or ecosystems by including the possibility of multiple stable states, non-linear causation, feedback, and self-organization (Holling 1973). Therefore, the idea of resilience being purely related to our capacity to absorb shocks and change while maintaining structure and function incorporates aspects of transformability and adaptation (Folke 2006). Understanding these conceptual roots and their evolution is especially relevant for tracing gaps in CI climate resilience. Despite the epistemological divergence of "engineering resilience" and "ecological resilience", these concepts are ontologically complementary in what concerns the paradigm shift from technocentric and hazard-centric approaches of CI protection to sociotechnical and socioecological approaches of CI resilience. This paradigm shift implies that resilient CI, within its social, ecological, and technological dimensions, process or contain information more and more effectively reflecting the ability to respond to environmental stimuli through self-organization, learning and reasoning; and through a dynamic balance of creative and conservative cross-scale cycles (Folke 2006; Cumming 2011). The advent of ecological resilience for CI climate risk management therefore demands process-based and iterative management policies, capable of incorporating what Holling calls "panarchical" cycles of stasis and change (syntropic or entropic change[1]). Panarchical cycles are not necessarily sequential, fixed or single cycles; but nested within each other with the ability of interacting across different spatial and temporal scales (Holling 2001; Folke 2006). The idea of panarchy expanded the resilience scope to "human-natural systems"—i.e., socioecological systems—as a heuristic, non-linear approach to understanding adaptive cycles[2] in the human-natural interface (Walker and Cooper 2011). It is not difficult therefore to understand why the operationalization of ecological resilience has been prominent in socioecological systems studies (Cumming 2011; Solé and Bascompte 2006).

Finally, complexity sciences have been most relevant to the socioecological aspect of CI systems as they harness empirical evidence from biology and ecology fields. The challenge remains in translating the concepts of *function* from ecosystems and ecology perspectives to the social or sociotechnical perspective. What is considered as a cycle of stasis and a cycle of entropic or syntropic change for a sociotechnical or socioecological system can present conflicting rationales and rises questions of "whose system" is being considered (Smith and Stirling 2008). Again, this brings questions of stakeholder engagement and collaborative governance to the forefront of complex system resilience goals. Notably in the case of CI it raises the question: "critical for whom?" (Steele et al. 2017). Some argue that keeping such concepts vague serves a useful purpose to give space for the development of the scientific community in the emerging field of complex systems without getting stuck in semantics (Cumming 2011). Considering the historical and quantitative issues tied to the institutionalization of CI, the vagueness in the definition of *function* is not only a passive space for reflection, but it flags the need for active, iterative, and critical examinations of what is being assumed as *critical*; notably in the context of DRR and climate risk management, where resilience has become a status quo policy goal in the last two decades (Boin et al. 2010) and assumed as a desirable outcome.

[1]According to McHarg (1981, p. 111) all living systems are oscillating between two states: syntropic and entropic. "Syntrophic-fitness" describes "certain energetic transactions that produce matter and energy at a higher level of order than at the onset", photosynthesis being a classical example. "Entropy is the tendency for energy and matter to degrade from higher to lower levels of order in any energetic transaction […]. Entropic-misfitness would be represented by an organism, unable to find a fit environment and/or unable to adapt that environment and/or unable to adapt itself".

[2]These cycles include rapid successional growth, followed by the conservation state of equilibrium which are both to be followed by a phase of collapse and then spontaneous organization leading to a new growth cycle.

18.3.2 Inputs of Disaster Risk Theories and Social Construct of Risk to Qualify CI Functions

Failures of massive international infrastructure investments in jump-starting economic development have already justified criticism of infrastructure as being purely grandiose technical and physical capital as understood in the 1960s and 1970s (Carse 2017). Concomitantly, the emergence of sociotechnical theories in safety science (systems engineering) and social science approaches of organizational behavior (Perrow 1972; Mitroff 1983; Wildavsky 1988; Roberts 1989) and of science and technology in the 1970s and 1980s (Latour 2005; Hanmer and Abram 2017) helped frame infrastructure beyond hardware. As sociotechnical systems, infrastructures represent complex interactions between humans, organizations, institutions, and technology (Amir and Kant 2018). The relevance of sociotechnical frameworks for resilience studies has been highlighted as a powerful approach in engineering systems for "joint optimization" to cope with uncertainties of disturbances under "un-designed" and "non-linear relationships" between human and technical factors (Walker 2015). Amir and Kant (2018) harness further contributions from science and technology studies by emphasizing the "transformability" character of sociotechnical resilience through informal relations, sociomaterial structures, and anticipatory practices. These are considered as key mechanisms of resilience that enable "sociotechnical systems to shift from one configuration to another in response to shock and disruption".

The view of infrastructure as sociotechnical systems developed our understanding of modern and systemic vulnerabilities derived from the inseparableness of technology, society, and nature (Edwards 2002). Our knowledge of these systemic vulnerabilities, of infrastructure influence on planetary carbon metabolism, and of global environmental change issues, was incorporated in the international policy agenda with milestone guidelines from the United Nations (UN) and the Intergovernmental Panel on Climate Change (IPCC) between the late 1970s and late 1980s.[3] Concomitantly, in social sciences the pivotal concepts of "tight couplings" and "risk society" were being produced, respectively, by Charles Perrow in 1984 and by Ulrich Beck in 1986. Fundamentally from similar scholarships, these terms are rarely associated in CI literature but were clearly described new forms of risk as a consequence of heightened interdependencies and modern society while pointing to important contributions to CI risk management (Tierney 2007; Gephart et al. 2009).

"Tight couplings" is the term used to characterize interactions within infrastructure systems and between infrastructure and human systems with processes that happen very fast and cannot be isolated in case of failure. Tight couplings lead to

[3] International policy milestones include the 1979 first World Climate Conference in Geneva where scientist of over 50 countries held the unanimous agreement on the urgency of acting on climate mitigation (Rich 2019), the United Nations Brundtland report coining "sustainability" as a normative concept within the social, economic, and environmental dimensions; the UN 1988 resolution to create the Intergovernmental Panel on Climate Change (IPCC) and the UN declaration of the decade for disaster risk reduction in 1989 and the Framework for Action for 1990–2000.

unavoidable system failures. Often used in safety and reliability and system engineering disciplines, "tightly coupled systems" refers to complex interactions "of unfamiliar sequences, and either not visible or not immediately comprehensible" (Perrow 1984). Perrow's organizational theory of the production of risk describes unexpected failures, or "normal accidents" of technological systems that can quickly lead to catastrophic risk. This path between "seamless accidents" and catastrophic risk is characteristic of higher interdependencies between society and technology through infrastructure systems and sociopolitical contexts of organizations (Roberts 1990; Rinaldi et al. 2001; Haimes 2012). Although not commonly explored in CI literature, Perrow's sociotechnical perspective on modern risk also contributes to the identification of populations with no power in the decision of their exposure to hazards. This population is also referred as "third" and "fourth-level victims" of modern catastrophes such as innocent bystanders or future generations that are exposed to far reaching effects of technological hazards but excluded from discussions that affect them. As opposed to "first" and "second level victims", such as operators of the system and users or suppliers of the system respectively, they do not choose to participate in the system but still share the risk. This concept of shared risk also brings therefore, stakeholder engagement (Gaillard and Mercer 2013; Alpaslan et al. 2009), collaborative governance (Meuleman 2018; Gieseke 2020), and ethical issues such as questions of fair exchange of risk (Hansson 2013), to the forefront of CI climate resilience problems.

From the socio-cultural perspective, Beck's "risk society" presents a theory of "man-made" risk as "unintended consequence of modernity" (Wimmer and Quandt 2006). Beck also argued that the novelty of contemporary risk is in the scale of spatial distribution and in the ability to affect multiple generations (Bulkeley 2001). "Risk society" corroborates with theories of social production of risk which emphasize the intrinsic characteristic of risk, originating within society itself (Hewitt 2000; Tierney 2014). This *intrinsic* characteristic of risk has been theorized and applied in (DRR)[4] as the vulnerability paradigm (Burton et al. 1968; Haas and White 1975; Hewitt 1983; Cardona 2005; Blaikie et al. 1994; Cutter 1996). It has been noted that the advent of the vulnerability paradigm in DRR in the 1980s, relates to the social construct of risk; the authors add it also corroborates the sociotechnical approach of CI resilience.

Hellström (2007) highlights the relevance of the vulnerability paradigm to assess the recursive progression of vulnerability in CI. The DRR's classic "pressure-and-release" (PAR) vulnerability model, first published in 1994 by Wisner et al. (2003), was designed to identify the progression of vulnerability looking through the links of "root causes" and "dynamic pressures" that create "fragile livelihoods" or "unsafe location conditions" which, when intersected in time and space with hazards, provide a chain of explanation to disaster risk. By understating the links that feed into dynamic pressures of vulnerability we can also develop methods to "reverse-engineer" the process and introduce coping or resilience capacities and "release the pressures". PAR functions as process-tracing inquiry system to understand risk by emphasizing

[4]Disaster risk reduction theory views risk as a product of hazard and vulnerability (Wisner et al. 2012).

the vulnerability component, thus expanding on socioeconomic, ideological, cultural, political, and historical processes.

The relevance of the PAR model is emphasized by the sociotechnical system framework Hellström applies to information communication technology networks. The vulnerability paradigm demonstrates the shortcomings of technocratic approach to risk. Likewise, traditional CI protection focuses on facing hazards with war-like strategies, such as engineering-centered guidelines for "infrastructure *defence*" developed under the International Decade for Natural Disaster Reduction (United Nations 1999). Issues of overemphasizing hardening or structural measures have been thoroughly critiqued within the vulnerability paradigm as they fix on infrastructure "hardware" and containment of hazards with reactive command-and-control strategies (Burton et al. 1968). Taken in isolation, these techno-and hazard-centric approaches have been proven to (1) function as "band-aid" solutions that fail to understand the dynamic pressures and roots causes of disasters (Wisner et al. 2003); and (2) become very disconnected from everyday development practices thus counter-productive to holistic disaster planning and increase quality of life overall (Gaillard 2010; Cutter et al. 2013; Turner et al. 2003). Furthermore, focus on infrastructure hardware, or "technical" dimensions, has resulted, sometimes tragically, on an over-reliance on design and building standards for implementing resilience strategies in multiple sectors. This overreliance, especially in the case of long-lived and capital intensive infrastructure investment, is no longer considered as a reasonable assumption in engineering and architecture practices that are compelled to plan in view of deep climate uncertainties (Walker et al. 2013; CNRA 2018).

CI resilience studies informed by complex adaptive systems theory point to the gap of properly addressing the "panarchical" nature of CI systems collapse and its social, ecological, and technological interlinks. Pescaroli and Alexander (2016) explain that the interconnectedness and interdependence of CI significantly increase the potential for cascading failures but they build on promoting the investigation of CI vulnerability pathways to better understand cascading disasters root causes and unpredictability (similar to the PAR vulnerability model). These vulnerability pathways, beyond the physical infrastructure fragilities, address cross-scale vulnerabilities rising from CI services ambivalent relationship with socioecological systems, i.e., providing resource and damage simultaneously. There is a clear distinction between what is narrowly defined as a sequence of events with linear cause-effect relationships, and the "toppling domino" effect of CI cascading failure and "panarchical collapses". Pescaroli and Alexander suggest a paradigm shift necessary to interpret CI complex adaptive systems where interdependencies and interlinks represent local and global dynamics that together form panarchical cycles. When a cascading event is triggered it progresses through unresolved vulnerabilities concentrated in these interdependencies. The "alignment" of unsolved vulnerabilities will result in non-linear amplification of disaster impacts through complex social, ecological, and technological chains creating panarchical collapses. Corroborating with the vulnerability paradigm, this calls for a shift from hazard-centric triggered cascade pathway inquiries to vulnerability cascade pathway inquiries. The emphasis here is on understanding "worst case *amplification* scenarios" instead of "worst case scenarios based on initial triggers" (Pescaroli and Alexander 2016; Hellström 2007).

In Sect. 18.4 we illustrate the ingression of CI resilience in DRR and climate change policies at the international and U.S. levels, with special focus on California's climate change impact assessment system and pioneering infrastructure climate adaptation standards and regulations.

18.4 Resilience Ingression in DRR and Climate Change Policies in the U.S. and Its Implication for Future CI Research

Rinaldi et al. (2001) bridged ecological resilience theories and CI resilience, defining both as "complex collections of interacting components in which change often occurs as a result of learning processes; that is, they are complex adaptive systems". From this bridge, Rinaldi et al. pioneered fundamental concepts of CI interdependence which have been widely used in infrastructure network engineering and spatial models. After the turn of the twenty-first century, the concept of resilience, originating in systems theory and ecology in the 1950s and 1970s, becomes the premise for DRR, climate change adaptation, and CI science-policy guidelines. Nevertheless, the alignment of ecological resilience discourse with CI network risk management and the vulnerability paradigm is still underway. In this section, we provide a brief description on how CI resilience goals emerged in international, U.S. and regional policies within climate adaptation, DRR and CI planning documents. We specifically underline the recent preoccupation with cascading failure of CI as well as cascading risk due to CI exposure to climate disasters in DRR and climate change documents. This coincides with the incorporation of ecological resilience to infrastructure risk management. Two important challenges and future trajectories for CI resilience can be inferred from these policies studies: creating and reinforcing climate-cognizant infrastructure design and standards, and collective organizational challenges of regional or large-scale infrastructure systems governance to address climate adaptation within collaborative frameworks.

At the international level, it was not until the 2005–2015 UN Hyogo Framework Action that resilience, based on persistence (robustness and reliability), preparedness (learning capacity and redundancy), adaptability (being flexible), and transformability (being innovative) (Davoudi et al. 2013), was officially incorporated in DRR policy guidelines (Cutter et al. 2013; UNISDR 2005). The relevance of understanding cascading risk or "sequential effects", through social and spatial scales was only officially addressed in the current U.N. Sendai Framework priorities for action (UNISDR 2015). This latest international policy guideline for DRR also explicitly addresses, for the first time, complex supply chains as drivers of vulnerability (I.6. p. 10), and identifies resilience of new or existing "critical infrastructure" and "critical facilities" as a priority at national and local scales (IV.30. c., and 33.c). The *criticality* of CI here is inferred as "life-saving and essential services" during and after disasters such as water, transportation, telecommunications educational, and health facilities. Within climate risk management in international policy, the IPCC only targeted CI interdependencies and cascading failures as key resilience challenges for the first time in the latest cycle of the Assessment Reports (AR5) (IPCC 2014).

At the U.S. level, the establishment of climate science-policy guidelines for infrastructure adaptation is centralized in the National Climate Change Assessments (NCA), first published in 2000 (NAST 2000), and which is charged with interpreting global climate risk at the U.S. scale and reporting it to the Executive and Legislative powers. In the early 2000s climate change science started to slowly penetrate national security agenda in the U.S. Military intelligence reports, to dismiss uncertainties of the climate change "scientific debate" over the potential consequences of extreme weather on food, water, and energy security (Schwartz and Randall 2003); and in which famously acknowledging climate change as "threat multiplier" to underlying geopolitical instabilities due to natural hazards or natural resource accessibility issues (Gerson and Goodman 2007). This military recognition of climate change as a security threat inspired metaphorical descriptions of U.S. mitigation and adaptive efforts as "war against climate change", which, some argue, represented a milestone in the communication of urgency and risk of climate resilience (Flusberg et al. 2017).

At the regional level, California's Climate Change Assessment (CCA) was the first state-level guideline to iteratively downscale climate models, improving their accuracy and promoting state-relevant infrastructure adaptation policies in the U.S. Nevertheless, it was not until 2009, at the second NCA and California's CCA that the focus by both was with climate mitigation (emissions reduction) and with adaptation policies, including infrastructure-sector-specific strategies (Franco et al. 2011; U.S. Global Change Research Program 2009). The main outcome of California's 2nd CCA was the 2009 Climate Adaptation Strategy that supported new climate adaptation laws, notably Assembly Bill AB-1482: Climate Adaptation Strategy introduced in February 27, 2015 and establishing the Strategic Growth Council leadership to continually update the state's climate adaptation strategy, aligning it with California's Five-Year Infrastructure Plan in coordination with other agencies. Later, California AB-2800 "Climate Change: Infrastructure Planning", approved September 24, 2016, was introduced which required California state agencies to incorporate current and future climate change impacts specifically when planning, designing, building, operating, maintaining, and investing in state infrastructure. The advent of the resilience discourse in DRR and climate risk international policies has only recently trickled down to national and local infrastructure resilience policies. In other words, the incorporation of climate risk into infrastructure planning is novel and the institutional alignment of DRR, climate risk management, and CI resilience is embryonic. Nevertheless, policies such as AB2800 are pioneering where they promote "climate-safe infrastructure" guidelines which simultaneously address issues of climate mitigation and adaptation emphasizing social equity discourse and governance as a major enabler of resilient infrastructure (CSIWG 2018).

From the CI planning perspective, DHS's association of *resilience* to security of CI was one of the main changes of the theoretical discourse of the latest National Infrastructure Protection Plan from 2013, a contrast to the former version that focused narrowly on infrastructure *protection.* The resilience concept is conceived as the "ability to prepare and adapt to changing conditions and withstand and recover rapidly from disruptions […] Resilient infrastructure assets, systems and networks must also be robust, agile and adaptable." (DHS 2013). Although this integration mostly focuses on technical domain infrastructure systems, addressing physical and cyber aspects of CI, it benchmarks increasing influence of ecological resilience within CI management and planning (Zio 2016; Ayyub 2014; Chang et al. 2014; Forzieri et al. 2016; Hosseini et al. 2016; Maliszewski and Perrings 2012; Lewis 2019). Furthermore, sociotechnical dimensions of CI are integrated in this latest national policy through the reinforcement of collaborative platforms referred as the "National Partnership Structure". Notably promoting public–private sector and cross-sector coordinating structures for government agencies and private sector operators and owners across jurisdictions and administrative boundaries. Yet, in light of recent disasters and updates on climate change risk, assessments of CI resilience challenges demonstrate limitations of current governance structures to find balance between prescriptive and voluntary collaborative mechanisms for effective partnerships (Northeastern University 2018; OECD 2019). The effectiveness of these CI partnerships and their complex governance structures remains obscure. Further research of these governance structures through organizational and policy network analysis (Knoke 2016; Bond and Harrigan 2014; Maldonado 2017) would contribute to our understanding of CI sociotechnical interlinks as well as network-level CI stakeholders which should strive to include Perrow's third-level victims.

18.5 Conclusion and Discussion

Considering the substantial research that flags the militarized legacy of infrastructure planning and its counterproductive effects on DRR, we lack understanding of how this conceptual approach has informed quantitative assessments of infrastructure criticality. The relationships between institutional CI definitions and current metrics of criticality are rarely addressed, in other words consensus of criticality is assumed and this assumption is visible in quantitative assessments, where explicit discussions of infrastructure network functionality and spatial boundaries are rare or non-existent. Therefore, despite considerable progress in our ability to model and quantitatively assess network-level criticality through local (i.e., centrality metrics) and global network parameters (i.e., cascading thresholds, giant connected component), *functions* of CI networks are often implicit premises and the effects of *spatial boundaries* are omitted. Investigating how the institutional legacy informs these assumptions is an important next step to realign criticality metrics to CI resilience goals, especially in the context of climate change risk. This realignment responds to current policy premises of ecological resilience for CI, viewing them as complex adaptive systems, with social, ecological, and technological dimensions.

When modeling CI, we call for a reexamination of infrastructure networks *functions* and spatial *boundaries* to harness our conceptual advancements on CI resilience within disaster risk theories and complexity sciences. Figure 18.1 presents a summary of the gaps from different literature bodies covered in this article as well as their complementary contributions to the paradigm shift defining CI as sociotechnical and socioecological systems for climate resilience.

Rethinking infrastructure network *criticality* through an ecological resilience lens provides insights to prevent maladaptation and to proactively address climate change lock-in issues (Markolf et al. 2018; Lewis and Ernstson 2017). Climate lock-in mechanisms stem from our reliance on hydrocarbon-dependent infrastructures are well researched and established, but they are also embedded in the amplification of vulnerabilities due to CI exposure to climate threats. With consideration of the militarization, securitization, techno-and hazard-centric critique of CI, some argue that the resilience premise is necessary to ultimately re-invent our relationship with vital infrastructures (Wakefield 2018; Gupta 2018). This new relationship could rise with the recognition of co-benefits from green infrastructures and ecosystem services (Raven et al. 2018; Pultrone 2018); the evidence of increased efficient, effective, and equitable use of natural resources through decentralization and short lead times of alternative energy infrastructure (Lovins and Lovins 1982; Alstone et al. 2015; Ottinger 2013), and the emancipation of grass-roots infrastructural stewardship (Lamond and Everett 2019).

Most common applications of ecological resilience within socioecological systems apply graph theory to ecosystem conservation or socioecological sustainability. Some examples are studies that enhance conservation management through governance networks (Minor and Urban 2008; Bodin and Crona 2009; Cumming et al. 2010; Rayfield et al. 2011; Bodin and Tengö 2012; Mills et al. 2014) and transition management for sustainable natural resource regimes (Bodin and Crona 2009; Pahl-Wostl et al. 2010; Chaffin et al. 2016; Bodin 2017; Stein et al. 2011). Fewer examples apply concepts of ecological resilience and graph theory to DRR and climate change risk management, some of these have developed pioneering methods to improve collaborative governance for climate adaptation (Hamilton and Lubell 2018); and incorporate social network analysis to improve flood risk management (Ceddia et al. 2017; Calliari et al. 2019), wildfire risk management (Ager et al. 2018; Hamilton et al. 2019); and model organizational networks to improve situational awareness for disaster response and emergency management (Comfort 1999; Nowell et al. 2018; Bodin and Nohrstedt 2016). The dominating risk assessment framework in DRR focuses on direct exposure, or calculating risk based on spatial overlay of people, goods, and hazard. The bridge between DRR and complexity sciences still has much room to improve our understanding of indirect exposure and vulnerability from a systems perspective, i.e., better assessing elements that are not in hazardous areas but still prone to suffer damage due to their inherent interconnectedness to complex socioecological and sociotechnical systems, which are in turn, directly exposed. Even though this concept of indirect exposure (or network-level vulnerability) stemming

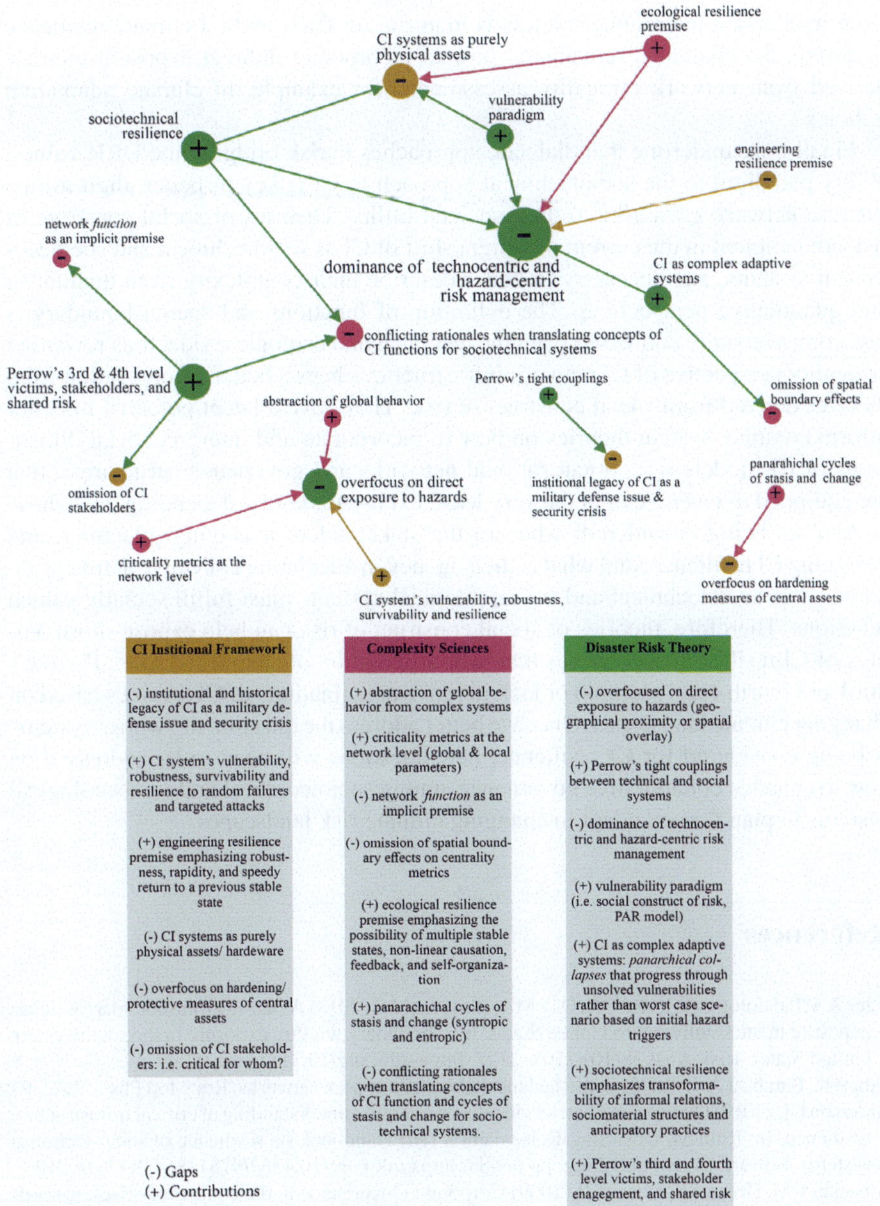

Fig. 18.1 Summary of gaps and complementary contributions for the paradigm shift defining CI as sociotechnical and socioecological systems for climate resilience

from heightened interconnectedness is maturing in DRR and CI climate resilience literature, the challenge remains on how to incorporate indirect exposure metrics, derived from network criticality assessments, for example, to climate adaptation policies.

Finally, we underline that dialectic approaches to risk bridging the DRR vulnerability paradigm to the sociotechnical approach to CI is key to better align infrastructure network *criticality* to their societal utility. Theories of social construct of risk are incipient in the current paradigm shift of CI as sociotechnical and socioecological systems, and necessary to better describe their complexity from qualitative and quantitative perspectives. The definition of functions and spatial boundary is less controversially addressed from the technical and ecological side, thus revisiting criticality perspective of CI network infrastructures begs a better incorporation from theories derived from social construct of risk. They have a latent potential to better inform complex system theories on how to incorporate and interpret social dimensions of CI models (i.e., organizational networks and governance structures), that are equipped to leverage or at the very least, expose trade-offs depending on "whose system" is being considered; who are the stakeholders managing, planning, and designing CI resilience; and what is their agency in alleviating collective action problems. CI as sociotechnical and socioecological systems must fulfill socially valued functions. Therefore, theories of social construct of risk can help expose shortcomings of CI resilience approaches that lack articulation of shared risk (i.e., Perrow's third and fourth-level victims) or lack iterative examinations of CI services based on changing climate landscapes. Once we better address the question of "whose system" is being considered for CI resilience, new questions will need to be addressed on how to enhance collaborative governance among sociotechnical and socioecological systems to plan for and adapt to changing climate risk landscapes.

References

Ager AA, Palaiologou P, Evers CR, Day MA, Barros AMG (2018) Assessing transboundary wildfire exposure in the southwestern United States: transboundary wildfire exposure in the southwestern United States. Risk Anal 38(10):2105–2127. https://doi.org/10.1111/risa.12999

Albert R, Barabási A-L (2002) Statistical mechanics of complex networks. Rev Mod Phys 74:47–97

Alderson DL (2019) Overcoming barriers to greater scientific understanding of critical infrastructure resilience. In: Ruth M, Goessling-Reisemann S (ed) Handbook on resilience of socio-technical systems. Edward Elgar Publishing, pp 66–88. https://doi.org/10.4337/9781786439376.00010

Alpaslan CM, Green SE, Mitroff II (2009) Corporate Governance in the context of crises: towards a stakeholder theory of crisis management. J Conting Crisis Manag 17(1):38–49. https://doi.org/10.1111/j.1468-5973.2009.00555.x

Alstone P, Gershenson D, Kammen DM (2015) Decentralized energy systems for clean electricity access. Nat Clim Chang 5(4):305–314. https://doi.org/10.1038/nclimate2512

Altshuler Y, Puzis R, Elovici Y, Beckhor S, Pentland A (2015) On the rationality and optimality of transportation networks defense: a network centrality approach. In: Securing transportation systems. Wiley-Blackwell, pp 35–63. https://doi.org/10.1002/9781119078203.ch3

Amir S, Kant V (2018) Sociotechnical resilience: a preliminary concept: sociotechnical resilience. Risk Anal 38(1):8–16. https://doi.org/10.1111/risa.12816

Amoaning-Yankson S, Amekudzi-Kennedy A (2017) Transportation system resilience: opportunities to expand from principally technical to sociotechnical approaches. Transp Res Rec J Transp Res Board 2604(1):28–36. https://doi.org/10.3141/2604-04

Ayyub BM (2014) Systems resilience for multihazard environments: definition, metrics, and valuation for decision making: systems resilience for multihazard environments. Risk Anal 34(2):340–355. https://doi.org/10.1111/risa.12093

Barabási A-L (2012) The network takeover. Nat Phys 8(1):14–16. https://doi.org/10.1038/nphys2188

Barabási A-L (2016) Network science. Cambridge University Press, Cambridge, UK

Barabasi A-L, Albert R (1999) Emergence of scaling in random networks. 286:5

Barnes TJ, Wilson MW (2014) Big data, social physics, and spatial analysis: the early years. Big Data Soc 1(1):205395171453536. https://doi.org/10.1177/2053951714535365

Barthélemy M (2011) Spatial networks. Phys Rep 499(1–3):1–101. https://doi.org/10.1016/j.physrep.2010.11.002

Von Bertalanffy L (1950) An outline of general system theory. Br J Philos Sci I(2):134–165. https://doi.org/10.1093/bjps/I.2.134

Blaikie P, Cannon R, Davis I, Wisner B (1994) At risk. Routledge, New York, NY

Boccaletti S, Bianconi G, Criado R, del Genio CI, Gómez-Gardeñes J, Romance M, Sendiña-Nadal I, Wang Z, Zanin M (2014) The structure and dynamics of multilayer networks. Phys Rep 544(1):1–122. https://doi.org/10.1016/j.physrep.2014.07.001

Bodin Ö, Nohrstedt D (2016) Formation and performance of collaborative disaster management networks: evidence from a swedish wildfire response. Glob Environ Chang 41(November):183–194. https://doi.org/10.1016/j.gloenvcha.2016.10.004

Bodin Ö (2017) Collaborative environmental governance: achieving collective action in social-ecological systems. Science 357(6352):eaan1114. https://doi.org/10.1126/science.aan1114

Bodin Ö, Crona BI (2009) The role of social networks in natural resource governance: what relational patterns make a difference? Glob Environ Chang 19(3):366–374. https://doi.org/10.1016/j.gloenvcha.2009.05.002

Bodin Ö, Tengö M (2012) Disentangling intangible social-ecological systems. Glob Environ Chang. no. Adding Insult to Injury: Climate Change, Social Stratification, and the Inequities of Intervention. https://doi.org/10.1016/j.gloenvcha.2012.01.005, Database: ScienceDirect

Boin A, Comfort LK, Demchak CC (2010) The rise of resilience. In: Designing resilience: preparing for extreme events. University of Pittsburgh Press, p. 13. http://www.jstor.org/stable/10.2307/j.ctt5hjq0c

Bollobás B (1998) Fundamentals of graph theory. In: Modern graph theory. Springer, New York, NY, pp 1–37

Bond M, Harrigan N (2014) Political dimensions of corporate connections. In: Scott J, Carrington P (eds) The SAGE handbook of social network analysis. SAGE Publications Ltd., London, UK, pp 196–209. https://doi.org/10.4135/9781446294413.n14

Bono F, Gutiérrez E (2011) A network-based analysis of the impact of structural damage on urban accessibility following a disaster: the case of the seismically damaged port Au prince and carrefour urban road networks. J Transp Geogr 19(6):1443–1455. https://doi.org/10.1016/j.jtrangeo.2011.08.002

Brandes U, Borgatti SP, Freeman LC (2016) Maintaining the duality of closeness and betweenness centrality. Soc Netw 44(January):153–159. https://doi.org/10.1016/j.socnet.2015.08.003

Bulkeley H (2001) Governing climate change: the politics of risk society? Trans Inst Br Geogr 26(4):430–447. https://doi.org/10.1111/1475-5661.00033

Burton I, Robert K, White G (1968) The human ecology of extreme geophysical events. Natl Hazard Res 1–40

Cagley E (1964) A general network simulation model. National Resource Evaluation Center, Technical Report 31. U.S. Army Engineer Mathematical Computational Agency

Calliari E, Michetti M, Farnia L, Ramieri E (2019) A network approach for moving from planning to implementation in climate change adaptation: evidence from Southern Mexico. Environ Sci Policy 93(March):146–157. https://doi.org/10.1016/j.envsci.2018.11.025
Cardona OD (2005) The need for rethinking the concepts of vulnerability and risk from a holistic perspective: a necessary review and criticism for effective risk management, p 17
Carse A (2017) Keyword: Infrastructure. How a Humble Fench term shaped the modern world, pp 27–39
Ceddia MG, Christopoulos D, Hernandez Y, Zepharovich E (2017) Assessing adaptive capacity through governance networks: the elaboration of the flood risk management plan in Austria. Environ Sci Policy 77(November):140–146. https://doi.org/10.1016/j.envsci.2017.08.014
Chaffin BC, Garmestani AS, Gosnell H, Craig RK (2016) Institutional networks and adaptive water governance in the Klamath river Basin, USA. Environ Sci Policy 57(March):112–121. https://doi.org/10.1016/j.envsci.2015.11.008
Chang SE, McDaniels T, Fox J, Dhariwal R, Longstaff H (2014) Toward disaster-resilient cities: characterizing resilience of infrastructure systems with expert judgments: toward disaster-resilient cities. Risk Anal 34(3):416–434. https://doi.org/10.1111/risa.12133
CNRA (2018) Paying it forward: the path toward climate-safe infrastructure in California: executive summary. California Climate-Safe Infrastructure Working Group. California Natural Resources Agency. https://resources.ca.gov/CNRALegacyFiles/docs/climate/ab2800/AB2800_Climate-SafeInfrastructure_FinalWithAppendices.pdf
Collier SJ, Lakoff A (2008) The vulnerability of vital systems: how 'critical infrastructure' became a security problem. In: Dunn M, Kristensen KS (eds) The politics of securing the homeland: critical infrastructure, risk and securitisation. Routledge, New York, pp 17–39
Collier SJ, Lakoff A (2015) Vital systems security: reflexive biopolitics and the government of emergency. Adey P, Anderson B, Graham S (eds). Theory Cult Soc 32(2):19–51. https://doi.org/10.1177/0263276413510050
Comfort LK (1999) Shared risk. Complex systems in seismic response, 1st edn. University of Pittsburgh Press, Pittsburg
Comfort LK (2017) Organizational reliability in practice: searching to resilience in communities at risk. In: Organizational reliability: a guide for research and practice, Chapter 11. Stanford University Press, Redwood City, CA
Comfort LK (2019) The dynamics of risk: changing technologies and collective actions in seismic events. Princeton University Press.
CSIWG (2018) Paying it forward: the path toward climate-safe infrastructure in California
Cumming GS (2011) Spatial resilience in social-ecological systems. Springer, Dordrecht
Cumming GS, Bodin Ö, Ernstson H, Elmqvist T (2010) Network analysis in conservation biogeography: challenges and opportunities: challenges and opportunities in network analysis. Divers Distrib 16(3):414–425. https://doi.org/10.1111/j.1472-4642.2010.00651.x
Cutter SL (1996) Vulnerability to environmental hazards. Prog Hum Geogr 20(4):529–539. https://doi.org/10.1177/030913259602000407
Cutter SL, Ahearn JA, Amadei B, Crawford P, Eide EA, Galloway GE, Goodchild MF et al (2013) Disaster resilience: a national imperative. Environ Sci Policy Sustain Dev 55(2):25–29. https://doi.org/10.1080/00139157.2013.768076
Davoudi S, Brooks E, Mehmood A (2013) Evolutionary resilience and strategies for climate adaptation. Plann Pract Res 28(3):307–322. https://doi.org/10.1080/02697459.2013.787695
DHS (2013) NIPP 2013: partnering for critical infrastructure security and resilience. Department of Homeland Security
Dunn S, Wilkinson S, Ford A (2016) Spatial structure and evolution of infrastructure networks. Sustain Cities Soc 27(November):23–31. https://doi.org/10.1016/j.scs.2016.08.011
Edwards PN (2002) Infrastructure and modernity: force, time and social organization in the history of sociotechnical systems. In: Brey P, Rip A, Feenberg A (eds) Technology and modernity: the empirical turn. MIT Press, Cambridge, MA, p 35

Etkin D (2016) Disaster theory: an interdisciplinary approach to concepts and causes. Butterworth-Heinemann, Waltham, MA

Flusberg SJ, Matlock T, Thibodeau PH (2017) Metaphors for the war (or race) against climate change. Environ Commun 11(6):769–783. https://doi.org/10.1080/17524032.2017.1289111

Folke C (2006) Resilience: the emergence of a perspective for social-ecological systems analyses. Glob Environ Chang 16(3):253–267. https://doi.org/10.1016/j.gloenvcha.2006.04.002

Forino G, Bonati S, Calandra LM (2018) Governance of risk, hazards and disasters: trends in theory and practice. Routledge, London and New York. https://ebookcentral.proquest.com/lib/berkeley-ebooks/detail.action?docID=5259931

Forzieri G, Marin Herrera MA, Silva FB, Bianchi A, Lavalle C, Feyen L, European Commission, Joint Research Centre (2016) Resilience of large investments and critical infrastructures in Europe to climate change. Publications Office, Luxembourg. https://bookshop.europa.eu/uri?target=EUB:NOTICE:LBNA27906:EN:HTML

Franco G, Cayan DR, Moser S, Hanemann M, Jones M-A (2011) Second California assessment: integrated climate change impacts assessment of natural and managed systems. Guest editorial. Clim Chang 109(S1):1–19. https://doi.org/10.1007/s10584-011-0318-z

Freeman LC (1978) Centrality in social networks conceptual clarification. Soc Netw 1(3):215–239. https://doi.org/10.1016/0378-8733(78)90021-7

Gaillard JC (2010) Vulnerability, capacity and resilience: perspectives for climate and development policy. J Int Dev 22(2):218–232. https://doi.org/10.1002/jid.1675

Gaillard JC, Mercer J (2013) From knowledge to action: bridging gaps in disaster risk reduction. Prog Hum Geogr 37(1):93–114. https://doi.org/10.1177/0309132512446717

Gao J, Barzel B, Barabási A-L (2016) Universal resilience patterns in complex networks. Nature 530(7590):307–312. https://doi.org/10.1038/nature16948

Gao J, Buldyrev SV, Eugene Stanley H, Havlin S (2011) Networks formed from interdependent networks. Nat Phys 8(1):40–48. https://doi.org/10.1038/nphys2180

George Mason University (2007) Critical thinking... Moving from infrastrcuture protection to infrastructure resilience. CIP program discussion paper series. Critical Infrastructure Protection Program, Arlignton, VA

Gephart RP, Van Maanen J, Oberlechner T (2009) Organizations and risk in late modernity. Organ Stud 30(2–3):141–155. https://doi.org/10.1177/0170840608101474

Gerson NL, Goodman S (2007) National security and the threat of climate change. The CNA Corporation, The Military Advisory Board. https://www.cna.org/cna_files/pdf/National%20Security%20and%20the%20Threat%20of%20Climate%20Change.pdf

Gieseke TM (2020) Collaborative environmental governance frameworks; a practical guide, p 211

Gorman SP (2005) Networks, security and complexity. The role of public policy in critical infrastructure protection. Edward Elgar Publishing, Northhampton, MA, USA

Gorman SP, Schintler L, Kulkarni R, Stough R (2004) The revenge of distance: vulnerability analysis of critical information infrastructure. J Conting Crisis Manag 12(2):48–63. https://doi.org/10.1111/j.0966-0879.2004.00435.x

Graham S (2011) Disrupted cities: infrastructure disruptions as the achilles heel of urbanized societies, p 16

Grubesic TH, Matisziw TC, Murray AT, Snediker D (2008) Comparative approaches for assessing network vulnerability. Int Reg Sci Rev 31(1):88–112. https://doi.org/10.1177/0160017607308679

Gupta A (2018) The future in ruins: thoughts on the temporality of infrastructure. In: The promise of infrastructure, pp 62–79. https://libproxy.berkeley.edu/login?qurl=https%3a%2f%2fsearch.ebscohost.com%2flogin.aspx%3fdirect%3dtrue%26db%3dnlebk%26AN%3d1848521%26site%3deds-live&ebv=EB&ppid=pp_62

Haas JE, White GF (1975) Assessment of research on natural hazards. MIT Press, Boston, MA

Haimes YY (2012) Strategic preparedness for recovery from catastrophic risks to communities and infrastructure systems of systems: recovery from catastrophic risks to communities. Risk Anal 32(11):1834–1845. https://doi.org/10.1111/j.1539-6924.2012.01930.x

Hamilton M, Fischer AP, Ager A (2019) A social-ecological network approach for understanding wildfire risk governance. Glob Environ Chang 54(January):113–123. https://doi.org/10.1016/j.gloenvcha.2018.11.007

Hamilton M, Lubell M (2018) Collaborative governance of climate change adaptation across spatial and institutional scales: adaptation across spatial and institutional scales. Policy Stud J 46(2):222–247. https://doi.org/10.1111/psj.12224

Hanmer C, Abram S (2017) Actors, networks, and translation hubs: gas central heating as a rapid socio-technical transition in the United Kingdom. Energy Res Soc Sci 34(December):176–183. https://doi.org/10.1016/j.erss.2017.03.017

Hansson S (2013) The ethics of risk. Ethical analysis in an uncertain world. Palgrave Macmillan, New York, NY

Hasan S, Foliente G (2015) Modeling infrastructure system interdependencies and socioeconomic impacts of failure in extreme events: emerging R&D challenges. Nat Hazards 78(3):2143–2168. https://doi.org/10.1007/s11069-015-1814-7

Hellström T (2007) Critical infrastructure and systemic vulnerability: towards a planning framework. Saf Sci 45(3):415–430. https://doi.org/10.1016/j.ssci.2006.07.007

Hewitt K (1983) Interpretations of calamity from the viewpoint of human ecology. Geogr Rev 74(2):226. https://doi.org/10.2307/214106

Hewitt K (2000) Safe place or 'catastrophic society'? Perspectives on hazards and disasters in Canada. Can Geogr 4(44):325–341

Holling CS (1973) Resilience and stability of ecological systems. Annu Rev Ecol Syst 4(1):1–23. https://doi.org/10.1146/annurev.es.04.110173.000245

Holling CS (2001) Understanding the complexity of economic, ecological, and social systems. Ecosystems 4(5):390–405. https://doi.org/10.1007/s10021-001-0101-5

Holmgren AJ (2006) Using graph models to analyze the vulnerability of electric power networks. Risk Anal 26(4):955–969. https://doi.org/10.1111/j.1539-6924.2006.00791.x

Hosseini S, Barker K, Ramirez-Marquez JE (2016) A review of definitions and measures of system resilience. Reliab Eng Syst Saf 145(January):47–61. https://doi.org/10.1016/j.ress.2015.08.006

IPCC (2014). Climate change 2014: impacts, adaptation, and vulnerability. Part A: Global and sectoral aspects. In: Field CB, Barros DJ, Dokken KJ, Mach MD, Mastrandrea TE, Bilir M, Chatterjee KL et al (eds)Controbutions of working Group II to the 5th assessment report of the intergovernmental panel on climate change. Cambridge University Press, New York, NY

Kasmalkar IG, Serafin KA, Miao Y, Avery Bick I, Ortolano L, Ouyang D, Suckale J (2020) When floods hit the road: resilience to flood-related traffic disruption in the San Francisco Bay area and beyond. Sci Adv 6(32):eaba2423. https://doi.org/10.1126/sciadv.aba2423

Knoke D (2016) Policy networks. In: Scott J, Carrington P (eds) The SAGE handbook of social network analysis. SAGE Publications Ltd., London, UK, pp 210–222. https://doi.org/10.4135/9781446294413

Lamond J, Everett G (2019) Sustainable blue-green infrastructure: a social practice approach to understanding community preferences and stewardship. Landsc Urban Plann 191(November):103639. https://doi.org/10.1016/j.landurbplan.2019.103639

Latour B (2005) Reassembling the social: an introduction to actor-network theory. Oxford University Press

Lazari A (2014) European critical infrastructure protection. Springer International Publishing, Cham. https://doi.org/10.1007/978-3-319-07497-9

Lei TE, Der V, Bekebrede G, Nikolic I (2010) Critical infrastructures: a review from a complex adaptive systems perspective. Int J Crit Infrastruct 6(4):380. https://doi.org/10.1504/IJCIS.2010.037454

Lewis JA, Ernstson H (2017) Contesting the coast: ecosystems as infrastructure in the Mississippi river delta. Progr Plann. https://doi.org/10.1016/j.progress.2017.10.003

Lewis LP (2019) Critical infrastructure interdependency analysis: operationalising resilience strategies

Lewis TG (2006) Critical infrastructure protection in homeland security: defending a networked nation. Wiley

Li W, Milburn L-A (2016) The evolution of geodesign as a design and planning tool. Landsc Urban Plann 156(December):5–8. https://doi.org/10.1016/j.landurbplan.2016.09.009

Lovins AB, Hunter Lovins L (1982) Brittle power: energy strategy for national security. Brick House Pub. Co., Andover, Mass

Maldonado JK (2017) The practical and policy relevance of social network analysis for disaster response, recovery and adaptation. In: Jones EC, Faas AJ (eds) Social network analysis of disaster response, recovery and adaptation. Elsevier, pp 255–267

Maliszewski PJ, Perrings C (2012) Factors in the resilience of electrical power distribution infrastructures. Appl Geogr 32(2):668–679. https://doi.org/10.1016/j.apgeog.2011.08.001

Markolf SA, Chester MV, Eisenberg DA, Iwaniec DM, Davidson CI, Zimmerman R, Miller TR, Ruddell BL, Chang H (2018) Interdependent infrastructure as linked social, ecological, and technological systems (SETSs) to address lock-in and enhance resilience. Earth's Future, December, 2018EF000926. https://doi.org/10.1029/2018EF000926

Matisziw TC, Murray AT, Grubesic TH (2009) Exploring the vulnerability of network infrastructure to disruption. Ann Reg Sci 43(2):307–321. https://doi.org/10.1007/s00168-008-0235-x

McHarg IL (1981) Human ecological planning at Pennsylvania. Landsc Plann 8(2):109–120. https://doi.org/10.1016/0304-3924(81)90029-0

Meuleman L (2018) Metagovernance for sustainability: a framework for implementing the sustainable development goals, 1st edn. Routledge. https://doi.org/10.4324/9781351250603

Miller HJ (2004) Tobler's first law and spatial analysis. Ann Assoc Am Geogr 94(2):284–289. https://doi.org/10.1111/j.1467-8306.2004.09402005.x

Miller RA (2009) There's infrastructure and ...critical infrastructure. Int J Crit Infrastruct Prot 2(1–2):3–4. https://doi.org/10.1016/j.ijcip.2009.02.003

Mills M, Álvarez-Romero JG, Vance-Borland K, Cohen P, Pressey RL, Guerrero AM, Ernstson H (2014) Linking regional planning and local action: towards using social network analysis in systematic conservation planning. Biol Cons 169(January):6–13. https://doi.org/10.1016/j.biocon.2013.10.015

Minor ES, Urban DL (2008) A graph-theory framework for evaluating landscape connectivity and conservation planning. Conserv Biol 22(2):297–307. https://doi.org/10.1111/j.1523-1739.2007.00871.x

Mitroff I (1983) Stakeholders of the organizational mind. Jossey-Bass, San Francisco

Monstadt J (2009) Conceptualizing the political ecology of urban infrastructures: insights from technology and urban studies. Environ Plann A Econ Space 41(8):1924–1942. https://doi.org/10.1068/a4145

Moteff J, Parfomak P (2004) Critical infrastructure and key assets: definition and identification. Order Code RL32631. Congressional Research Service

Murray AT, Matisziw TC, Grubesic TH (2008) A methodological overview of network vulnerability analysis. Growth Chang 39(4):573–592

NAS, National Research Council, Committee on Network Science for Future Army Applications, Board on Army Science and Technology, and Division of Engineering and Physical Science (2005) Network science. National Academies Press, Washington, D.C. https://doi.org/10.17226/11516

NAST (2000) Climate change impacts on the United States. The potential consequences of climate variability and change. National Assessment Synthesis Team. https://downloads.globalchange.gov/nca/nca-2000-report-overview.pdf

Nelson A, Lindbergh S, Stephenson L, Halpern J, Espinet-Alegre X, Arroyo FA, Gonzalez M. forthcoming. Coupling natural hazard estimates with road network analysis to assess vulnerability and risk: a case study of freetown. Tranportation Research Reccords.

Northeastern University (2018) Resilience governance for critical infrastructure interdependencies. A practical model for regional critical infrastructure resilience. Global Resilience Institute

Nowell B, Steelman T, Velez A-L, Yang Z (2018) The structure of effective governance of disaster response networks: insights from the field. Am Rev Publ Adm 48(7):699–715. https://doi.org/10.1177/0275074017724225

Nuss P, Graedel TE, Alonso E, Carroll A (2016) Mapping supply chain risk by network analysis of product platforms. Sustain Mater Technol 10(December):14–22. https://doi.org/10.1016/j.susmat.2016.10.002

OECD (2019) Good governance for critical infrastructure resilience. OECD Reviews of Risk Management Policies, OECD. https://doi.org/10.1787/02f0e5a0-en

Osleeb JP (1979) An evaluation of the strategic petroleum reserve program of the united states department of energy*. Prof Geogr 31(4):393–399. https://doi.org/10.1111/j.0033-0124.1979.00393.x

Ottinger G (2013) The winds of change: environmental justice in energy transitions. Sci Cult 22(2):222–229. https://doi.org/10.1080/09505431.2013.786996

Ouyang M (2014) Review on modeling and simulation of interdependent critical infrastructure systems. Reliab Eng Syst Saf 121(January):43–60. https://doi.org/10.1016/j.ress.2013.06.040

Pahl-Wostl C, Holtz G, Kastens B, Knieper C (2010) Analyzing complex water governance regimes: the management and transition framework. Environ Sci Policy 13(7):571–581. https://doi.org/10.1016/j.envsci.2010.08.006

Peck H (2005) Drivers of supply chain vulnerability: an integrated framework. Int J Phys Distrib Logist Manag 35(4):210–232. https://doi.org/10.1108/09600030510599904

Perrow C (1972) Complex organizations. A critical essay, 3rd edn. Newberry Award Records

Perrow C (1984) Normal accidents. Living with high-risk technologies, 2nd edn. Princeton University Press, New Jersey. https://search.alexanderstreet.com/view/work/bibliographic_entity%7Cbibliographic_details%7C2138933?utm_source=aspresolver&utm_medium=MARC&utm_campaign=AlexanderStreet#page/6/mode/1/chapter/bibliographic_entity%7Cdocument%7C2138939

Pescaroli G, Alexander D (2016) Critical infrastructure, panarchies and the vulnerability paths of cascading disasters. Nat Hazards 82(1):175–192. https://doi.org/10.1007/s11069-016-2186-3

Pultrone G (2018) The ecological challenge as an opportunity and input for innovative strategies of integrated planning. In: Smart innovation, systems and technologies. New metropolitan perspectives, vol 101. Springer, pp 691–698. https://link.springer.com/chapter/10.1007/978-3-319-92102-0_75

Raven J, Stone B, Mills G, Towers J, Katzschner L, Leone MF, Gaborit P et al (2018) Urban planning and urban design. In: Rosenzweig C, Solecki WD, Romero-Lankao P, Mehrotra S, Dhakal S, Ali Ibrahim S (eds) Climate change and cities, 1st ed, Cambridge University Press, pp 139–172. https://doi.org/10.1017/9781316563878.012

Rayfield B, Fortin M-J, Fall A (2011) Connectivity for conservation: a framework to classify network measures. Ecology 92(4):847–858. https://doi.org/10.1890/09-2190.1

Rich N (2019) Losing earth. A recent history. MCD Farrar, Straus and Giroux, New York

Rinaldi SM, Peerenboom JP, Kelly TK (2001) Identifying, understanding and analyzing critical infrastructure interdependencies. Inst Electr Electron Eng Control Syst Mag 21(6):11–25. https://doi.org/10.1109/37.969131

Roberts KH (1989) New challenges in organizational research: high reliability organizations. Ind Crisis Q 3(2):111–125

Roberts KH (1990) Managing high reliability organizations. Calif Manag Rev 32(4):101–113. https://doi.org/10.2307/41166631

Roe E, Schulman PR (2016) Reliability and risk. The challenge of managing interconnected infrastructures. Stanford University Press, Stanford, CA

Roukny T, Bersini H, Pirotte H, Caldarelli G, Battiston S (2013) Default cascades in complex networks: topology and systemic risk. Sci Rep 3(1). https://doi.org/10.1038/srep02759

Schulman P, Roe E, van Eeten M, de Bruijne M (2004) High reliability and the management of critical infrastructures. J Conting Crisis Manag 12(1):14–28

Schwartz P, Randall D (2003) An abrupt climate change scenario and its implications for United States national security. Defense Technical Information Center, Fort Belvoir, VA. https://doi.org/10.21236/ADA469325
Scott J (2014) Social physics and social networks. In: Scott J, Carrington P (eds) The SAGE handbook of social network analysis. SAGE Publications Ltd., London, UK, pp 55–66. https://doi.org/10.4135/9781446294413.n5
Shen S, Feng Xi, Peng ZR (2016) A framework to analyze vulnerability of critical infrastructure to climate change: the case of a coastal community in Florida. Nat Hazards 84(1):589–609. https://doi.org/10.1007/s11069-016-2442-6
Smith A, Stirling A (2008) Social-ecological resilience and socio-technical transitions: critical issues for sustainability governance. Working Paper, STEPS Centre, University of Sussex, Brighton
Solé RV, Bascompte J (2006) Self-organization in complex ecosystems. Princeton University Press, Princeton. https://doi.org/10.1515/9781400842933
Steele W, Hussey K, Dovers S (2017) What's critical about critical infrastructure? Urban Policy Res 35(1):74–86. https://doi.org/10.1080/08111146.2017.1282857
Stein C, Ernstson H, Barron J (2011) A social network approach to analyzing water governance: the case of the Mkindo Catchment, Tanzania. Phys Chem Earth, Parts A/B/C 36(14–15):1085–1092. https://doi.org/10.1016/j.pce.2011.07.083
Taylor MAP (2003) Concepts of network vulnerability and applications to the identification of critical elements of transport infrastructure, p 15
Thayer SB, Shaner WW (1960) The effects of nuclear attacks on the petroleum industry. Office of civil and defense mobilization. Stanford Research Institute, Menlo Park, California
Theoharidou M, Kotzanikolaou P, Gritzalis D (2010) A multi-layer criticality assessment methodology based on interdependencies. Comput Secur 29(6):643–658. https://doi.org/10.1016/j.cose.2010.02.003
Tierney KJ (2007) From the margins to the mainstream? Disaster research at the crossroads. Ann Rev Sociol 33(1):503–525. https://doi.org/10.1146/annurev.soc.33.040406.131743
Tierney KJ (2014) A different perspective. The social production of risk. In: The social roots of risk: producing disasters, promoting resilience. Stanford Univeristy Press, pp 31–49
Turner BL, Kasperson RE, Matson PA, McCarthy JJ, Corell RW, Christensen L, Eckley N et al (2003) A framework for vulnerability analysis in sustainability science. Proc Natl Acad Sci 100(14):8074–8079
UNISDR (2005) Hyogor framework for action 2005–2015: building resilience of nations and communities to disasters. In: UN 2nd conference on disaster risk reduction. Kobe, Japan: United Nations Office for Disaster Risk Reduction (UNISDR). https://www.unisdr.org/we/inform/publications/8720
United Nations (1999) Coping study on disaster resilient infrastructure. 12107. Secretariat for the International Decade for Natural Disaster Reduction for the IDNDR Programme Forum
U.S. Global Change Research Program (ed) (2009) Global climate change impacts in the United States: a state of knowledge report. Cambridge University Press, Cambridge [England]; New York. https://www.globalchange.gov/browse/reports/global-climate-change-impacts-united-states
Vatn J, Hokstad P, Utne IB (2012) Defining concepts and categorizing interdependencies. In: Risk and interdependencies if critical infrastructures: a guideline for analysis. Springer series in reliability engineering, pp 13–22
Vespignani A (2010) Complex networks: the fragility of interdependency. Nature 464(7291):984
Vitali S, Glattfelder JB, Battiston S (2011) The network of global corporate control. PLoS ONE 6(10):6
Wakefield S (2018) Infrastructures of liberal life: from modernity and progress to resilience and ruins. Geogr Compass 12(7):e12377. https://doi.org/10.1111/gec3.12377
Walker G (2015) Come back sociotechnical systems theory, all is forgiven .... Civil Eng Environ Syst 32(1–2):170–179. https://doi.org/10.1080/10286608.2015.1024112

Walker J, Cooper M (2011) Genealogies of resilience: from systems ecology to the political economy of crisis adaptation. Secur Dialogue 42(2):143–160. https://doi.org/10.1177/0967010611399616

Walker W, Haasnoot M, Kwakkel J (2013) Adapt or perish: a review of planning approaches for adaptation under deep uncertainty. Sustainability 5(3):955–979. https://doi.org/10.3390/su5030955

Watts DJ, Strogatz SH (1998) Collective dynamics of 'small-world' networks. Nature 393:3

Wildavsky A (1988) Searching for safety. Social Philosophy and Policy Center, Transaction Books, New Brunswick, N.J.

Wimmer J, Quandt T (2006) Living in the risk society: an interview with Ulrich Beck. Journal Stud 7(2):336–347. https://doi.org/10.1080/14616700600645461

Wisner B, Blaikie P, Cannon T, Davis I (2003) At risk: natural hazards, people's vulnerability, and disasters, 2nd edn. Routledge, New York, NY

Wisner B, Gaillard JC, Kelman I (2012) The Routledge handbook of hazards and disaster risk reduction. Routledge, Oxon, OX

Yadav N, Chatterjee S, Ganguly AR (2020) Resilience of urban transport network-of-networks under intense flood hazards exacerbated by targeted attacks. Sci Rep 10(1):10350. https://doi.org/10.1038/s41598-020-66049-y

Zhang X, Li H (2018) Urban resilience and urban sustainability: what we know and what do not know? Cities 72(February):141–148. https://doi.org/10.1016/j.cities.2017.08.009

Zhuge H, Zhang J (2009) Topological centrality and its applications. ArXiv:0902.1911 [Cs], February. https://arxiv.org/abs/0902.1911

Zio E (2016) Challenges in the vulnerability and risk analysis of critical infrastructures. Reliab Eng Syst Saf 152(August):137–150. https://doi.org/10.1016/j.ress.2016.02.009

Zio E, Sansavini G (2011) Component criticality in failure cascade processes of network systems: component criticality in failure cascade processes. Risk Anal 31(8):1196–1210. https://doi.org/10.1111/j.1539-6924.2011.01584.x

Correction to: Integrated Research on Disaster Risks

Riyanti Djalante, Mizan B. F. Bisri, and Rajib Shaw

Correction to: R. Djalante et al. (eds.), *Integrated Research on Disaster Risks*, Disaster Risk Reduction, https://doi.org/10.1007/978-3-030-55563-4

The original version of the book was inadvertently published with an incorrect layout in Chapter 7, and an incorrect version of Chapter 18. The layout of Chapter 7 has been corrected. The correct version of Chapter 18 has been included with one page reduced in the book and the title corrected in the Table of Contents.

The updated versions of these chapters can be found at
https://doi.org/10.1007/978-3-030-55563-4_7
https://doi.org/10.1007/978-3-030-55563-4_18

R. Djalante et al. (eds.), *Integrated Research on Disaster Risks*, Disaster Risk Reduction,
https://doi.org/10.1007/978-3-030-55563-4_20

The manufacturer's authorised representative in the EU is Springer Nature Customer Service Centre GmbH, Europaplatz 3, 69115 Heidelberg, Germany. If you have any concerns regarding our products, please contact ProductSafety@springernature.com

Printed and bound by CPI Group (UK) Ltd, Croydon, CR0 4YY
23/09/2026
02221544-0010